The Warfare between Science and Religion

The Warfare between Science and Religion

The Idea That Wouldn't Die

Edited by
JEFF HARDIN,
RONALD L. NUMBERS,
and RONALD A. BINZLEY

Johns Hopkins University Press
Baltimore

© 2018 Johns Hopkins University Press
All rights reserved. Published 2018
Printed in the United States of America on acid-free paper
2 4 6 8 9 7 5 3 1

Johns Hopkins University Press
2715 North Charles Street
Baltimore, Maryland 21218-4363
www.press.jhu.edu

Library of Congress Cataloging-in-Publication Data

Names: Hardin, Jeff, editor.
Title: The warfare between science and religion : the idea that wouldn't die / edited by Jeff Hardin, Ronald L. Numbers, and Ronald A. Binzley.
Description: Baltimore : Johns Hopkins University Press, 2018. | Includes bibliographical references and index.
Identifiers: LCCN 2017054175 | ISBN 9781421426181 (paperback : alk. paper) | ISBN 9781421426198 (electronic) | ISBN 1421426188 (paperback : alk. paper) | ISBN 1421426196 (electronic)
Subjects: LCSH: Religion and science.
Classification: LCC BL240.3 .I34 2018 | DDC 201/.65—dc23
LC record available at https://lccn.loc.gov/2017054175

A catalog record for this book is available from the British Library.

Special discounts are available for bulk purchases of this book.
For more information, please contact Special Sales at 410-516-6936 or specialsales@press.jhu.edu.

Johns Hopkins University Press uses environmentally friendly book materials, including recycled text paper that is composed of at least 30 percent post-consumer waste, whenever possible.

CONTENTS

Acknowledgments *vii*

Introduction 1
MARK A. NOLL AND DAVID N. LIVINGSTONE

1 The Warfare Thesis 6
LAWRENCE M. PRINCIPE

2 The Galileo Affair 27
MAURICE A. FINOCCHIARO

3 Rumors of War 46
MONTE HARRELL HAMPTON

4 The Victorians: Tyndall and Draper 65
BERNARD LIGHTMAN

5 Continental Europe 84
FREDERICK GREGORY

6 Roman Catholics 103
DAVID MISLIN

7 Eastern Orthodox Christians 123
EFTHYMIOS NICOLAIDIS

8 Liberal Protestants 143
JON H. ROBERTS

9 Protestant Evangelicals 163
BRADLEY J. GUNDLACH

10 Jews 184
NOAH EFRON

11 Muslims 203
M. ALPER YALÇINKAYA

12 New Atheists 220
RONALD L. NUMBERS AND JEFF HARDIN

13 Neo-Harmonists 239
PETER HARRISON

14 Historians 258
JOHN HEDLEY BROOKE

15 Scientists 279
ELAINE HOWARD ECKLUND AND CHRISTOPHER P. SCHEITLE

16 Social Scientists 302
THOMAS H. AECHTNER

17 The View on the Street 324
JOHN H. EVANS

Contributors 341
Index 345

ACKNOWLEDGMENTS

This volume would not exist without the generous support of the Issachar Fund, especially its president, Kurt Berends, and its event coordinator, Sara Merrilees, as well as the Department of Zoology (now the Department of Integrative Biology), University of Wisconsin–Madison. On May 14–16, 2015, the Issachar Fund and the Department of Zoology co-sponsored an international three-day conference on the so-called warfare thesis—that is, the belief that an inevitable and irreconcilable conflict exists between science and religion. The gathering brought together more than thirty distinguished scholars—primarily historians and sociologists—to critique the papers that would eventually become the chapters of this book. In addition to the chapter authors, expert participants included Charles L. Cohen (University of Wisconsin–Madison), Edward B. Davis (Messiah College), Paul Erickson (Wesleyan University), Joan Fujimura (University of Wisconsin–Madison), Albert Gunther (University of Wisconsin–Madison), Salman Hameed (Hampshire College), Jonathan Hill (Calvin College), Nicholas Jacobson (University of Wisconsin–Madison), G. Blair Nelson (Benedictine University), Scott Prinster (University of Wisconsin–Madison), Robert J. Richards (University of Chicago), Christopher M. Rios (Baylor University), Michael H. Shank (University of Wisconsin–Madison), Lawrence Shapiro (University of Wisconsin–Madison), Elliott Sober (University of Wisconsin–Madison), John Stenhouse (University of Otago), Rodney L. Stiling (Seattle Pacific University), Marc Swetlitz (Naperville, Illinois), and Stephen P. Weldon (University of Oklahoma). To each of them we are very thankful.

We would also like to thank Jon Roberts for allowing us to borrow the title from his previously published essay "'The Idea That Wouldn't Die': The

Introduction

MARK A. NOLL AND DAVID N. LIVINGSTONE

In 1996, Steven Shapin introduced a book inaugurating a new series on the history of science from the University of Chicago Press with a captivating opening sentence: "There was no such thing as the Scientific Revolution, and this is a book about it."[1] Based on the careful presentation of evidence in that book, this apparently outlandish statement proved surprisingly persuasive. Although Shapin's sentence has already been too often borrowed for too many purposes, one more rip-off seems justified for the persuasive demonstrations of the present volume: "There has never been systemic warfare between science and theology, and this is a book that explains why the notion nonetheless lives on."

Like a stubbornly adaptive virus, the idea of inevitable struggle between science and religion is proving hard to eradicate, as Michael Reiss, an evolutionary biologist, discovered to his bitter cost in 2008. On September 16, he was forced to resign his position as education director for the Royal Society over observations he had made about how science teachers should deal with creationism. Reiss's critics, who sprang immediately into action, were animated by the presumption of inexorable conflict between science and religion. Notwithstanding all the outstanding work by a generation of historians dismantling the "conflict model," their revisionist accounts have scarcely made a dent on leading public intellectuals. In a letter to the president of the Society demanding that Reiss step down, the Nobel Prize winner Sir Richard Roberts of New England Biolabs quipped: "We gather Professor Reiss is a clergyman, which in itself is very worrisome. Who on earth thought that he would be an appropriate Director of Education, who could be expected to answer questions about the differences between science and religion in a scientific, reasoned way?"[2] In similar vein, another Nobel laureate, Sir Harold Kroto, commenting on the whole issue for the *New Scientist*, observed:

"There is no way that an ordained minister—for whom unverified dogma must represent a major, if not the major, pillar in their lives can present free-thinking, doubt-based scientific philosophy honestly or disinterestedly."[3]

Perhaps one of the reasons why the scholarly subversion of the conflict thesis is passing the public by is that the idea of perpetual warfare between science and religion serves the interests of partisans rather well. It can be deliberately used to excite controversy. As Geoffrey Cantor reminds us, Arthur Keith used it in the 1920s "to infuriate the clergy."[4] So too did Thomas Henry Huxley with his rhetorical gibe that it was only "old ladies, of both sexes, [who] consider [*On the Origin of Species*] a decidedly dangerous book."[5] Comparable taunts in our own day obviously boost sales, make for feisty radio and television, stimulate Internet excitement, and provide colorful courtroom drama. Of course this does not mean that the whole issue can be reduced to just these considerations. But, like all myths, its survival against the historical odds suggests that the conflict fable continues to perform work that quite a few combatants find useful in today's culture.

It certainly means that many share an inclination to find conflict where scant conflict exists. Take the case of the highly distinguished philosopher Jerry Fodor, a self-declared secular humanist. In 2010, he brought out a book with fellow cognitive scientist Massimo Piatelli-Palmarini with the daring title *What Darwin Got Wrong*. In fact, Fodor had been thinking about Darwinism for several years and had been expressing some doubts about the explanatory power of natural selection in places like the *London Review of Books* and, rather more technically, in the journal *Mind and Language*. The details need not detain us. To be clear, Fodor has no problem with evolution and insists that it is "a mechanical process through and through."[6] What he wants to say is that natural selection cannot be the mechanism of adaptive evolution that enthusiastic Darwinians claim it is.

The reaction was stunning. Recall that Fodor has no religious axe to grind; he and his coauthor describe themselves as "card-carrying, signed-up, dyed-in-the-wool, no-holds-barred atheists."[7] That looks pretty clear. But you would not think so from the reviews. Reviewer after reviewer cast their book into the cauldron of science-corrupted-by-religion. Some worried that it would give succor to advocates of Intelligent Design; others that it was reminiscent of the Huxley-Wilberforce confrontation; one even described Fodor as a "creationist" and claimed that he sounded "like Christoph Schönborn, Catholic archbishop of Vienna, the chap duped by the Intelligent

Design folks."[8] Naturally enough, Fodor did not take these responses too kindly. "Well none of that is remotely our view," he observed. "There's not a scintilla of text in our book (or elsewhere) to support the accusation of creeping theism. . . . Short of trial by fire, water, or the House Un American Activities Committee, what must one do to prove one's bona fides?"[9]

Another dreary episode of the same rush to judgment took place in late 2012 when Thomas Nagel and Alvin Plantinga exchanged nuanced, sophisticated appreciations for books that each had written on the question of meaning in the natural world.[10] Although the atheist Nagel and the theist Plantinga found much to criticize in the other's work, they also found much to commend in their common dissatisfaction with explanations for how self-consciousness that could appreciate and evaluate might arise in a universe understood strictly as matter in motion. Despite the high level of their exchange, obloquy descended on Nagel for being willing to say anything positive about anything even remotely threatening to "science."

Denunciations of this variety cannot be understood outside the time and place of their making. Fears about an increasingly militant global fundamentalism, concerns about the integrity of the science curriculum in schools, and apprehensions over pushy Intelligent Design activists are prominent on the intellectual horizon of twenty-first-century scientific America. These anxieties provide the backdrop against which culture wars over science need to be interpreted.

Nothing new attends this turn of events of course. Science has always been a located enterprise, and—a fortiori—conflicts between champions of science and religious observers have always "taken place" at particular times in particular spaces. The fate of Darwin's theory in New Zealand, for example, cannot be grasped without attending to the politics of Māori dispossession and settler colonialism. The scientific rejection of evolution in the American South was part and parcel of a cultural resistance to northern abolitionism. The suspicion that Russian naturalists entertained about the Malthusian dimensions of natural selection and their promotion of evolution by mutual aid was all-of-a-piece with their valorizing of the peasant commune in the challenging landscapes of the Siberian north.[11]

In much the same way, the opposition to the new science by Calvinist clergy that broke out in Belfast in the 1870s cannot be divorced either from the aggressive anticlericalism exhibited by John Tyndall in his infamous presidential address to the British Association or from the thorny question of who should control university education in a profoundly divided sectarian society. By the same token, the substantial acceptance of evolution by Scottish lowland Calvinists at the same time owed much to the fact that their defensive energies were directed toward a different threat—the radical biblical criticism emanating from Germany. In the American South, disgust at Darwinism had much to do with a long-standing abhorrence of a variety of modernizing trends—polygenist anthropology, radical French republicanism, and creeping emancipation—which ran counter to the biblical literalism on which their culture had long rested. In cases like these, what on the surface looked like a warfare between science and religion turned out to have more to do with cultural anxieties of one kind or another.[12]

A full academic lifetime ago, the authors of this introduction ventured to express opinions on how relationships of science and theology might best be expressed. A review-essay by Noll titled "The War Is Over" and a monograph by Livingstone, *Darwin's Forgotten Defenders*, suggested that while such relationships have certainly involved tension, complexity, ambiguity, irony, negotiation, compromise, maneuvering, accommodation, false starts, rethinking, and incomprehension, "warfare" has only rarely and then only momentarily been the right word. In documenting the surprisingly large number of theologically traditional thinkers who supported, to one degree or another, Darwin's theory of natural selection, Livingstone opined that only "misinformation and half-baked history" could justify that bellicose expression.[13] For his part, Noll confidently affirmed, "The war between science and religion is over. While not everyone yet knows this, it is nonetheless true."[14] His evidence included a growing number of scientifically informed Christian authors and a host of deeply researched books that seemingly punctured the warfare trope once and for all, including especially James Moore's *The Post-Darwinian Controversies: A Study of the Protestant Struggle to Come to Terms with Darwin in Great Britain and America, 1870–1900* and the volume edited by David Lindberg and Ronald Numbers, *God and Nature: Historical Essays on the Encounter between Christianity and Science*.[15]

Little did we know. Yet despite the persistence of the idea that would not die, the book that lies before you once again shows how very, very inadequate this particular idea was, is, and probably always will be. Its pages are graced with another effort by Ronald Numbers to nuance the story along with a sampling (but only a sampling) of the now numberless scholars who seek a better way.

Dispensing with the warfare metaphor by no means resolves the admittedly difficult questions for those who think seriously about the natural world, about God, and about the relation of the two. Books like this one do, however, clear the smoke of a battle that has never really existed so that meaningful work can proceed.

NOTES

1. Steven Shapin, *The Scientific Revolution* (Chicago: University of Chicago Press, 1996).
2. Quoted in Priya Shetty and Andy Coghlan, "Royal Society Fellows Turn on Director over Creationism," *New Scientist*, September 16, 2008, http://www.newscientist.com/article/dn14744-royal-society-fellows-turn-on-director-over-creationism.html.
3. Quoted in Shetty and Coghlan, "Royal Society Fellows Turn on Director."
4. Geoffrey Cantor, "What Shall We Do with the 'Conflict Thesis'?," in Thomas Dixon, Geoffrey Cantor, and Stephen Pumfrey, eds., *Science and Religion: New Historical Perspectives* (Cambridge: Cambridge University Press, 2010), 294.
5. Thomas Henry Huxley, *Man's Place in Nature and Other Essays* (London: Dent, 1911), 301.
6. Jerry Fodor and Massimo Piatelli-Palmarini, *What Darwin Got Wrong* (London: Profile Books, 2010), xv.
7. Fodor and Piatelli-Palmarini, *What Darwin Got Wrong*, xv.
8. Daniel Dennett, "Fun and Games in Fantasyland," *Mind and Language* 23, no. 1 (2008): 25–31, on 26–27.
9. Jerry Fodor, "From the Darwin Wars." We are grateful to Professor Fodor for allowing us to see this unpublished piece.
10. Thomas Nagel, "A Philosopher Defends Religion," review of *Where the Conflict Really Lies: Science, Religion, and Naturalism* by Alvin Plantinga, *New York Review of Books*, September 27, 2012; Alvin Plantinga, "Why Darwinist Materialism Is Wrong," review of *Mind and Cosmos: Why the Materialist Neo-Darwinian Conception of Nature Is Almost Certainly False* by Thomas Nagel, *New Republic*, November 16, 2012.
11. For the importance of historical context to the understanding of science, see David N. Livingstone, *Putting Science in Its Place: Geographies of Scientific Knowledge* (Chicago: University of Chicago Press, 2003).
12. See David N. Livingstone, *Dealing with Darwin: Place, Politics, and Rhetoric in Religious Engagements with Evolution* (Baltimore, MD: Johns Hopkins University Press, 2014).
13. David N. Livingstone, *Darwin's Forgotten Defenders: The Encounter between Evangelical Theology and Evolutionary Thought* (Edinburgh: Scottish Academic Press; Grand Rapids, MI: Eerdmans, 1987), 1.
14. Mark Noll, "The War Is Over," *Reformed Journal*, August 1986, 4.
15. James R. Moore, *The Post-Darwinian Controversies: A Study of the Protestant Struggle to Come to Terms with Darwin in Great Britain and America, 1870–1900* (Cambridge: Cambridge University Press, 1979); David Lindberg and Ronald Numbers, eds., *God and Nature: Historical Essays on the Encounter between Christianity and Science* (Berkeley: University of California Press, 1986).

CHAPTER ONE

The Warfare Thesis

LAWRENCE M. PRINCIPE

Although rejected by every serious historian of science today, the "conflict" or "warfare" model for the historical interaction of science and religion remains not only widespread but naturalized as a "fact" in popular culture. Like a pernicious biological species introduced into a new environment, the conflict thesis has proven difficult to eradicate despite the continued efforts of the scholars best qualified for the task. Historians identify two late-nineteenth-century books as the chief vectors of the conflict thesis: John William Draper's *History of the Conflict between Religion and Science* (1874) and Andrew Dickson White's *A History of the Warfare of Science with Theology in Christendom* (1896). The melodramatized "history" and the dubious "facts" upon which these books rest are easy to point out and refute, and many historians have done so.[1] Others have explored the social, intellectual, and religious contexts that allowed the formulations of Draper and White to establish themselves so firmly despite their obvious historical and intellectual failings.[2] Somewhat less studied in detail are the *origins* of Draper's and White's notions. Was there a significant "prehistory" to the conflict/warfare thesis? What ideas did Draper and White draw on, and how do their accounts fit with the rest of their work? Draper and White are so routinely cited together as the originators of the warfare/conflict thesis that the significant differences in their attitudes and motivations are easily overlooked. Like the historical relationship between science and religion, the background to the theses of Draper and White turns out to be more complex (and thus more interesting) when closely examined in context.

Prehistory of the Conflict Thesis

A few limited claims that "religion" (broadly understood) opposed "science" (equally broadly understood) do predate the nineteenth century, but these claims differ essentially in character from the theses of Draper and White. They are invariably far more specific, restricted to particular events, groups, or persons rather than generalized diachronically or extended to religion or science as a whole. Early comments regarding the Galileo affair illustrate this divergent character. For example, the young Robert Boyle (1627–91) was in Italy when Galileo died in 1642. Upon his return to England, the twenty-one-year-old Boyle rather precociously began writing his autobiography, wherein he recalled that Galileo's "ingenious Opinions, perhaps because they could not be so otherwise, were confuted by a Decree from Rome; the Pope it seems presuming (and that justly) that the Infallibility of his Chaire extended to determine points in philosophy as in Religion; & loath to have the stability of that Earth question'd, in which he had established his Kingdome."[3]

Boyle voices here a typical English Protestant contempt for Catholicism, expressed in sarcastic references to papal authority and worldliness. His aborted autobiography contains other similar disapprobations of Catholic practices and morals (having no relation to science) as would have been expected of the young son of a Protestant noble family, perhaps especially one seated in Ireland.[4] As such, Galileo functions rhetorically as a convenient source of anti-Catholic polemic rather than as a general statement about science and religion, or even about science and Catholicism. In reality, Boyle was convinced of the mutual benefits that science and religion offer one another, and he wrote extensively on this issue, most notably in his *Christian Virtuoso* (1690). The book's subtitle neatly sums up Boyle's theme: "That by being addicted to Experimental Philosophy, a Man is rather Assisted, than Indisposed, to be a *Good Christian*."[5] John Milton also deployed Galileo's condemnation to criticize a parliamentary censorship law, thereby implicitly linking the affair to the suppression of free philosophical dissemination of ideas. Yet once again, this was a specific event applied rhetorically in a narrow context rather than any broad statement about science and religion generally.[6]

One potential eighteenth-century locus for the emergence of the warfare model is the influential "Discours préliminaire" by Jean le Rond d'Alembert (1717–83), written in 1751 for the *Encyclopédie* that he coedited with Denis Diderot (1713–84). The "Discours" sketches out the progress of *philosophie* (which includes the sciences) from the "barbarous" Middle Ages to d'Alembert's "enlightened" eighteenth century. Following a biting critique of Renaissance humanists for their addiction to antiquity, d'Alembert alludes to the abuses of "some theologians" who "feared or appeared to fear the blows that blind reason might deliver upon Christianity" or who "opposed the advancement of philosophy" in order to forestall criticism of personal opinions that they endeavored to elevate into dogmas. To these unworthy theologians, he adds others "just as dangerous" who desired to use religion "to enlighten us about the system of the world, that is to say about those matters that the Omnipotent expressly left to our disputes." D'Alembert then names "theological despotism" as the reason that the Inquisition "condemned a celebrated astronomer for having maintained the earth's motion and declared him a heretic." He concludes by lamenting how "the abuse of spiritual authority, joined with the temporal, forced reason into silence, and nearly forbade the human race to think," ending at last with a military metaphor, writing that "poorly instructed or ill-intentioned adversaries openly waged war upon philosophy [*faisaient ouvertement la guerre à la philosophie*]."[7]

Some articles of the *Encyclopédie* can be interpreted to exhibit a negative reading of the interaction of science and religion, although such expressions are not common and are more attributable to conventional Enlightenment rhetoric regarding the "darkness" of the Middle Ages and Scholasticism. To be sure, the eighteenth-century *philosophes*' negative, dismissive attitude toward (indeed lack of understanding of) the Middle Ages provided a *predisposition* toward the later conflict thesis. But d'Alembert's comments, like those of the *Encyclopédie*'s individual articles, are narrowly directed at a few specific incidents. His "Discours," for example, maintains the need for revealed religion "that instructs us about so many diverse matters" as a "supplement to natural knowledge."[8] It presents revealed theology as more valuable and rich than natural theology, since it "draws from sacred history a much more perfect knowledge of that Supreme Being" and defines the subject explicitly as "reason applied to revealed facts."[9] D'Alembert even underscores that he wishes to criticize only *some* theologians, not "the entire respectable and highly enlightened body" of them. (This qualifying sentence was deleted from the 1764 edition and thereafter, perhaps indicating a move to a stronger position.) Thus while it is possible that the "Discours" and *Encyclopédie* may have supplied some general notions or inclinations toward the construction of a warfare model—d'Alembert's choice of the word "war" might be the earliest such usage—these eighteenth-century claims remain far more restricted than those later expressed by Draper or White.

Andrew Dickson White

Historians often treat Draper first and White second, following the chronology of their major works on the subject—Draper's 1874 *Conflict* and White's 1896 *Warfare*—yet White's opening volley actually preceded Draper's, although White continued to expand his attacks over a longer period. Interestingly, both Draper and White first formulated and published their views within the same short interval—from 1869 to 1873.

White first aired his warfare rhetoric on December 17, 1869, in a lecture titled "The Battle-Fields of Science." This lecture was delivered as the opening installment of a "winter series" of lectures presented at Cooper Union in New York City under the aegis of the American Institute of the City of New York.[10] The text was published the next day in the *New York Daily Tribune* (whose editor, Horace Greeley, was both president of the American Institute and a trustee of Cornell University where White was president).[11] Shorter accounts of the lecture appeared in other publications (e.g., *Scientific American*), and White repeated the lecture at Boston, New Haven, Ann Arbor, and as a Phi Beta Kappa lecture at Brown University.[12] In somewhat expanded form and replete with footnotes, it was published as "The Warfare of Science" in *Popular Science Monthly* in early 1876, and then as a slim volume later that year, which was republished a few months later in London (with a preface by John Tyndall).[13] White fell relatively silent on the matter for much of the following decade, although the book was translated and published in several languages. Shortly after resigning as president of Cornell in June 1885, White again took up the cudgels, and from 1885 to 1895 provided *Popular Science Monthly* with a further *twenty-five* articles, all titled "New Chapters in the Warfare of Science." He cobbled these articles together in 1896, over twenty-five years after his opening salvo, into the ponderous two-volume *A History of the Warfare of Science with Theology in Christendom.*

Modern historians have regularly seen the *Warfare* as a response to criticism from clergy and denominational colleges aimed at White during his work to found Cornell University as a nonsectarian institution. White acknowledges as much in his 1905 autobiography and in the preface to the 1896 *Warfare.* Indeed, his 1869 lecture ends with a rather peevish allusion to these events.

> . . . [I]n concluding, I might allude to another battle-field in our own land and time. I might show how an attempt to meet the great want of this State for an

> institution providing scientific and modern instruction has been met with loud outcries from many excellent men who fear injury thereby to religion. . . . I might show how it has been denounced from many pulpits, and in many sectarian journals. . . . I might show how, as the battle has waxed hotter, the honored founder of the institution . . . has been charged with "swindling the colleges of the State,". . . . I will not weary you with so recent a chapter in the history of the great warfare extending through the centuries.[14]

However, White's depiction of his offensive as a response to ignorant fears and declamations by backward clergy against enlightened education masks the real story. In fact, his characterization of the relationship between science and religion as an epic and immortal battle has made his explanation for the origin of his *Warfare* seem more plausible than it ever should have done. The golden apple of discord that set White's warfare in motion was actually the Morrill Land-Grant Act of 1862. This law gave large tracts of federal land to each state, which could be sold for financing colleges that would teach (among other things) agriculture and engineering. Choosing among many claims from competing colleges, the New York legislature voted in 1863 that New York's share of the funds would go to People's College in Havana (now Montour Falls), an innovative if yet embryonic institution intending to specialize in the technical subjects the Morrill Act was meant to support.[15] In order to receive the funds, People's College had to meet certain conditions within three years. After a year, the struggling college had made little progress in this direction, and a new idea—proposed by Senator Ezra Cornell—was to split the proceeds from sale of the land between People's and the (equally financially shaky) State Agricultural College at Ovid, one of the former competitors for the funds. At just this time, White was elected to the New York Senate and made head of its education committee. He argued forcefully against dividing the money and killed the bill, proposing instead that funds not be given to any existing college, but used to found a new one. After further negotiations, White—eventually in concert with Ezra Cornell—introduced a bill to this effect. This bill also specified that the new institution's board of trustees would not have a majority "of any one religious sect, or of no religious sect" and that positions would be open to all regardless of religious affiliation or lack thereof.[16] Unsurprisingly, once the Land Grant funds seemed up for grabs again, colleges across the state—all of them denominational, as was then the norm for American colleges—lobbied for their share of the funds and opposed White's bill. After violent and contentious debate, legal maneuvering, and backroom horse trading, the White-Cornell bill passed the State Assembly in 1865 and was signed by the governor.

Following such fractious dealings, it is not difficult to see why the disgruntled losers would continue to voice their criticisms, and thereafter harbor (and express) suspicions of impropriety when one senator who introduced and fought for a bill to create a new university left the Senate to become its first president, and the other had it named after him and obtained exclusive control over the sale of the federal land.[17] To what extent religious sectarianism really played any significant role in the customarily contentious matter of squabbling over federal funds is open to debate. It seems extremely unlikely that White was accurate in attributing it all to reactionary sectarian zealotry against the nonsectarian nature of Cornell University. Such emotionally loaded attacks may have been little more than convenient clothing for financial and personal grievances and competition over students, and in fact much of the criticism focused on wholly nonreligious themes like financial and administrative improprieties.[18] White's return to the subject in 1885 after a decade-long hiatus may have been provoked by renewed criticism of Cornell University and perhaps most of all by his repeated clashes with Henry Sage, chairman of the Cornell board of trustees, regarding the religious character of the institution.[19]

There is yet more historical context. White's opposition to sectarianism certainly predated the establishment of Cornell. It may have derived from Ezra Cornell himself, who had been expelled from the Quakers for marrying outside the sect, and it followed closely the views of Henry P. Tappan (1805–81), first president of the University of Michigan, who had hired White as a professor in 1857, and whom White acknowledged as a major formative influence. Tappan envisioned a new sort of American college akin to German models and strikingly similar to what White would later desire for Cornell, and he found himself frequently opposed by the older, smaller (and of course religiously based) colleges of Michigan when trying to secure funding from the legislature.[20]

The military metaphor that White applied so consistently (and melodramatically) to the relationship between science and religion already appeared in his earlier writings on other topics. Judging from the title, one might guess that his 1866 Yale lecture, "The Most Bitter Foe of Nations and the Way to Its Permanent Overthrow," whose topic he described as a "sacred struggle and battle of so many hundred years," concerned sectarian religion as well.

Nevertheless, White's "most bitter foe" in 1866 was aristocracy. Yet this lecture's military language, its tone and concatenation of dubious historical "facts," recur in strikingly similar form in the later *Warfare of Science*. Absent however is any criticism of religion—quite the opposite, White argues here that Italy's future lies in balancing power between church and state, and claims that "ecclesiasticism" in Spain shielded the lower classes from the aristocracy.[21] The same militaristic language occurs in White's account of the legislation to found Cornell. He called his bill a "signal for war," described the other senators and colleges as the "enemy," and dramatized the whole process as a succession of "struggles," "new dangers," and "fresh enemies."[22] White's pen seems to have been ever ready with martial metaphors.

White's warfare model for science and religion thus resulted from a range of personal issues and experiences, and *not* from historical evidence. He first confected the warfare model as a response to personal criticism, and then retailored or cherry-picked historical events to suit it. Indeed, an early response to the 1896 version now appears to have been substantially correct in stating that "after all its growth in size, the same judgement must be passed upon the completed work: it is still an 'indiscriminate and uncritical agglomeration of facts,' brought together for the support of a thoroughly one-sided and fatally misleading proposition . . . a piece of special pleading like this was not worth writing."[23] Some of White's friends remarked that his customarily tart replies to criticism were overkill—"using a triphammer to crush a midget," as one expressed it.[24] What remains difficult to fathom is how White, a president of the American Historical Association, justified to himself his grotesquely unhistorical (mis)use of sources—from resorting to the romanticized fictions of Washington Irving for "facts" about Columbus to the strategic truncation of a quotation from St. Augustine to make the African Doctor appear to say something exactly opposite to what he meant.[25] Did he believe, lawyer- and politician-like, that historical accuracy could be legitimately subjugated to the goal of winning the argument at hand? Whatever the cause, and despite the repeatedly noted failures of his evidence and conceptions, White's battle cry has continued to characterize and straightjacket public understanding of the science-religion issue to the present.

John William Draper

John William Draper's involvement in the science-religion issue was of shorter duration than White's, and although its background also included personal experiences, it grew out of a much more uniform and clearly enunciated philosophical position. Draper built up this position, which he believed unified and explained all of history, in lengthy works that preceded his 1874 *Conflict between Religion and Science.* The *Conflict* has not generally been fully situated in the context of these earlier works; looking at the totality of Draper's oeuvre provides crucial insight on his best-known work.

Draper was born in England, the son of a Methodist minister; he immigrated to the United States in 1832 to take up a job as a professor, first at Hampton-Sydney College in Virginia and later, in 1839, as professor of chemistry and then of physiology at New York University. In 1876, he became the first president of the American Chemical Society.[26] His numerous scientific papers during the 1830s dealt primarily with chemistry. In the 1840s, his attentions shifted to plant and animal physiology. Although he showed some early interest in history, it was only in the late 1850s that he turned almost exclusively to historical topics. His major publications in this area were the *History of the Intellectual Development of Europe* (1863) and *A History of the American Civil War* (1867–70); between these two works he also published *Future Civil Policy of America* (1865), which grew out of a series of invited lectures. At first glance, Draper's wide-ranging publications might seem eclectic, but they form a coherent philosophical trajectory. The *Conflict* is both an extension of this trajectory and, in important ways, a deviation from it.

Draper's thought can be characterized as an obsession with law, and his sequential interests represent an expanding campaign of hegemonic reductionism. The beginning of the sequence appears in Draper's first sustained foray beyond chemistry. His 1844 *Treatise on the Forces Which Produce the Organization of Plants* proposes that the growth of plants "unquestionably depends upon the laws of physics" and thus refutes the possibility of any guiding vital force: "Vegetable and Animal Physiology are to have their foundations laid on Chemistry and Natural Philosophy, the only basis which can elevate them from their present deplorable condition to that of true sciences." Taking the obedience of celestial motions to a single common law where "no invisible or extraneous hand ever intervenes" as his model, Draper asserts that all of nature is likewise controlled by uniform law. He does not rule out the Deity, but while he states that the succession of terrestrial life-forms springs "from the operations of the same Intelligent Mind," it is clear that Draper's conception of this "Mind" is that of a disinterested primordial lawgiver, not the providential God of revealed religion.[27] Strikingly, hints of Draper's future extension of his metaphysical assumptions are already present here in 1844, for he claims that the life even of human "individuals and races is completely determined by external conditions." A supposed universal "law of progress and of evolvement" governs everything, bringing about, as one example, the extermination of Native Americans: "the race, like each individual of it, submits in silence to an irreversible doom."[28] Such notions are far indeed from the expressed subject of the book—the effects of light and water on plant growth—but accurately delineate the direction of Draper's authorship over the next thirty years, suggesting that he had this progression already in mind during the early 1840s.

The same notions inform Draper's 1856 textbook on human physiology. While the book's intended theme is the parts and functions of the human body, a constant refrain throughout is how external influences, particularly climate, cause "metamorphoses" of the human body, and thence of human societies. The human body and human society thus both have physiologies manipulated by natural laws. What makes a sudden appearance here is the explicit influence of the positivism of Auguste Comte. Claiming that all systems of metaphysics have fallen into disarray, Draper asserts the need for a new guide for human thought and declares that this "guide is Positive Science."[29]

Draper's newfound devotion to positivism offered a blueprint for future authorship: "according to the methods of Positive Philosophy, there are but two classes of facts which can be admitted into our discussions respecting man, those furnished by his structure and function, and those gathered from his historical career."[30] *Human Physiology* expounds the first, and Draper's next two books would expound the second.[31] Here lies the rationale for Draper's shift from scientific to historical writing. In the second edition of *Human Physiology* (1858) Draper mentions that he has "nearly completed a volume that will serve as a companion to this, in which . . . the laws which preside over the career of nations [are] established." The guiding principle of that book—published in 1863 as *Intellectual Development of Europe*—is the "analogy between [man's] advance from infancy through childhood, youth, manhood, to old age and his progress through the stages of civilization."[32] This notion appears to adapt Comte's "Law of Three Stages" (theological-philosophical-scientific/positive) into what might be called Draper's "Law of Five Stages," where the inescapable influence of a uniform law—predominantly the effect of climate on physiology—governs the development of human individuals, human societies, and the human race as a whole.

Intellectual Development takes the reader through European history from ancient Greece to the nineteenth century. The single argument is that but one lesson is to be learned "from inquiries respecting the origin, maintenance, distribution, and extinction of animals and plants, their balancing against each other; from the variations of aspect and form of an individual man as determined by climate; from his social state, whether in repose or motion; from the secular variations of his opinions, and the gradual dominion of reason over society: this lesson is, that the government of the world is accomplished by immutable law."[33]

His later work on the American Civil War applies the latest chapter of human history to the same purpose, namely, to show how that conflict "exemplified the great truth that societies advance in a preordained and inevitable course."[34] The war was thus, according to Draper, an inescapable effect of physiological factors resulting from the operation of a universal natural law. The differing climates of North and South caused the metamorphosis of their respective inhabitants into different "types" that came, unavoidably, into conflict for supremacy. Future peace therefore requires a constant mixing of northerners and southerners to avoid the reemergence of divergent types—"that is the true method for combatting climate effects." Had there been more North-South travel in the 1850s, Draper concludes, "the civil war would not have occurred." Draper thus moves from explanations of the past to the promotion of social policies for the future.[35]

Because Draper was a stranger to the footnote and infrequently cites any authority by name, his sources are difficult to identify. One often wishes to know where he obtained the innumerable facts and figures he so confidently asserts, and from which he draws such dazzlingly arbitrary, and often contradictory, conclusions. For his general outlook, Comte was clearly a major influence; indeed, the final stage of Draper's journey fulfills Comte's desire for a "science of sociology" to govern societies. But Draper was not simply a Comtean positivist. Comte's Law of Three Stages restricted itself to describing the development of human society through the theological, philosophical/metaphysical, and positive/scientific phases. For Comte, this progression occurred through society's successive acquisition of knowledge and understanding. Draper is far more radical. Human actions count for little or nothing; we are all subjected to inescapable laws external to ourselves that predestine both individual and collective actions and outcomes. Likewise, Comte's ideas envisioned overall progress, but for Draper, the degradation of societies appears with equal emphasis, analogous to the old age and death of human beings. Most significantly, Draper's universal "Law of Development" applies equally to *everything*—from inanimate physical, chemical, and astronomical systems, through plants and animals, to human individuals, societies, and all of civilization as a whole. This astonishing reductionism points toward further sources. In terms of the development and decay of life-forms and the physiological influence of climate, he is clearly following the eighteenth-century theories of Buffon and Lamarck. His addiction to uniform law greatly resembles the natural philosophy of Robert Chambers's popular *Vestiges of a Natural History of Creation* (1844), which posits a similar guiding law of development of increasingly advanced systems and life-forms over the course of Earth's history. Indeed, Chambers summarizes the message of his book in terms similar to Draper's: "We have fixed mechanical laws at one end of the system of nature. If we turn to the mind and morals of man, we find that we have equally fixed laws at the other."[36] In terms of historical method, Draper's notions are closely aligned with those of Henry Thomas Buckle (1821–62), whose 1857 *History of Civilization in England* attempted to found a "science of history" wherein human societies are governed by fixed natural laws, not by human choices, free will being an illusion.[37] Nevertheless, Draper is not merely an imitator of Buckle, for the former expressed many of these ideas well before the latter published anything—yet Buckle undoubtedly provided further inspiration to Draper's program.

Two other midcentury authors might be expected to have exerted considerable influence on Draper. The first is Charles Darwin. This is all the more to be expected since Draper was present at the fabled 1860 Oxford debate over evolution between Bishop Wilberforce and Thomas Henry Huxley. Prior to their face-off, Draper read a summary of his physiological conception of history that would form the basis of *Intellectual History*; it was not well received.[38] Yet Darwinian evolution plays no role in Draper's evolutionary scheme. Natural selection is absent; there is no hint of a mechanism for the operation of Draper's universal law of development—human physiology simply "changes to get into correspondence with" external environments.[39] For example, Draper explains that the "red-haired, blue-eyed" northern Europeans described by Roman authors have largely disappeared not because of interbreeding or extermination, but rather, because "the red-haired man has himself been slowly changing to get into correspondence with the conditions that have been introduced through the gradual spread of civilization—

purely physical conditions with which the darker man was more in unison." His reference is to better clothing and heated housing, which meant exposure to higher average temperature, akin to that found naturally further south, where (of course) people have darker complexions. When *Intellectual History* was published in 1863, four years after *On the Origin of Species*, Draper nodded toward Darwin's work only to note that since 1858 he had made no changes to his book's "discussion of several scientific questions, such as that of the origin of species, which have recently attracted public attention so strongly." While certainly a preemptive expression of intellectual independence, it is also true that no Darwinian ideas appear in Draper.[40] A second plausible influence is Herbert Spencer who likewise envisioned an all-encompassing, evolutionary natural law, and endeavored to derive a "Synthetic Philosophy" from scientific principles, especially the "conservation of force." But Spencer published an outline of his ideas only in 1858, by which time Draper's ideas were already well-formed. Undoubtedly, once Spencer's views were published, they presumably served to support Draper's ideas. Chambers's *Vestiges* may represent a common inspiration for both, and Spencer, Buckle, and Draper may all draw on ideas fashionable in the 1850s for which it is difficult to assign a discrete origin.

Draper's oeuvre thus appears as a chain of largely consistent, but constantly expanding, argumentation running from the early 1840s until 1870. How then does the 1874 *Conflict* fit into this progression? Unsurprisingly, Draper's "Law of Development" undergirds the work. His obsession with law leads him already in the preface to apply what sounds like a modification of Boyle's Gas Law to the science-religion dynamic, summarizing the history of science as a "narrative of the conflict of two contending powers, the expansive force of the human intellect on one side, and the compression arising from traditionary faith and human interests on the other." He then contrasts "artistic" and "scientific" modes of writing history, averring that he will follow the latter, which shows the concatenation of "facts" and "sternly impresses us with a conviction of the irresistible dominion of law, and the insignificance of human exertions." In tones reminiscent of Dickens's Thomas Gradgrind, Draper wants a book with "every page . . . glistening with facts," and then promises a "clear and impartial statement" promising "not to advocate the views and pretensions of either party."[41] What follows is neither factual nor impartial.

Some of the book's historical material is cribbed from *Intellectual History*, although there are significant differences, most notably in the book's melodramatic—often hysterical—tone. The *Conflict* has long been recognized as a specifically *anti-Catholic* rant. While Draper considers Protestantism the "twin-sister" of science (thus any persecution of science by Protestants comes solely from an "incomplete emancipation from Catholicism") and Islam to be the "Southern Reformation" carried out by a people ever benevolent to science and progress, Catholicism is the brutal enemy of science, progress, and civilization. "Roman Christianity and Science are recognized by their respective adherents as being absolutely incompatible; they cannot exist together . . . mankind must make its choice—it cannot have both."[42] Any attempt at a "scientific" or impartial analysis of history is swept away by a torrent of inflammatory rhetoric built on cherry-picked, contorted, or simply fabricated "facts." This attitude—akin to the fulminations of the slightly earlier Know-Nothings—differs substantially in tone from Draper's earlier historical works, where scattered criticisms of Catholicism are often softened by words of praise for papal government and for monastic intellectual and humanitarian achievements.[43]

The causes for this shift are traceable to three diverse factors. The first, and most diffuse, was the growing American panic over Catholic immigrants rapidly filling the nation's cities. Draper referred repeatedly to the "insidious agency" of immigration, the dangers of a "hybrid population" (which he identifies as the cause for the collapse of ancient Rome through the pollution of the "pure Roman race" into an "adulterated festering mass") and the need to address the problem. He remarks specifically that Irish Catholic immigrants kept regions where they settled "in a lower intellectual state."[44] The exact extent to which such concerns engendered the rabidity of *Conflict* remains open to debate, but the extraordinary success and continued popularity of the book owed much to the well-documented anti-Catholic panic of late-nineteenth-century America. Second, Draper's family history unquestionably augmented his anti-Catholic sentiments. His father's conversion to Methodism caused his ostracism by the rest of the Draper family, which was Catholic. Draper's sister Elizabeth—who had emigrated with him from England—converted *to* Catholicism while in America, which, together with an incident involving a devotional book at the time of the death of one of Draper's sons, led to her being thrown out of the Draper household where she had been living.[45] Third, conciliar and papal declarations in 1864 and 1870 that responded to philosophical and political ideas that Catholic authorities considered contrary to religion undeniably helped drive Draper over the edge. In 1864, Pius IX issued a "Syllabus of Errors" listing philosophical positions considered erroneous and contrary to religion. These theses included such things as absolute rationalism and the denial of God's action in the world—two things deeply embedded in Draper's metaphysics. In 1870, the First Vatican Council spoke out against materialism, defined the notion of papal infallibility, reasserted the role of faith in human life and understanding, defended Christian education, and emphasized the role religion should play in governing and guiding society. Draper surely saw such statements as directly opposed to his devotion to positivism and as challenges to his deterministic "Universal Law of Development."

Additionally, by asserting the teaching role of the church and denying that scientific reasoning could answer *all* questions, these declarations probably affected Draper viscerally by placing limits on scientists' authority to speak on *all* matters whatsoever. Of course, as James Moore points out in detail, Draper completely ignores the historical context surrounding these events (for example, Vatican I as the conservatives' response to ultramontanism) and Pius IX's notable embrace of new science and technology, and concludes that Draper's response, if made to midcentury pronouncements alone, was a "serious overreaction." One also wonders to what extent Draper imagined his attacks could "finish off" the Catholic Church, then deeply shaken by the contemporaneous seizure of Rome and papal lands, the virtual imprisonment of the pope, and the forced secularization of many of its Italian institutions.[46]

It would be a serious mistake to see Draper as simply "antireligious"—a "scientist" in favor of scientific ideas. Draper's *Conflict* implicitly and repeatedly promotes a religion of his own devising, a theology opposed to Christianity—and to Roman Catholic Christianity in particular. His theology has several clear features, all of which derive from his starting metaphysical assumption of the all-encompassing, immutable nature of law. Draper's religion is overtly anti-Trinitarian; he claims that the doctrine of the Trinity was introduced from Egyptian paganism (as was veneration of the Virgin). Hence he labels Islam as the "Southern Reformation," for supposedly introducing monotheism in opposition to Christianity, which he implicitly casts as polytheistic.[47] Given his obsession with the unswerving operation of universal law, miracles are forbidden (presumably including that of the Incarnation, hence perhaps a foundation of his anti-Trinitarianism). By forbidding God from ever intervening in the world, an act that would upset the metaphysics of inviolate law, Draper not only creates an entirely nonprovidential God, but effectively deifies his supposed "universal law." For this reason he praises the "fatalism of the Arabs" as insightful resignation to the absolutism of impersonal law. Draper's great hero and religious teacher is Averroes (or rather a caricature of Averroes). He defends the Averroistic doctrine of emanation and absorption of the soul, which he supports with appeals to the "conservation of force," a link that surely indicates a dependence on Spencer.[48] He condenses virtually the entire activity of the Spanish Inquisition to suppressing Averroistic ideas—which he equates with scientific ideas—and notes that "orthodox" Muslims and Jews collaborated in this suppression, thus bringing all three religions under his criticism. Notably, he fulminates repeatedly against Vatican I, specifically for condemning what he sees as specifically Averroistic doctrines—pantheism, emanation and absorption, and so forth.[49] In short, Draper champions a deterministic, pantheistic, pseudo-Averroistic religion as the one true faith, and does not hesitate to condemn in the roundest terms any—for example, the Catholic Church—who dare object to it. In the end, his book is not so much about a conflict between "science" and "religion," but rather about a conflict between Draper's religion and all others.

A few comments in regard to the book's reception help better contextualize it by showing that, regardless of the subsequent popularity of the warfare/conflict thesis, not all readers were impressed, and not only those writing for religious journals. *Scribner's Monthly*, for example, concluded that "a more hasty, pretentious, incorrect work, claiming the title of 'history,' has seldom fallen into our hands." The modern historian will agree with the reviewer's assessment that the work is "swarming with statements, either positively erroneous, or put in such a form as to be misleading."[50] Some of the most trenchant criticism comes from John Fiske, who notes that *Conflict* is essentially "a repetition *en petit*" of the earlier *Intellectual Development*, and that book was little more than "a succession of pleasant though thread-bare anecdotes" with a "great quantity of superficial and second-hand information." After noting Draper's strangely one-sided adulation of Islam, Fiske observes that *Conflict* depends on a "very superficial conception" of religion, for "the word 'religion' is to him no more than a symbol that stands for unenlightened bigotry or narrow-minded unwillingness to look facts in the face."[51] Fiske had previously reviewed Draper's other books in the *North American Review*, eloquently revealing their flaws of both fact and interpretation. But perhaps most interesting in terms of understanding the future popularity of the *Conflict*, are his comments about Draper's *Civic Policy*. While he notes that Draper is "haunted by the idea that there is a particular fatality in climates" and that "his scientific facts are . . . too frequently obscured by paradoxes, and his historical facts by doubtful theories," he still expects the work will be well-received. He explains this contradictory judgment by saying that "Draper, like Mr. Spencer, is a popular writer, and interests us by nearly the same means which have heretofore entertained us in treatises on Natural Theology. The interest is nearly the same, whether the lesson be on Divine Providence or on the force of an inscrutable and irresistible fate. The main interest is in the facts of science and the narratives of history."[52]

Common Ground: Edward L. Youmans

Although Draper and White arrived at their theses by different routes, their final products bear some significant similarities, and they share some sources of inspiration. White cites the influence of Buckle and Spencer—authors clearly also of significance for Draper—and mentions Draper in the same breath, although certainly referring to the historical works that predated the *Conflict*.[53] A more significant commonality was the means by which their ideas were diffused. Draper was commissioned to write *Conflict* by the science popularizer Edward L. Youmans (1821–87) as part of his International Scientific Series. The preface to *Conflict* appeared before the book's publication as an article titled "The Great Conflict" in *Popular Science Monthly*, the same magazine that featured the first publication and later serialization of White's *Warfare* articles, and which, significantly, was founded and run by Youmans.[54] White's final 1896 *Warfare* was published by D. Appleton & Company, the same publishing house that produced Youmans's International Scientific Series and that was run by Youmans's friend William H. Appleton, who frequently took Youmans's advice on what to publish. Youmans thus emerges as a key player in the diffusion and popularization of the myth of the warfare of science and religion.

A glance over the articles published in Youmans' *Popular Science Monthly* reveals a consistent view on science, society, and religion akin to the notions spread by Draper and White. The magazine's very first article was by Herbert Spencer, another friend of Youmans, followed by twelve further articles on the "Study of Sociology," and many other contributions from Spencer. Articles on science and religion, agnosticism, evolution, and materials from or about Huxley, Tyndall, and of course White and Draper were standard fare for the magazine throughout Youmans's editorship.[55] Youmans often used the pulpit of his "Editor's Table," featured in every issue, to express views completely in line with those of Draper and White. Nor did he hesitate to publish there his own glowing review of Draper's *Conflict*, the volume he himself commissioned and published.[56] He also printed a spirited response to the book's critics. After noting with a publisher's pecuniary delight that the critics have "execrated it into about thrice the circulation that it would otherwise have had," Youmans chose to refute specifically criticism coming from "the outside sects—Jews, Unitarians, and Catholics, whom the orthodox repudiate as beyond the pale of Christianity, as knowing nothing of true religion." To their objection that there is no true conflict between science and religion, he replied with a stark binary opposition: either there is conflict or there is harmony, and since there exist books about reconciling science and religion, there must not be harmony, ergo, there is conflict, and ergo, Draper's book is a good and a needed one.[57] He publicized White's *Warfare* articles in the magazine in a piece he titled "The Conflict of Ages," where he also vigorously asserted his belief in an inherent and continuous conflict between science and religion.[58] What proportion of Youmans' advocacy consists of honestly held opinions and what proportion represents a publisher desirous of hawking his wares is matter for discussion; it seems unlikely that the latter motivation was entirely lacking. What Youmans's strong words force us to contemplate is how much of the conflict metaphor was of Draper's own invention, and how much was fed to him by Youmans. In any event, the emergence and the popularity of the warfare/conflict metaphor cannot be fully understood without the inclusion of Youmans's significant role in commissioning, publishing, translating, disseminating, and championing it.

Conclusion

The closer consideration of the background to Draper's and White's publications serves not only to contextualize them, but also to illuminate their authors' motivations and trajectories. In White's case, we find a person stung by criticism (and not just over sectarianism) regarding his actions in the founding of Cornell University and lashing out with an almost obsessive "over-the-top" response that quickly passed from defensive to offensive. Draper's case is more interesting, since his attacks form part of a coherent philosophical perspective developed over the course of thirty years, partly following Comte, partly striking out in his own directions. Significantly, his *Conflict* espouses not just a "religion of science" but in fact a religion of his own devising, replete with its own theology and metaphysics. The exercise of examining the roots of the *Warfare* and *Conflict* naturally renders the historical situation more complex, further ruling out the oversimplification of narratives about a supposed polarity between science and religion. This result is in turn an advantageous development in undercutting the erroneous notion that the warfare/conflict metaphor is an accurate or acceptable description of the interactions of science and religion.

NOTES

1. One survey of Draper and White appears in James R. Moore, *The Post-Darwinian Controversies: A Study of the Protestant Struggle to Come to Terms with Darwin in Great Britain and America, 1870–1900* (Cambridge: Cambridge University Press, 1979), 19–40; refutations of their claims appear in essays in Ronald L. Numbers, ed., *Galileo Goes to Jail and Other Myths about Science and Religion* (Cambridge, MA : Harvard University Press, 2009); and David C. Lindberg and Ronald L. Numbers, eds., *God and Nature: Historical Essays on the Encounter between Christianity and Science* (Berkeley: University of California Press, 1986).

2. For example, Frank M. Turner, "The Late Victorian Conflict of Science and Religion as an Event in Nineteenth-Century Intellectual and Cultural History," in Thomas Dixon, Geoffrey Cantor, and Stephen Pumfrey, eds., *Science and Religion: New Historical Perspectives* (Cambridge: Cambridge University Press, 2010), 87–110; Turner, "Victorian Conflict between Science and Religion: A Professional Dimension," *Isis* 69 (1978): 356–76; Bernard Lightman, "Victorian Sciences and Religions: Discordant Harmonies," *Osiris* 16 (2001): 343–66; and David Wilson, "Victorian Science and Religion," *History of Science* 15 (1977): 52–77.

3. Robert Boyle, *An Account of Philaretus* (1649) in Michael Hunter, *Robert Boyle by Himself and His Friends* (London: Pickering, 1994), 1–22, on 19.

4. See for example, Boyle, *Account of Philaretus*, 20–22.

5. Michael Hunter, *Robert Boyle: Between God and Science* (New Haven, CT: Yale University Press, 2009), 201–4; Hunter, "Science and Heterodoxy: An Early Modern Problem Reconsidered," in *Science and the Shape of Orthodoxy* (Woodbridge, UK: Boydell, 1995), 225–44; Jan W. Wojcik, *Robert Boyle and the Limits of Reason* (Cambridge: Cambridge University Press, 1997).

6. Maurice Finocchiaro, *Retrying the Galileo Affair* (Berkeley: University of California Press, 2005), 76–77.

7. Jean le Rond d'Alembert, *Discours préliminaire de l'Encyclopédie*, ed. Michel Malherbe (Paris: Vrin, 2000), 125; d'Alembert, *Preliminary Discourse to the Encyclopedia of Diderot*, trans. Richard N. Schwab with Walter E. Rex (Chicago: University of Chicago Press, 1995), 3–140. On d'Alembert's treatment of Galileo, see Finocchiaro, *Retrying the Galileo Affair*, 120–25.

8. D'Alembert, *Discours préliminaire*, 95.

9. D'Alembert, *Discours préliminaire*, 113.

10. Andrew Dickson White, "The Battle-Fields of Science," *Annual Report of the American Institute of the City of New York, 1869–70* (Albany, NY: Argus, 1870), 199–218.

11. "First of the Course of Scientific Lectures—Prof. White on 'The Battlefields of Science,'" *New York Daily Tribune*, December 18, 1869, 4.

12. Andrew Dickson White, *The Warfare of Science* (New York: Appleton, 1876), 5. The review in *Scientific American* praises the lecture but notes that White could barely be heard at the back of the room and that many attendees left shortly after it began (*Scientific*

American 22 [1 January 1870]: 8). The section on anatomy and medicine reappeared also in the *North American Journal of Homeopathy* 18 (1870): 421–23.

13. Andrew Dickson White, "The Warfare of Science," *Popular Science Monthly* 8 (February 1876): 385–409 and (March 1876): 553–70.

14. White, "Battle-Fields," 217–18; cf. *Warfare* (1876), 144–45 where the reference to Ezra Cornell is omitted, as is the earlier claim that Cornell's "Christian trustees and professors [are] earnestly devoted to building up Christian civilization."

15. Walter P. Rogers, "The People's College Movement in New York State," *New York History* 26 (1945): 415–46, esp. 434–46.

16. New York State Senate Bill 145, sec. 2 and 4, February 7, 1865.

17. The details of the founding of Cornell and the attendant criticism (from White's perspective) are given in his *Autobiography*, 2 vols. (New York: Century, 1905), 1:100–106, 294–323, 330ff.; see also George M. Marsden, *The Soul of the American University: From Protestant Establishment to Established Nonbelief* (Oxford: Oxford University Press, 1994), 113–22.

18. For the notion that religious controversy was a smoke screen for financial interests and for criticisms not based on denominationalism, see Samuel D. Halliday, *History of the Federal Land-Grant of July 2, 1862* (Ithaca, NY, 1905), esp. 16, 18.

19. The bibliography in White, *Autobiography*, 2:575–82, on 578 cites "*Letter Defending the Cornell University from Sundry Sectarian Attacks*, Elmira, December 17, 1884," but I have not been able to locate a copy of this putative publication. It may be related to White, *Address Delivered to the Students of the Cornell University, Friday 4 May 1883, in Reply to Certain Attacks upon the Institution* (Ithaca, [NY], 1883), which is not listed in White's bibliography; I have not yet seen the latter pamphlet. For clashes with Sage, see Glenn C. Altschuler, *Andrew D. White—Educator, Historian, Diplomat* (Ithaca, NY: Cornell University Press, 1979), 95–97, 133–45. For further causes, see Moore, *Post-Darwinian Controversies*, 38–40.

20. White, *Autobiography*, 1:271, 279–81; Altschuler, *Andrew D. White*, 80; Laurence Veysey, *Emergence of the American University* (Chicago: University of Chicago Press, 1965).

21. Andrew Dickson White, "The Most Bitter Foe of Nations and the Way to Its Permanent Overthrow" (New Haven, CT, 1866), 1, 8, 13.

22. White, *Autobiography*, 1:300, 316, and passim.

23. *Presbyterian and Reformed Review* 9 (1898): 510–12; the quotation within this passage was Princeton president James McCosh's remark regarding the 1876 version. Altschuler, *Andrew D. White*, 202–16 contains a useful sampling of responses.

24. Quoted in Altschuler, *Andrew D. White*, 95.

25. White, *Warfare* (1876), 19, 99–100.

26. For biographies of Draper, see Donald Fleming, *John William Draper and the Religion of Science* (Philadelphia: University of Pennsylvania Press, 1960); and George F. Barker, *Memoir of John William Draper, 1811–1882* ([Washington, DC?]: [National Academy of Sciences?], [1886?]).

27. John William Draper, *Treatise on the Forces Which Produce the Organization of Plants* (New York, 1844), v, vii, 1–4.

28. Draper, *Organization of Plants*, 7–8. In an earlier reference to native Americans, human action, not impersonal law, operates; see "Lecture: The Last of a Course of Lectures Delivered during the Years 1836–1837," *Southern Literary Messenger* 3 (1837): 693–98, where he refers to how "the Anglo-Saxon, the son of freedom, . . . has driven the red man from these forests. . . . Born the champion of freedom—the protector of science—from all points on the surface of the earth he is exercising a silent, but prodigious influence on the destinies of man. . . . [H]e is, as it were, the heart of the universe" (695–96).

29. John William Draper, *Human Physiology* (New York, 1856), v. Fleming, *John William Draper*, 49, suggests that Draper's acquaintance with Comte came indirectly through digests, rather than from reading Comte himself.

30. Draper, *Human Physiology*, vi.

31. The historical approach already fills the last two chapters of *Human Physiology*, 563–637.

32. Draper, *Human Physiology*, 2nd ed. (New York, 1858), viii.

33. John William Draper, *A History of the Intellectual Development of Europe* (New York, 1863), 15.

34. John William Draper, *A History of the American Civil War*, 3 vols. (New York, 1867–70), 1:iii.

35. John William Draper, *Thoughts on the Future Civil Policy of America* (New York, 1865), 83–87.

36. [Robert Chambers], *Explanations: A Sequel to Vestiges of a Natural History of Creation* (New York, 1846), 17; see also 19–20. Chambers moves from the individual to society using ideas of the Belgian statistician Quetelet; Draper does the same in *Human Physiology*, 15–16.

37. On Buckle, see Ian Heskith, *The Science of History in Victorian Britain* (London: Pickering and Chatto, 2011).

38. Despite Draper's later involvement with the science-religion issue, he never commented on the Wilberforce-Huxley debate and devoted almost none of the *Conflict* to Darwinian evolution.

39. Draper, *Human Physiology*, 591.

40. Draper, *Intellectual Development of Europe*, iv. In the preface, Draper cites his presence at the Oxford meeting, but only to mention his own lecture. On Draper's attendance at the meeting, see James C. Ungureanu, "A Yankee at Oxford: John William Draper at the British Association for the Advancement of Science at Oxford, 30 June 1860," *Notes and Records of the Royal Society of London* 70 (2016): 135–50. In 1877, Draper lectured on evolution to Unitarian ministers; he covered Darwin in less than half a page and criticized natural selection by saying that nature never selects, but only obeys law. He spent more time on Lamarck's superiority to Cuvier; "Dr. Draper's Lecture on Evolution," *Popular Science Monthly* 12 (1877): 175–92.

41. John William Draper, *History of the Conflict between Religion and Science* (New York, 1874), vi, ix, xi–xii.

42. Draper, *Conflict*, 68ff., 353, 363–64.

43. E.g., Draper, *Civic Policy*, 316.

44. Draper, *Civic Policy*, 110–12, 150–51, 165, 241; *Civil War*, 182. In the latter work, Draper also claims that immigration retards the natural process of physiological metamorphosis that can be attributed to climate, leading to increased mortality (98, 178).

45. Fleming, *John William Draper*, 31, 143–44. The incident involving Elizabeth occurred in 1853, so it would be too early to be, on its own, the major trigger for Draper's vitriol of the early 1870s.

46. Moore, *Post-Darwinian Controversies*, 25–29. It should be noted that Pius IX reestablished the Accademia dei Lincei, a scientific academy to which, ironically enough, Draper would be elected.

47. Draper, *Conflict*, 47–48, 53–56, 68, 84.

48. Draper, *Conflict*, 106–7, 126–28, 138–40, 210, 358.

49. Draper, *Conflict*, e.g., xiv, 141–51, 357–58.

50. *Scribner's Monthly* 9 (March 1875): 635–37.

51. [John Fiske], "Draper's Science and Religion," *Nation* 21 (November 11, 1875): 343–45. Identification of the anonymous reviewer is given in Fleming, *John William Draper*, 194.

52. John Fiske, *The North American Review* 101 (October 1865): 589–97, on p. 597. His review of Draper's *Civil War* is 105 (October 1867): 664–70, and his review of *Intellectual Development of Europe* is 109 (July 1869): 197–230.

53. White, *Autobiography*, 42. Draper's son John Christopher gave the fourth lecture in the series at Cooper Union that White opened with his "Battle-Fields of Science."

54. John William Draper, "The Great Conflict," *Popular Science Monthly* 6 (December 1874): 227–32.

55. See Bernard Lightman, "Spencer's American Disciples: Fiske, Youmans, and the Appropriation of the System," in Bernard Lightman, ed., *Global Spencerism: The Communication and Appropriation of a British Evolutionist* (Leiden: Brill, 2015), 123–48; and also William E. Leverette Jr., "E. L. Youmans' Crusade for Scientific Autonomy and Respectability," *American Quarterly* 17 (1965):12–32; Charles M. Haar, "E. L. Youmans: A Chapter in the Diffusion of Science in America," *Journal of the History of Ideas* 9 (1948): 193–213; and John Fiske, "Edward Livingston Youmans," *Popular Science Monthly* 37 (1890): 1–18.

56. Edward L. Youmans, "Editor's Table," *Popular Science Monthly* 6 (1875): 371–72.

57. Edward L. Youmans, "Draper and His Critics," *Popular Science Monthly* 7 (1875): 230–33. See his lengthy commendation of the *Conflict* in "The Conflict of Religion and Science," *Popular Science Monthly* 6 (1875): 361–64.

58. Edward L. Youmans, "The Conflict of Ages," *Popular Science Monthly* 8 (1876): 493–94.

CHAPTER TWO

The Galileo Affair

MAURICE A. FINOCCHIARO

Galileo's Trial

In 1633, at the conclusion of one of history's most famous trials, the Roman Inquisition found Galileo Galilei (1564–1642) guilty of "vehement suspicion of heresy."[1] This was a specific category of religious crime intermediate in seriousness between formal heresy and mild suspicion of heresy. He had committed this alleged crime in a book that defended Nicolaus Copernicus's (1473–1543) hypothesis of the earth's motion and denied the scientific authority of scripture. The book, *Dialogue on the Two Chief World Systems, Ptolemaic and Copernican*, had been published in 1632.

The verdict was accompanied by several penalties: Galileo had to immediately recite an "abjuration" of his erroneous beliefs; the *Dialogue* was banned; he was condemned to house arrest until the end of his life; and he had to recite the seven penitential psalms once a week for three years.

This condemnation was the climax of a series of events starting in 1543, when Copernicus published his epoch-making book, *On the Revolutions of the Heavenly Spheres*. This book advanced an argument for the idea that the earth moves, with daily axial rotation and yearly heliocentric revolution; the argument amounted to showing that the known facts about the motions of heavenly bodies could be explained better in terms of the geokinetic hypothesis, as compared with the geostatic, geocentric view.

Although novel and significant, Copernicus's argument was inconclusive. Moreover, the idea faced many powerful objections. In summary, the earth's motion seemed epistemologically absurd because it contradicted direct sense experience. It seemed empirically false because it had astronomical consequences that were not observed, such as the phases of Venus and the annual parallax of fixed stars. It seemed mechanically impossible because it

Copernican view; and an intelligent layperson, who wants to make up his mind after a critical scrutiny of the evidence. Galileo tried his best to carry out the evaluation fairly and validly. The arguments for the earth's motion turned out to be better than those against it. This was at worst an *implicit* defense of the *probability* of Copernicanism.

Galileo's gamble was that friendly church officials would not blame him for this type of discussion, because they would recognize that the defense was not explicit and absolute, and that he had acted within the spirit of Bellarmine's warning. His effort misfired not because it was foolhardy, but because in 1632 the special injunction came to the surface, and from its point of view any discussion of the earth's motion by Galileo was prohibited, whether or not it amounted to a defense. The book's publication thus led to a trial during which he denied receiving the special injunction, but admitted receiving and violating Bellarmine's warning, insisting that his violation was unintentional. This admission spared Galileo more drastic punishments, such as being burned at the stake.

While the 1633 condemnation ended the original controversy, it started a new one that continues to our own day. The subsequent Galileo affair concerns the facts, causes, responsibilities, implications, and lessons of the original. Some of the later developments involve actions by the church, such as the partial unbanning of Galileo's *Dialogue* and Copernican books in general by Pope Benedict XIV (in 1740–58); the repeal of the condemnation of the Copernican doctrine in 1820–35; the implicit theological vindication of Galileo's biblical hermeneutics by Pope Leo XIII's encyclical *Providentissimus Deus* (1893); and the attempted rehabilitation of Galileo by Pope Saint John Paul II (in 1979–92).

Original Affair: Conflict, Harmony, and Deep Structure

As traditionally interpreted, Galileo's trial epitomizes the conflict between science and religion. In other words, it has become the paradigmatic or iconic case of this conflict. In rhetorical terms, the trial has become a so-called synecdoche, i.e., a figure of speech that mentions a part to represent the whole. This interpretation has been advanced not only by relatively injudicious writers who have recently been widely discredited (for example, the nineteenth-century figures John Draper, a chemist and historical philosopher, and Andrew Dickson White, the first president of Cornell University), but also by such scientific, philosophical, and cultural icons as Albert Einstein, Bertrand Russell, and Karl Popper.[3]

However, the interpretation of the affair as conflict is just the beginning of the story; it should come as no surprise that the situation is more complicated. To begin with, at the opposite extreme, there is the revisionist thesis that the trial really shows the *harmony* between science and religion. This interpretation does call for some elaboration.

The most significant advocate of the harmony thesis is Pope John Paul II. In a speech to the Pontifical Academy of Sciences at the 1979 commemoration of the centennial of Einstein's birth, the pope expressed his regret for Galileo's suffering "at the hands of men and organisms of the Church," and quoted the Second Vatican Council's general condemnation of such interferences with freedom of speech and thought. He went on to give his support for new and deeper studies of the affair, conducted with a "loyal recognition of wrongs from whatever side they come." In his remarks, the pope also stressed that Galileo himself believed that religion and science cannot contradict each other; that his justification of this belief was essentially identical to that given by the Second Vatican Council; that he conducted his scientific research in the spirit of piety and divine worship which the same council recommended as exemplary; and that he formulated important epistemological norms about the relationship between science and scripture, which the church later recognized as correct. The pope summarized his own interpretation thus: "[I]n this affair the agreements between religion and science are more numerous and above all more important than the incomprehensions which led to the bitter and painful conflict that continued in the course of the following centuries."[4]

This speech was not an accidental pronouncement; it began a process that lasted thirteen years and was not concluded until 1992. Popular media and some scholars described this process as a "rehabilitation" of Galileo and an admission of error on the part of the church, which is a questionable characterization. A few critics have gone so far as to accuse the church of exploiting Galileo for its own self-rehabilitation and of constructing a new myth about him. Whatever the truth of such accusations, there is no question that this latest Vatican reexamination of Galileo's trial was, among other things, an attempt to elaborate the notion of harmony between science and religion.[5]

Let me now attempt some evaluation, beginning with the strengthening of John Paul's harmony thesis by reconstructing his argument more explicitly. The harmony interpretation starts by making a distinction between the Catholic religion as such on the one hand, and men and institutions of the church on the other. It then goes on to say that the injustices, errors, and abuses were committed by men and institutions for which they and not the church are responsible; thus, the conflict was between a scientist and some churchmen. Regarding the relationship between science and religion, the correct view is the one elaborated by Galileo himself, which the church later adopted as its own. That view says that God revealed himself to humanity in two ways, through his word and through his work. His word—namely, Holy Scripture—aims to give us information which we cannot discover by studying and examining his work, the physical universe and our interior selves. But to find out what his work is like, we need to observe it by using those parts of it which are our bodily senses and by reasoning about it with that other aspect of his work which is our mind. In short, scripture is only an authority on questions of faith and morals, not on scientific factual questions about physical reality.[6] In Galileo's trial, a key difficulty was the misunderstanding of these principles by the churchmen in power. Once these principles are clarified, as Galileo himself ironically contributed to doing, the conflict between science and religion evaporates and subsists only in the imagination of people who do not know better.

However, these considerations, although helpful, cannot be the end of our analysis. Again, the situation is more complicated. Let's focus on the Inquisition's sentence that concluded the 1633 trial. The sentence condemned Galileo for believing two things, namely, that the earth moves and that scripture is not a scientific authority. Thus, the controversy involved at least two issues. One was the scientific issue of the location and behavior of the earth in physical reality; the other was the philosophical, methodological, theological, and hermeneutical question of the relationship between astronomical science and scripture. This second issue involved a disagreement between those (like Galileo) who held, and those (like the Inquisitors) who denied, that it is proper to defend the truth of a physical theory contrary to scripture. That is, if in this controversy we take Copernicanism to represent science and scripture to represent religion, then Galileo was the one claiming that there is no real incompatibility between the two (because scripture does not aim to convey scientific information), whereas the Inquisition was the one claiming that the apparent conflict between Copernicanism and scripture was real. It follows that there was an irreducible conflictual element in Galileo's trial, between those like Galileo, who believed that there is no conflict between scripture and science, and those like his Inquisitors, who believed that there is a conflict. The irony of the situation is that it was the victim who held the more fundamentally correct view. However, insofar as that nonconflictual view is the more nearly correct one, then the content of that view suggests a minor harmonious element in the trial.

Furthermore, although the conflict was real at the level of what I would call the surface structure of the situation, there is a deeper aspect that needs to be taken into account. That is, Galileo was not the only one who held that there was no conflict. Many of those who agreed with him on this question of principle were themselves churchmen. For example, such was the receptive correspondent of the letter in which Galileo refuted the biblical argument against Copernicanism and which started the Inquisition proceedings, namely Castelli, a Benedictine monk; such also was the author explicitly condemned in the Index's anti-Copernican decree (for arguing that Copernicanism is compatible with scripture), namely, Foscarini, a Carmelite friar; and such was the author of the first published apology of Galileo, namely, Tommaso Campanella (1568–1639), a Dominican friar.[7]

In other words, in Galileo's time, there was a division within the church between those who did and those who did not accept the scientific authority of scripture. A similar split existed in scientific circles. A further division existed with regard to the other main issue of Galileo's trial, the physical proposition of the earth's motion. Thus, rather than having an ecclesiastic monolith on one side clashing with a scientific monolith on the other, the real conflict was between two attitudes that crisscrossed both.[8]

What shall we call these two sides? How shall we conceive of them? I believe the most fruitful way of describing them is to label them conservatives or traditionalists on one side and progressives or innovators on the other. The deep-structural conflict was between these two groups. In this sense, Galileo's trial illustrates the conflict between conservation and innovation and involves an episode where the conservatives happened to win one particular battle. This conflict is one that operates in such other domains of human culture as politics, art, economy, and technology. It cannot be eliminated on pain of stopping cultural development; it is a moving force of human history.[9]

However, this is not to say that the outcome is predictable, predetermined, or inevitable. Moreover, I don't want to give the impression that the task of detecting the conservative and innovative elements is trivial. There are problems with this task. For example, historical agents do not come with these labels attached to them; and even if they did, the scholarly challenge would be to ascertain whether such descriptions are accurate. Furthermore, innovators often defend their novelties by arguing that they are rooted in tradition; Galileo did this in his *Letter to Christina*, by basing his hermeneutical principles partly on the views of St. Augustine and other church fathers. Conversely, conservatives often oppose innovations by arguing that the alleged novelties are really old ideas discarded long ago; this was a common argument of the anti-Copernicans, who claimed that the earth's motion was a Pythagorean idea that had been refuted by Aristotle (384-322 BC) and Ptolemy (about AD 100-178). Finally, conservation and innovation are relative concepts whose application depends on the selection of a relevant historical period; thus, Galileo's opponents argued in part that his hermeneutical ideas contradicted the newest church policies stemming from the Council of Trent (1545–63), and so represented a conservative throwback to medieval traditions.

These problems do not imply that the dialectic between conservation and innovation is a useless idea. Rather, they reinforce the need to study it concretely and contextually, without losing sight of its intended function: to help us explain the facts of the surface structure.

Subsequent Affair: Conflict, Myth, and Deep Structure

However, again, the story, or rather the analysis, cannot end here, not yet. For, in Galileo's case, there are also complications stemming from what happened subsequently. After Galileo's condemnation in 1633, the interpretation and evaluation of the trial became a cause célèbre in its own right, which continues to our day. Even those who advocate the harmonious account of the original trial do not deny that the key feature of the *subsequent* Galileo affair is indeed a conflict between science and religion. In fact, as we saw earlier, Pope John Paul believed that the lesson from Galileo's original trial is the harmony between science and religion, and he wanted to stress this lesson in order to put an end to the subsequent, very real, but presumably unjustified, science-versus-religion conflict. Thus, we need to look more closely at the subsequent controversy.

Let's begin by adding some nuances to the view of the subsequent affair as the story of a science-versus-religion conflict, namely, the story that in the past four centuries the church has had to face criticism and difficulties because of the condemnation of Galileo. The first refinement is the following.

It may very well be true, as Pope John Paul and others have advocated, that such difficulties are a thing of the past and there is no longer any good reason for them to continue. According to their argument, during the

Enlightenment the view emerged that Galileo's trial embodied the inherent incompatibility between science and religion, and later this view became widely accepted.[10] But this view was the result of inadequate historical knowledge about the trial and of ideological biases. For example, it overlooks the pro- and anti-Galilean split within Catholicism at the time of the original affair, described earlier. It also overlooks the crucial fact that despite the practical opposition Galileo experienced, at the reflective level he himself believed in the harmony between science and religion. And the view also presupposes the Platonist principle that science and religion are eternal unchanging self-subsisting entities which by definition have a nature that places them at war with each other, rather than being historical dynamic entities that are sometimes at war and sometimes in harmony.

However, although this pro-harmony argument is important (for a full understanding of the *original affair*), it does not undermine the essential correctness of the *subsequent conflict*, for several reasons. First, the view of Galileo's trial as epitomizing the conflict between science and religion was not an Enlightenment invention, but started being developed immediately after the 1633 condemnation; this happened when (1633–36) an international group of liberal-minded secularists translated into Latin and published in Strasbourg Galileo's banned *Dialogue* and the incriminating *Letter to Christina*.[11] Second, even if the conflictual view of Galileo's trial were incorrect, and a thing of the past that should now be replaced by the harmony view, it would be naive to deny the reality of the historical fact that for about four centuries such an incorrect view has been the most popular interpretation of the episode. Third, although the Platonist, static conception of science and religion may be inadequate, and so science and religion are not necessarily and always in conflict, and so there are many episodes when they were in harmony, Galileo's case may very well be one of those where science and religion happened to be in conflict. In fact, and this is a fourth point, as we saw earlier, Galileo's trial does exhibit such a conflict if we take science to be represented by Copernicanism and religion to be represented by scripture, and if we focus on the disagreement between Galileo and his opponents over whether Copernicanism is contrary to scripture.

Next, we need to add another qualification, involving a contrast between the original and the subsequent Galileo affairs. First let's understand that the conflict between science and religion is a striking feature of both the original and the subsequent Galileo affair: in the original episode it takes the form of Copernicanism versus scripture; in the subsequent controversy, it takes the form that Galileo's trial was widely perceived to epitomize the conflict between science and religion. The important difference involves the deep structure that underlies them. With regard to the original affair, as we have already seen, that deep structure is the conflict between conservation and innovation, which generated the split within the church (and within Protestantism, and within astronomy and natural philosophy) about Copernican astronomy vis-à-vis scriptural interpretation. Regarding the subsequent controversy, the deep structure lies in the conflict between myths and facts, that is, the phenomenon of the rise, evolution, and fall of cultural myths; for the trial became a great occasion for mythologizing not only (more obviously) on the part of anticlerical and pro-Galilean elements, but also (somewhat reactively) on the part of pro-clerical and anti-Galilean forces.

In fact, mythmakers on both sides have been busy for four centuries. Let me mention some of the more iconic, emblematic, and quotable myths.

On the anticlerical side, in the *Areopagitica* (1644) John Milton, while recalling his visit to Tuscany in 1638–39, expressed one of his most memorable impressions thus: "[T]here it was that I found and visited the famous Galileo grown old, a prisoner to the Inquisition, for thinking in astronomy otherwise than the Franciscan and Dominican licensers thought."[12]

In the *Essay on the Customs and Spirit of Nations* (1753), Voltaire opined that "in 1616, a congregation of theologians declared Copernicus's opinion . . . 'not only heretical in the faith, but also absurd in philosophy.' This judgment against a truth later proved in so many ways is clear testimony of the force of prejudice. It should teach those who have nothing but power . . . not to interfere by deciding what is not within their jurisdiction. Then in 1633, Galileo was condemned by the same tribunal . . . and he was obliged to recant on his knees. In truth, his sentence was . . . disgraceful to the reason of the judges of Rome."[13]

In 1841, in a book widely circulated in French, Italian, and German, Guglielmo Libri concluded his account by saying: "The persecution of Galileo was odious and cruel, more odious and more cruel than if the victim had been made to perish during torture. For . . . they forbade him to make discoveries. . . . This ill-fated vengeance . . . deprived humanity of the new truths which his sublime mind might have discovered. To restrain genius; to frighten thinkers; to hinder the progress of philosophy; that is what Galileo's persecutors tried to do. It is a stain which they will never wash away."[14]

Finally, in 1953, Einstein, writing the foreword to an English translation of the book (*Dialogue*) that occasioned Galileo's condemnation, expressed this judgment: "[A] man is here revealed who possesses the passionate will, the intelligence, and the courage to stand up as the representative of rational thinking against the host of those who, relying on the ignorance of the people and the indolence of teachers in priest's and scholar's garb, maintain and defend their position of authority."[15]

Yet, on the other side of the question, there is no dearth of anti-Galilean mythmaking. Soon after the condemnation, there was an attempt to discredit Galileo's ideas by taking his abjuration at face value. A good example is provided by Alexander Ross in 1646, in the context of a controversy with John Wilkins over how many astronomers followed Copernicus. Ross believed that "of these five you muster up for your defense, there was one, even the chiefest, and of longest experience, to wit, Galileus, who fell off from you; being both ashamed, and sorry that he had been so long bewitched with so ridiculous an opinion."[16]

Later, in 1784 in France, Jacques Mallet du Pan started the myth that "Galileo was persecuted not at all insofar as he was a good astronomer, but insofar as he was a bad theologian."[17] The bad theology which Mallet misattributed to Galileo was the use of scripture to prove astronomical propositions—the opposite of what Galileo preached and practiced!

In an 1841 biographical work, David Brewster portrayed Galileo as a coward: "[W]hat excuses can we devise for the humiliating confession and abjuration of Galileo? . . . Galileo cowered under the fear of man, and his submission was the salvation of the church. The sword of the Inquisition descended on his prostrate neck; and though its stroke was not physical, yet it fell with a moral influence fatal to the character of its victim, and to the dignity of science."[18]

Then in 1908, the French historian, philosopher, and physicist Pierre Duhem (1861–1916) portrayed Galileo as a bad logician and epistemologist, claiming "that logic was on the side of Osiander, Bellarmine, and Urban VIII, and not on the side of Kepler and Galileo; that the former had understood the exact import of the experimental method; and that, in this regard, the latter were mistaken."[19]

Finally, in 1988, Paul Feyerabend (1924–94), an Austrian-American professor of philosophy at the University of California–Berkeley, claimed that "the Church at the time of Galileo not only kept closer to reason as defined then and, in part, even now; it also considered the ethical and social consequences of Galileo's views. Its indictment of Galileo was rational and only opportunism and a lack of perspective can demand a revision."[20]

What I claim is that, whereas the dialectic of conservation and innovation forms the deep structure underlying the science-religion conflict of the original affair, the just described two-sided mythmaking forms the deep structure underlying the science-religion conflict of the subsequent Galileo affair. Furthermore, although bilateral, it is obvious that such mythmaking is not otherwise symmetric, for the pro-Galilean and anticlerical myths tend to be aggressive, and the pro-clerical and anti-Galilean ones tend to be defensive. And this asymmetry can be contrasted with the conflict in the original affair, in which there was also an asymmetry, but in the opposite direction; that is, "science" (through Copernicus and Galileo) was the apparent victim, and "religion" (through the church) was the apparent aggressor. Additionally, besides the important substantive issues raised by the various views of Galileo's trial, the myths in question also deserve study from the point of view of the rise, evolution, and fall of cultural myths.

Of course, such a study must be balanced, judicious, bipartisan, and objective with respect to the pro- and anticlerical and pro- and anti-Galilean dichotomies. Admittedly, this is easier said than done, but I believe the task is not impossible. Moreover, it is absolutely essential to have a proper conception of what myths are; thus, it may be useful here to give such a definition, even though this merely scratches the surface of an iceberg.

First of all, I would distinguish among three different kinds of myths, in three different contexts. One type might be labeled "popular" myths, in the sense of involving primarily popular culture (the beliefs of the masses). Another kind might be labeled "religious" myths, insofar as they involve primarily doctrines and practices of organized religions. A third variety of myth might be called "cultural" myths, in the sense of "high" culture; these chiefly involve claims and beliefs in various branches of scholarship, such as history, philosophy, and natural science. Of course, if one attaches the same term ("myths") to all three kinds, one is assuming that all three have something in common, but that is not an issue that can or should be pursued here. Our main interest here is what I have labeled cultural myths.

Cultural myths are (historical, philosophical, or scientific) claims which are (1) actually false, but which are (2) widely believed to be true, because they (3) appear to be true, (4) based on some evidence, and which also (5) perform valuable functions for the cultural cohesiveness of social groups. Note that, to be a myth, it is not enough that some assertion be false or untenable; this is only one out of five conditions. The assertion must also be accepted as true by many people, that is, it must be a relatively common belief. Nor are these two conditions sufficient, for it is also necessary that the assertion have some semblance of truth, that it be apparently or partially true. Moreover, the belief must be based on some evidence, however partial, incomplete, inadequate, or misinterpreted. And the assertion must be such that believing it is culturally and socially useful, from the point of view of some particular group, perhaps by defining its identity or preserving its unity.[21]

Beyond Conflict and Complexity

Thus, as we have seen, the Galileo affair displays various conflicts between science and religion, but also various harmonies between them. Furthermore, it embodies complications, in the relationship between science and religion, as well as between conflict and harmony, between conservation and innovation, and between myths and facts. However, I do not want to give the impression that the affair merely or primarily supports what has been called the complexity thesis,[22] for I believe that complexity per se is methodologically unsatisfactory, and that my account possesses a simplicity that transcends complexity. To see this, let me now summarize my account and be explicit about how it recognizes but transcends both the simple conflicts and the complications of the affair.

To begin with, we must be explicit about formulating a very simple idea about the Galileo affair, namely a distinction between the *original* and the *subsequent* affair. That is, by the phrase "Galileo affair" one can mean the sequence of events which began in 1613 when he defended Copernicanism from the scriptural objection in his letter to Castelli, and which climaxed in 1633 when he was condemned by the Inquisition. However, by "Galileo affair" one can also mean the controversy about the facts, causes, responsibilities, issues, lessons, and implications of that trial, which began immediately after his condemnation and continues to our own day.

This distinction is important because there is no a priori reason why the science-religion relationship that characterizes one of these two affairs should also characterize the other. And even if the same relationship were to hold in both controversies, the supporting evidence in the two cases would have to be different; thus, it would be obviously invalid to argue that there was a conflict between science and religion in the original episode because there was such a conflict in the subsequent affair, or that there was no conflict in the subsequent affair because there was no such conflict in the original episode. Moreover, even if the relationship were conflictual in both cases, the particular form of the conflict may very well be different (as we have seen).

We must also make another distinction: between *surface structure* and *deep structure*. The surface structure of a controversy refers to those phenomena or characteristics that are observable or ascertainable most easily or directly, at least relatively so. The deep structure refers to the set of occurrences that are less easily observable, but nevertheless are demonstrably present, and, more importantly, whose presence enables us *to explain how and why* the surface structure comes about. However, the deep structure does *not*, and is not meant to, *explain away* the surface structure; the reality of the latter does not disappear once it is explained in terms of the deep structure. We may also say that the deep structure causes the surface structure, or that the surface structure is the effect of, or results from, the deep structure. However, here the cause and effect relationship is not deterministic, nomological, or necessary, but it is rather a looser connection—one that is historical and contingent.

This distinction is important partly because one wants to know not only what "really" happened, but also why it happened. That is, one wants to gain not only knowledge, in the sense of information, but also understanding. Another reason is that the deep structure enables us not only to take into account the complexity of the situation, but also to do so in a way that does not leave the complexity as an undigested or brute fact but transcends it into a greater simplicity.

Applying this framework, one deals first with the original affair and its surface structure. I have claimed that the original affair undeniably exhibits a conflict between science and religion. The chief grounds for this conflict are the key events of Galileo's trial. Here, the conflict takes the form of religion versus science, that is, religion attacking science: the scientist Galileo was persecuted, tried, and condemned by institutions and officials of the Catholic religion.

It is important to also point out that the affair has harmonious elements, mainly Galileo's own belief and cogent argument that there is no real conflict between Copernicanism and scripture. And it is also important to point out that the cogency of Galileo's argument eventually was accepted even by the church. However, such harmony does not destroy the conflict, which reappears in the historical context of the original affair, where the church held the opposite view: that there really was a conflict.

On the deep structure of the original affair, I have claimed that the just described conflict between Galilean science and Catholic religion is the result of a conflict between conservation and innovation. This deeper conflict may be seen by stressing the fact that many churchmen sided with Galileo, and some scientists sided with his opponents. And here I would add something I have not even mentioned thus far: some key figures individually embodied the conflict between conservation and innovation in their own thinking, attitudes, and actions. For example, Pope Paul V was mostly a conservative, but in 1616 did not declare Copernicanism a formal heresy and did not have Galileo tried and condemned; Cardinal Bellarmine, although primarily a conservative, was willing to tolerate Copernicanism as a hypothesis to explain observed facts and to make calculations and predictions; and Pope Urban VIII had innovative inclinations that encouraged Galileo to write the *Dialogue*, but eventually his conservatism prevailed and he had Galileo condemned. The science-religion conflict resulted from the conservation-innovation conflict insofar as the conservatives and conservative attitudes happened to prevail over the innovators and innovative attitudes in Galileo's trial.

Next, there is the surface structure of the subsequent affair. I have argued that it too consists of a conflict between science and religion, but taking the form of science versus religion. That is, for the past four centuries, the church has been under fire emanating from scientists, or alleged representatives of the scientific attitude, on account of its treatment of Galileo. Evidence for such a conflict is found in writings such as those mentioned earlier by Milton, Voltaire, Libri, and Einstein; these four examples are merely the tip of an iceberg of anticlerical and pro-Galilean criticism. The other main body of evidence consists of various apologetic attempts to defend the church and blame Galileo, such as the above-mentioned examples of Ross, Mallet du Pan, Brewster, Duhem, and Feyerabend; these particular examples are also the tip of an iceberg, although much smaller than the anticlerical one. Additionally, although I have not elaborated this, another piece of evidence substantiating this conflict is found in the sequence of actions taken by the church to retract or undo its prohibition or condemnation of Copernicanism, Galileo, and the principle limiting the scientific authority of scripture.

Finally, I have argued that the deep structure of this subsequent conflict between science and religion is to be found in the conflict between cultural myths and documented facts. For, if we scratch under the surface of the anticlerical criticism and of the pro-clerical apologetics, we find the phenomenon of mythmaking and mythologizing, that is, the rise, evolution, and fall of cultural myths. Accordingly, one should explore whether both sides engaged in their share of exaggerations, distortions, propaganda, and rhetoric; whether such literally false beliefs (of both types) are really impossible to discard; and whether they perform a necessary and useful social function.

In short, in my account, there is a religion-versus-science conflict in the original Galileo affair and a (reverse) science-versus-religion conflict in the subsequent Galileo affair. However, both conflicts in these terms (science and religion) occur at the level of the surface structure, because at a deeper level they result from conflicts of things other than science and religion: the dialectic between conservation and innovation for the case of the original affair, and the dialectic of myths and facts for the case of the subsequent conflict. This is meant to explain, but not explain away, the conflict between science and religion in the Galileo affair.

On this last point, an analogy may be helpful. Physics explains the phenomenon of heat based on the motion and kinetic energy of the molecules that make up physical bodies. The kinetic theory enables us to understand many observed facts about heat and make additional predictions of less easily observable phenomena. But this does not undermine the reality of heat. Similarly, the reality of the science-and-religion conflict in the Galileo affairs is not undermined by the conflicts between conservation and innovation and between myths and facts.

With regard to complexity and simplicity, I believe my framework is complex enough to enable us to take into account the complications of the phenomenon we are dealing with. However, I also believe that this framework is elegant enough to give it a kind of simplicity, which nevertheless avoids oversimplifications. Moreover, in my account, complexity is not introduced merely to silence one-sidedness (the one-sidedness of "the idea that refuses to die"), but also to generate a new challenge (the challenge of rendering complexity intelligible). One way of rendering it intelligible is to search for the deeper simplicity underlying it.

Conclusion

Given this account of the Galileo affair, we can begin to see why the idea of warfare between science and religion refuses to die. A main reason is that the idea contains a lot of truth, as the Galileo affair indicates. Of course, it is not true that there is an *inevitable* conflict between science and religion. Nor is it correct to say that the conflict is *inherent* in the nature of science and religion. Nor can it be claimed that their conflict is *perennial* or *ubiquitous*. However, there is no good reason why the conflict thesis should be formulated in such sensationalist, inflated, and exaggerated terms. The idea to which I am attributing considerable truth is a more nuanced and moderate thesis: that there is often actual conflict between science and religion, and almost always potential conflict between them.

A second reason for the prevalence of the conflict thesis is that critics often make it sound as if the truth of the matter is that there is harmony or concord between science and religion, meaning that science and religion are always or inherently in harmony or concord.[23] But this is clearly untenable; that is, the harmony thesis, when expressed in inflated and exaggerated terms, is no more tenable than the sensationalist conflict thesis.

Now, many scholars who reject the conflict thesis do not do so in order to defend the harmony thesis; rather, they admit that the harmony thesis is also problematic. Their point is that the relationship between science and religion is much more complicated than what is conveyed by the notions of conflict or harmony. That is, they hold the "complexity thesis."

However, my account also suggests that a third reason for the continued prevalence of the conflict thesis is that its critics often object to it on the basis of the complexity thesis. For although the complexity thesis is sound as far as it goes, it does not go very far; and it is ineffective in providing an alternative to the conflict thesis. A truly effective analysis needs to transcend complexity, by means of some deeper simplicity that explains it, while avoiding oversimplifications.

ACKNOWLEDGMENT

I would like to thank Ron Binzley, John Brooke, Peter Harrison, Michael Shank, and Ron Numbers for their helpful comments.

NOTES

1. See Maurice A. Finocchiaro, trans. and ed., *The Galileo Affair: A Documentary History* (Berkeley: University of California Press, 1989); Finocchiaro, *Retrying Galileo, 1633–1992* (Berkeley: University of California Press, 2005); Finocchiaro, *Defending Copernicus and Galileo: Critical Reasoning in the Two Affairs* (Dordrecht: Springer, 2010); Finocchiaro, *Routledge Guidebook to Galileo's "Dialogue"* (London: Routledge, 2013); and Finocchiaro, trans. and ed., *The Trial of Galileo: Essential Documents* (Indianapolis: Hackett, 2014).

2. For an elaboration, documentation, and defense of this interpretation, and a criticism of the more traditional accounts, see Finocchiaro, *Defending Copernicus and Galileo*, 37–64.

3. John W. Draper, *History of the Conflict between Religion and Science* (New York: Appleton, 1874), x–xi, 335, 364; Andrew D. White, *A History of the Warfare of Science with Theology in Christendom*, 2 vols. (New York: Appleton, 1896), 1:130–70; Bertrand Russell, *Religion and Science* (1935; New York: Oxford University Press, 1997), 31–43; Albert Einstein, "Foreword," in Galileo Galilei, *Dialogue concerning the Two Chief World Systems, Ptolemaic and Copernican*, trans. Stillman Drake (Berkeley: University of California Press, 1953), vi–xx; Karl R. Popper, *Conjectures and Refutations* (New York: Harper, 1963), 33–65.

4. John Paul II, "Deep Harmony Which Unites the Truths of Science with the Truths of Faith," *L'Osservatore Romano*, weekly ed. in English, November 26, 1979, 9–10; all quotations from 9.

5. John Paul II, "Faith Can Never Conflict with Reason," *L'Osservatore Romano*, weekly ed. in English, November 4, 1992, 1–2. Cf. Michael Sharratt, *Galileo, Decisive Innovator* (Cambridge, MA.: Blackwell, 1994), 209–22; Antonio Beltrán Marí, "'Una reflexión serena y objectiva,'" *Arbor* 160 (1998): 69–108; Michael Segre, "Light on the Galileo Case?," *Isis* 88 (1997): 484–504; Hermes H. Benítez, *Ensayos sobre ciencia y religión* (Santiago, Chile: Bravo y Allende, 1999), 85–110; Finocchiaro, *Retrying Galileo*, 338–58.

6. See Galileo's letters to Castelli and to Christina in Finocchiaro, *The Trial of Galileo*, 43–77; see also Finocchiaro, *Defending Copernicus and Galileo*, 65–96, 243–48.

7. Paolo A. Foscarini, "A Letter . . . concerning the Opinion of the Pythagoreans and Copernicans," in Richard J. Blackwell, trans. and ed., *Galileo, Bellarmine, and the Bible* (Notre Dame, IN: University of Notre Dame Press, 1991), 217–51; Tommaso Campanella, *A Defense of Galileo*, trans. and ed. Richard J. Blackwell (Notre Dame, IN: University of Notre Dame Press, 1994).

8. On the nonmonolithic character of the Catholic Church, see also Michael Segre, "Science at the Tuscan Court, 1642–1667," in S. Unguru, ed., *Physics, Cosmology and Astronomy, 1300–1700: Tension and Accommodation* (Dordrecht: Kluwer, 1991), 295–308; Rivka Feldhay, *Galileo and the Church: Political Inquisition or Critical Dialogue?* (Cambridge: Cambridge University Press, 1995); David C. Lindberg, "Galileo, the Church, and the Cosmos," in David C. Lindberg and Ronald L. Numbers, eds., *When Science and Christianity Meet* (Chicago: University of Chicago Press, 2003), 33–60, on 58.

9. The dialectic of conservation and innovation has been appreciated by various scholars in various contexts: in the history of science in general, by Thomas S. Kuhn, *The Essential Tension: Selected Studies in Scientific Tradition and Change* (Chicago: University of Chicago Press, 1977); in the post-Darwinian controversies in Britain and the United States, by James R. Moore, *The Post-Darwinian Controversies: A Study of the Protestant Struggle to Come to Terms with Darwin in Great Britain and America, 1870–1900* (Cambridge: Cambridge University Press, 1979), 80–103; and in the history of creationism, by Ronald L. Numbers, *The Creationists: The Evolution of Scientific Creationism* (Berkeley: University of California Press, 1993), xiv–xv.

10. Here the sequence would be: Jean le Rond d'Alembert, "Discours préliminaire," in Denis Diderot and Jean d'Alembert, eds., *Encyclopédie, ou Dictionnaire raisonné des sciences, des arts et des métiers*, 35 vols. (Paris, 1751–80), 1:i–xlv; Draper, *History of the Conflict* (1874); and White, *History of the Warfare* (1896). See Agostino Gemelli, "Scienza e fede nell'uomo Galilei," in Università Cattolica del Sacro Cuore, ed., *Nel terzo centenario della morte di Galileo Galilei* (Milan: Vita e Pensiero, 1942), 1–27; John Paul II, "Deep Harmony"; and John Paul II, "Faith Can Never Conflict with Reason." Cf. Finocchiaro, *Retrying Galileo*, 120–25, 275–80, 338–58.

11. See Finocchiaro, *Retrying Galileo*, 72–76.

12. John Milton, *Areopagitica* (1644), in Milton, *Complete Prose Works*, 8 vols. (New Haven, CT: Yale University Press, 1953–82), 2:485–570, on 538.

13. Voltaire, *Oeuvres complètes*, ed. Louis Moland, 52 vols. (Paris, 1877–83), 12:249.

14. Guglielmo Libri, *Essai sur la vie et les travaux de Galilée* (Paris, 1841), 46–47.

15. Einstein, "Foreword," in Galilei, *Dialogue concerning the Two Chief World Systems* (1953), vii.

16. Alexander Ross, *The New Planet No Planet, or the Earth No Wandring [sic] Star, Except in the Wandring [sic] Heads of Galileans* (London, 1646), 9; cf. John Wilkins, *A Discourse concerning a New Planet, Tending to Prove That 'Tis Probable Our Earth Is One of the Planets* (London, 1640).

17. Jacques Mallet du Pan, "Mensognes imprimées au sujet de la persécution de Galilée," *Mercure de France*, July 17, 1784, 121–30, on 122.

18. David Brewster, *The Martyrs of Science* (London, 1841), 93–95.

19. Pierre Duhem, *To Save the Phenomena*, trans. E. Doland and C. Maschler (Chicago: University of Chicago Press, 1969), 113.

20. Paul K. Feyerabend, *Against Method*, rev. ed. (London: Verso, 1988), 129.

21. For a general account of myths, see Paulo de Carvalho Neto, *The Concept of Folklore*, trans. J.M.P. Wilson (Coral Gables, FL: University of Miami Press, 1971). For applications to the Galileo affair, see Benítez, *Ensayos sobre ciencia y religión*, 85–110; Maurice A. Finocchiaro, "Galileo as a 'Bad Theologian': A Formative Myth about Galileo's Trial," *Studies in History and Philosophy of Science* 33 (2002): 753–91; Finocchiaro, "Myth 8: That Galileo Was Imprisoned and Tortured for Advocating Copernicanism," in Ronald L. Numbers, ed., *Galileo Goes to Jail and Other Myths about Science and Religion* (Cambridge, MA: Harvard University Press, 2009), 68–78, 249–52; Thomas S. Lessl, "The Galileo Legend as Scientific Folklore," *Quarterly Journal of Speech* 85 (1999): 146–68; David N. Livingstone, "Science, Religion, and the Cartographies of Complexity," *Historically Speaking: The Bulletin of the Historical Society*, 8, no. 5 (May/June 2007): 15–16, on 15. For a better understanding of the distinction between myths and false claims, I find useful the discussion of the distinction between fallacies and invalid arguments; cf. John Woods, *Errors of Reasoning: Naturalizing the Logic of Inference* (London: College Publications, 2013); and Maurice A. Finocchiaro, "Essay-Review of J. Woods's *Errors of Reasoning: Naturalizing the Logic of Inference*," *Argumentation* 28 (2014): 231–39.

22. John H. Brooke, *Science and Religion* (Cambridge: Cambridge University Press, 1991), 5, 8–10, 33, 42, 50–51; John H. Brooke and Geoffrey Cantor, *Reconstructing Nature: The Engagement of Science and Religion* (Edinburgh: T & T Clark, 1998), xi, 21, 66; John H. Brooke, "Science, Religion, and Historical Complexity," *Historically Speaking: The Bulletin of the Historical Society*, 8, no. 5 (May/June 2007): 10–13, 16–17; Bernard Lightman, "Victorian Sciences and Religions," in John H. Brooke, Margaret J. Osler, and Jitse M. van der Meer, eds., *Science in Theistic Contexts: Cognitive Dimensions* (Chicago: University of Chicago Press, 2001), 343–66; David C. Lindberg and Ronald L. Numbers, "Introduction," in Lindberg and Numbers, eds., *God and Nature: Historical Essays on the Encounter between Christianity and Science* (Berkeley: University of California Press, 1986), 6, 10, 14; and David C. Lindberg and Ronald L. Numbers, "Introduction," in *When Science and Christianity Meet*, 3.

23. One author who comes close to this is Alvin Plantinga, *Where the Conflict Really Lies* (New York: Oxford University Press, 2011).

CHAPTER THREE

Rumors of War

MONTE HARRELL HAMPTON

In his 1869 speech "The Battle-Fields of Science," Andrew Dickson White, the cofounder and first president of Cornell University, starkly portrayed the relationship of science and religion as "war," a war that not only exceeded in duration and ferocity history's bloodiest actual wars, but a war that was perpetrated by religion.[1] Except for his allowing scripture little part in the subject matter of his speech, one might wonder whether White believed he had been personally anointed to inaugurate fulfillment of the biblical prophecy, "ye shall hear of wars, and rumors of wars."[2] But if White's 1869 speech constituted the opening salvo on American soil of the martial rhetoric that would comprise the warfare thesis in the decades to follow, to what extent did it characterize the *actual* relationship of science and religion during the decades leading up to it?

Of course, to speak in such reified terms of "science" and "religion," as if these were immutable entities, raises problems. After all, the semantic range of each of these terms has hardly been stable across time. "Science" once meant something closer to "knowledge," and arrived at its more limited modern sense of the systematic study of natural phenomena only in the late eighteenth and early nineteenth centuries. The term "religion" dates from an only slightly earlier time, when the creedal controversies of the post-Reformation European Enlightenment pushed believers to redefine the essence of faith. Polemical conflict over what constituted true and false Christianity gave birth to the notion of "religion," increasingly understood as an objective set of doctrinal shibboleths, rather than an internal devotional piety.[3]

By the mid-nineteenth century, the phrase "science and religion," understood to signify the relationship between two distinct disciplines, had become a commonplace in the Anglophone world. As early as the 1820s, "science and religion" began to appear in the titles of books and articles, with the honor of priority probably going to Thomas Dick, whose popular *The Christian Philosopher; or The Connection of Science and Philosophy with Religion* (1823) may have been the first English-language book to contain such a phrase in its title. Dick and the many Americans who bought or used his book in their classrooms believed an important "connection" existed between science and religion. Eventually America's Protestant denominations would give institutional expression to this perceived connection by establishing numerous academic chairs dedicated to its study, among the first of which was the Perkins Professorship of Natural Science in Connexion with Revealed Religion, established in 1859 at the Presbyterian seminary in Columbia, South Carolina. When the chair's first occupant, James Woodrow, accepted the position, he joined a discourse shaped by two impulses that stretched back to the eighteenth century: Enlightenment rationalism and evangelical Christianity.[4]

To be sure, American discourse on science and Christianity in the three or four generations prior to Andrew D. White's 1869 speech varied with denominational dynamics, individual idiosyncrasies, rhetorical context, and perceived associations. Still, a discernible pattern emerges from this diversity. From the turn of the nineteenth century until 1869, science and religion in America generally embraced each other, but theirs was sometimes an awkward embrace, a guarded embrace. If many participants in the discourse concerning science and religion saw the two pursuits as bound together in an alliance, even a sacred alliance, it was also at times an uneasy alliance. Both aspects of this posture—the embrace as well as the guardedness—stemmed from the centrality in American Protestant Christianity of a particular kind of hermeneutics that one might term "rationalist biblicism."[5]

Enlightenment impulses had more readily merged with Protestant impulses in the Anglophone world than on the European continent, and in the early nineteenth century, this fusion was most pronounced among American Protestants, who generally believed being biblical *entailed* being rational. As citizens of the *novus ordo seclorum*, they were hardly immune to the Enlightenment obsession with empirical reason as the key to unlocking all of nature's truths. Moreover, the spread of a resurgent, populist version of deism in the 1790s, loudly proclaiming that unimpeded empirical reason discredited biblical revelation, had fostered even greater eagerness among them to show that their faith and scriptures met the evidentiary criteria of rationalism. Yet in constructing their apologetics, American Protestants increasingly conceptualized biblical truth in the very terms set down by rationalists. Scripture functioned as nature did, as a repository of discrete "facts," to be accessed and defended in terms of empirical reason.

From its inception in the early nineteenth century, this tendency among American Protestants to read, apply, and defend the Bible according to rationalist, empiricist assumptions was an apologetic approach that was both promising and problematic. It held promise because a rational faith stood to resonate throughout a new society building itself upon the rational discovery of "the Laws of Nature and of Nature's God." Its adherents affirmed that reason was the indispensable companion of revelation, and their robust rationalism invigorated in America slogans that claimed God's "two books," nature and scripture, could never conflict or that science was the "handmaid of theology." As reason came increasingly to be identified with natural science in the ensuing decades of the nineteenth century, most American Protestants expected science to dwell peaceably with the teachings of the Bible. On the ground, however, this expectation proved more problematic than the irenic slogans suggested. Rather than merely corroborate the sure claims of religion, science sometimes seemed to complicate or contravene them. When such complications arose, it could become difficult to tell whether the embrace of science and religion was altogether affectionate.

This mixed potential that science presented to American Protestants at the threshold of the nineteenth century was evident in Samuel Miller's two-volume *A Brief Retrospect of the Eighteenth Century*, published in 1803. In his ambitious tome, Miller, a Presbyterian minister whose accomplishments would include membership in the American Philosophical Society and a professorship at Princeton, summed up both the positive and negative aspects of the eighteenth century, and much of his assessment focused on developments within natural science. The previous century, he noted, had boldly examined received wisdom through the lens of reason, and such rationalistic analysis carried both beneficial and worrisome possibilities. It had both expanded and distorted truth.[6] In a revealing metaphor, he described the science produced by such investigations as "a fantastic patch-work, enriched with many beautiful and precious materials, but deformed by the mixture of many gaudy colours and false ornaments."[7] The "deformed," "gaudy," and "false" part of this patchwork referred to threats against the integrity and authority of the Bible launched from the realm of natural science. Relegating such science to a familiar biblical category that would become more and more populated as the nineteenth century advanced, Miller classified such claims as "science falsely so called." "It may be confidently pronounced," Miller contended, "that there never was an age in which so many deliberate and systematic attacks were made on Revealed Religion, through the medium of pretended science, as in the last."[8] Unfortunately, such feral forms of rationalism could not be safely consigned to the bygone century. Rather, as Miller faced the new one, he feared that such audacious predictions of the "triumph of Reason" remained "yet clamorous and obtrusive."[9]

Indeed, the clamor of a resurgent deism in the 1790s still rang in the ears of American Protestants, and its heralds, men like Thomas Paine, systematically pitted reason against revelation. Trumpeting an all-sufficient natural religion with no need for scripture, Paine argued in *Age of Reason* (1794–1807) that Moses did not write the Pentateuch, that Jesus was not divine, and that the Old Testament Proverbs were "not more wise . . . than those of the American Franklin." To those who would then object, "Are we to have no Word of God," Paine replied, "Yes; there is a Word of God; there is a revelation. THE WORD OF GOD IS THE CREATION WE BEHOLD."[10] American evangelicals adamantly rejected deism with its "one book" theology, opting instead for the traditional "two books" theology associated with Francis Bacon, the embodied ideal of American science and religion in the early nineteenth century. Concerned to protect the authority of the "book" of scripture, they regarded Paine's attempt to establish a rational natural religion as infidelity of the worst sort, yet they nonetheless adopted a kind of "ambient deism." Professing a rational natural *and supernatural* religion, they tended to answer questions of science and religion in the idiom of rationalism.[11]

The Embrace of Science and Christianity

Despite any concern they may have had about rationalism's negative potential, most American evangelicals would have agreed with Samuel Miller that properly executed science was the companion of scripture. "In no century more than the one just completed," asserted Miller, "was it ever rendered so apparent, that the information and the doctrines contained in the sacred volume perfectly harmonize with the most authentic discoveries, and the soundest principles of science." Not only did natural science corroborate biblical teachings, Smith said, but the practitioners of natural science were more often than not Christian clergy.[12]

Indeed, in the early decades of the nineteenth century, it could sometimes appear that the embrace of science and Christianity was so intimate that one could scarcely tell where the one began and the other ended. This was quite evident within geology, where some of the most eminent men of the new science were practicing Christians, even Christian ministers. Several of them, like William Buckland, both taught geology at universities and also maintained ecclesiastical responsibilities at prominent Anglican churches. Their writings, familiar to many Americans, affirmed the amenability of new geological findings to the language of Genesis. The very title of Buckland's inaugural address at Oxford, published in 1820 as *Vindiciae geologicae; or, The Connexion of Geology with Religion Explained*, was telling.[13] This book contended that observable geological phenomena on the surface of the earth comported with the biblical flood.

Buckland and other Genesis-geology reconcilers believed their reconciliation efforts necessary at least in part because the ranks of geologists, hardly a monolithic lot, included a few whose researches proceeded with little reference to scripture or supernatural activity. James Hutton's *Theory of the Earth* (1785) and Charles Lyell's *Principles of Geology* (1830–33), for instance, endorsed a rigorously empirical approach that would explain all past geologic phenomena by presently observable processes, which these works' authors believed were uniform across time. Such "uniformitarian" approaches not only required a much older history of the earth than traditionally accepted, but effectively dismissed the relevance of scripture for geological investigation. "The physical part of Geological inquiry ought to be conducted," Lyell said, "as if the Scriptures were not in existence."[14]

Over against uniformitarianism, theologian-scientists, such as Buckland, Adam Sedgwick, and William Conybeare, for instance, presided over a paradigm known as "catastrophism," which maintained that the current state of the earth's crust had resulted from a series of past cataclysms. Undeniably, this Oxbridge catastrophism, ascendant during the third decade of the nineteenth century, often required unconventional readings of scripture such as multiple cycles of catastrophe and creation, during each of which God had repopulated the earth with successively more advanced sets of creatures. Furthermore, some pious geologists began moving away from the effort to find specific correspondences between their research and the language of Genesis altogether, content to show only that geology did not contradict scripture or that it further reinforced natural theology. Indeed, Buckland himself in his 1836 *Bridgewater Treatise* would recant his earlier contention that geological evidence corroborated the Noachian deluge.[15] Still, these adjustments should not be confused with antagonism, for the earnest discussion of "Genesis and geology" had more to do with maintaining their accord than pitting them against each other.[16]

While the fervor to square geology with a scientifically literal interpretation of Genesis was by the 1830s waning among professionalizing geologists in Britain, many American geologists worked more robustly to maintain the relevance of biblical language to geological investigation. Few were more influential in this regard than Yale's Benjamin Silliman, who embodied for Americans the peaceful union of science and scripture. A devout Congregationalist, Silliman had been invited to fill Yale's chair of chemistry and natural history by the institution's eminent president, the evangelical Timothy Dwight. Silliman commanded respect not only for his piety but as a scientist as well, editing the influential *American Journal of Science* and serving as president of the American Association for the Advancement of Science.

To his popular American edition of Robert Bakewell's *Introduction to Geology* (1828), Silliman appended his own essay on the relationship between geology and Genesis, in which he set forth his view that the two could be readily harmonized if the "days" of the Genesis creation account were read figuratively as extended geologic periods rather than literal twenty-four hour days. To be sure, Silliman disclaimed the notion that the Bible was a science textbook; he disapproved of those whose geological data were passages of scripture rather than pieces of rock strata. Contrary to the traditional view that the earth was but a few thousand years old, Silliman believed the geologic column evinced its great antiquity. Still, a strong current of biblicism ran through Silliman's work. He included with his appended essay a detailed chart titled "Table of Coincidences between the Order of Events as Described in Genesis, and That Unfolded by Geological Investigation," which served notice that, in this Christian geologist's estimation, the book of God's word and the book of God's works stood in blissful harmony.[17]

Silliman's "day"-age construct was not the only harmonization scheme available to Americans eager to reconcile Genesis and geology. In works such as the oft-reprinted *Elementary Geology* (1840) and *The Religion of Geology* (1851), Edward Hitchcock, Silliman's former student, fellow Congregationalist, and professor of geology at Amherst College, also declared the relationship between geology and scripture to be one of peaceful accord. Unlike Silliman, however, Hitchcock endorsed what became variously known as the "ruin-and-restoration" or "gap" scheme. Advanced decades earlier by Scottish Presbyterian Thomas Chalmers and French comparative anatomist Georges Cuvier, this view accommodated the vast ages of living and dying petrified in the strata beneath humanity's feet by exploiting the "gap" between the initial creation of the world recounted in Genesis 1:1, and the actual creative acts of God, which commenced in Genesis 1:2.[18]

Many Americans warmly received such attempts at accommodating scripture to science, even when they required the adjustment of older interpretations of relevant biblical texts. Indeed, Methodist Thomas Ralston claimed in 1847 that "most intelligent Christians of the present day" allowed scientific findings to compel reinterpretations of biblical texts.[19] Few pressed the authority of the naked biblical text more stringently than the Disciples of Christ, but even their leader, Alexander Campbell, believed geology had shown that changes in the surface of the earth revealed evidence of its antiquity.[20] Joseph Le Conte, Presbyterian professor of science at South Carolina College, went so far as to call geology "the chief handmaid of religion among the sciences."[21]

The majority of nineteenth-century American Protestants accommodated their reading of the Bible not only to new findings in geology but also in astronomy.[22] In 1796, Pierre Simon Laplace had presented his nebular hypothesis, according to which the solar system had been produced naturally out of the whirling atmosphere of the primitive sun. Because Laplace had formulated his thesis without reference to supernatural activity, and because skeptics sometimes employed it in the service of naturalistic conclusions, the nebular notion did elicit some apprehension among Christians. But its endorsement by numerous Christian scientists generally rendered it safe for American believers. From Joseph Henry's favorable lectures at Princeton to Cincinnati astronomer Ormsby MacKnight Mitchell's regaling packed theaters with the wonders of the nebular hypothesis, the notion seemed to redound to the glory of God.[23] Accordingly, one Mississippi writer exulted that astronomers' discovery of "nebulae after nebulae" served only to enlarge the grandeur of God, giving greater force to "[scriptural] language, 'who by searching can find out the Almighty' . . . or 'what is man that thou art mindful of him.'"[24]

In addition to the widespread acceptance of mainstream geology and astronomy, American Protestants also readily accommodated the methodological naturalism that was becoming more and more common in the actual practice of science during the early nineteenth century.[25] This insistence on finding "second" or natural causes, rather than explaining natural phenomena by appeal to direct divine action, increasingly characterized American science. And Christians who were eager to accept science and yet retain the sovereignty of God over the cosmos simply affirmed that these natural laws were God's way of working in nature.

For most believers, acceptance of methodological naturalism did not mean that God could never interrupt natural causation with miraculous acts, but it did mean that God's *typical* interaction with the natural world occurred through more uniform means. Accordingly, in the 1859–60 school year, James Woodrow, the Presbyterian theologian-scientist who held the Perkins Professorship of Natural Science in Connexion with Revealed Religion at Columbia Theological Seminary in South Carolina, defended scientists' "belief in the uniformity of nature," as an "axiom founded on what the Bible reveals of God & his dealings with Nature." Such uniform laws, Woodrow taught, were simply the "laws which God has made for the government of the Universe."[26] In this light, the natural processes of geological and astronomical change had originated and were maintained by God. "That every event in the universe takes place according to fixed laws I am ready to admit," wrote the Christian geologist Edward Hitchcock. "For what is a natural law? Nothing more nor less than the uniform mode in which divine power acts." Christians could embrace the rule of natural law because it "by no means renders a present directing and energizing Deity unnecessary."[27]

The widespread acceptance of methodological naturalism was just one indicator of the degree to which the methods, tropes, and cachet of science had permeated American Protestantism. The embrace of science and religion was symbolized on the front page of the Baptist newspaper, the *Columbiana Star*, which featured the words "science" and "religion" juxtaposed around a star on its masthead.[28] "Religion and science," exclaimed Princeton theologian Charles Hodge, "are twin daughters of heaven."[29] Everywhere, it seemed, science and Christian theology worked side by side, or shared the same space.

In the early to mid-nineteenth century, Americans conceptualized both science and religion, as well as the relationship between the two, in terms of a particular redaction of Enlightenment rationalism, which was influenced by Scottish commonsense realism and associated in their minds with Francis Bacon. Indeed, Samuel Miller's guarded embrace of natural science, discussed earlier in this chapter, had assumed that all sound science functioned according to the methodology of Bacon, being built on the foundation of Scottish commonsense epistemology. In contrast with Aristotelian deduction, Francis Bacon had advocated in his *Novum organum* (1620) the use of inductive reasoning to uncover the facts, and then laws, of nature. The successes of recent natural science, Miller was certain, had resulted from adhering, as all true science must, to "Lord Bacon's plan of pursuing knowledge by observation, experiment, analysis, and induction." In addition, Miller identified Scottish commonsense philosophy as the epistemological foundation for sound Christian reasoning. It was, he said, the "tribunal" before which all philosophy must be tried.[30]

Of course antebellum America would prove to be no philosophical monolith. Transcendentalists and other romantics, for instance, would resist the ascendancy of Baconianism and Scottish common sense in decades to come, urging an epistemology more in tune with intuition than empiricist rationalism.[31] Nevertheless, Miller's turn-of-the-century endorsement of Baconianism and Scottish common sense foreshadowed their ascendancy in antebellum America.[32]

These twin pillars supported a particular empiricist-biblicist construct, which in turn shaped the way antebellum Americans related science and religion. The Baconian pillar required that true science reject deductive speculation, focusing only on objective "facts," from which source alone the laws of nature could be derived. For some, to be sure, Baconianism had more to do with maintaining boundaries than with the actual scientific practice. In an 1846 letter to Benjamin Silliman, for instance, Joseph Henry clarified that by "science" he meant "the knowledge of the laws of phenomena and not a mere collection of facts or a classification of objects which is properly denominated by Bacon Natural History." Still, for most Americans such qualifications could only tweak, not topple, the Baconian ideal of science.[33]

Most American Protestants insisted that this Baconian approach applied equally to the study of the Bible. Princeton's James W. Alexander spoke for many when he likened the biblical interpreter's task to that of the scientist: "The theologian should proceed," he said, precisely as the chemist or the botanist proceeds."[34] James S. Lamar echoed these hermeneutic assumptions, commonly employed within his Stone-Campbell circles, in his tellingly titled book, *The Organon of Scripture: Or, the Inductive Method of Biblical Interpretation* (1859). The scriptures, he averred, must be "studied and expounded upon the principles of the inductive method," because "thus interpreted they speak to us in a voice as certain and unmistakable as the language of nature heard in the experiments and observations of science."[35]

Besides Baconian methodology, the empiricist-biblicist framework rested on an American appropriation of Scottish commonsense realism. Introduced to America in the late eighteenth century by Princeton presidents John Witherspoon and Samuel Stanhope Smith, this philosophical school had achieved hegemony by the early decades of the nineteenth century. While Scottish commonsense philosophy was not monolithic, the American redaction provided an epistemic foundation for the widespread conviction that knowledge was universally accessible and stable, unaffected by the perspective of the knower. Like a mirror, the mind was supposed to do little more than reflect what was there—in nature or in scripture. This highly objectivist epistemology not only honored the traditional Enlightenment confidence in humanity's capacity to discover the fixed laws of nature, but it also dovetailed with Americans' increasingly robust faith in the rational powers of everyman, a faith that flourished with the democratization of early-nineteenth-century culture. For its adherents, the benefit of this empiricist-biblicist construct seemed clear: reliance on common sense and submission to the plain "facts" of nature and scripture insured that these two divinely authored books could continue their peaceful embrace.

When Rationalism Runs Amok

No matter how insistently antebellum Americans affirmed the union of science and religion in the decades prior to 1869, it was sometimes difficult to tell whether their embrace was altogether affectionate. Indeed, the potential for trouble had in many ways been built into the empiricist-biblicist paradigm from the start, and the tensions inherent within this conception became more and more apparent as the nineteenth century progressed. The paradigm used nature to defend the supernatural. But did, in fact, empirical science reliably evince the hand of God in nature? While many nineteenth-century Americans were able to domesticate methodological naturalism, others worried that the trend toward deemphasizing first causes might exile the Creator from his own creation. The appearance in 1844 of the anonymous *Vestiges of the Natural History of Creation* seemed to confirm this fear. It deployed geology, astronomy, phrenology, and Lamarckian evolution to advance an exhaustive developmental explanation for the whole cosmos, "from nebula to man."[36]

The troubling question of supernatural agency and natural law was inextricably related to an even more fundamental question, which *also* problematized the empiricist-biblicist construct: how should modern readers ascertain the meaning of ancient biblical texts, especially those pertaining to nature and natural history? The co-option of empirical rationalism often lent a unique cast to Americans' conception and appropriation of scripture. Under the influence of Baconian inductivism, snippets of biblical text were assumed to be of the same character as specimens of nature, such that biblical language conveyed truth in precisely the same way as the inanimate rocks and chemical compounds studied by men of science. Just as nature offered up its "phenomena" for the scientist's study, wrote James Alexander, so scripture presented its "propositions" for the theologian's study.[37] This propositional, empirical approach could render the original literary and historical context surrounding a given biblical text virtually irrelevant in shaping and constraining its meaning. When these biblical texts touched on nature or natural history, it was far too easy to find their meanings vis-à-vis the questions of a science-obsessed contemporary culture.

But did the Bible, supposedly composed of empirical "facts," return straightforward, univocal answers to the questions Americans were asking of it? The University of North Carolina geologist/minister Elisha Mitchell believed mistaken assumptions about the purpose of scripture lay behind Christians' disagreements over the merits of geology. Mitchell's 1825 exchange with Episcopal Bishop John Stark Ravenscroft exhibited his belief in the all-sufficiency of scripture. Mitchell rejected Ravenscroft's view that the interpretive traditions of the early church fathers should be normative for Christians, affirming instead that "the Scriptures are a sufficient rule to all who approach them with humble and honest hearts."[38] The exchange manifested Mitchell's stout confidence in the bedrock Protestant doctrines of *sola scriptura* and the perspicuity of scripture, as now metamorphosed under the American pressures of Baconianism and democratization: Christians needed nothing but the Bible, and the meaning of the Bible was clear.

Yet, all Christians, and presumably many who shared Mitchell's "Bible-only" convictions, did *not* derive the same meanings from the Bible when it came to questions of geology. What did scripture truly say, for instance, about the age of the earth, the age of humanity, and the extent of Noah's flood? In his *Elements of Geology* (1842), Mitchell criticized fellow believers who "raise an outcry against geology, as hostile to the truths of Revelation."[39] Such opponents were in fact replicating the mistakes of the misguided inquisitors of Galileo, who had wrongly assumed that God's word meant to "communicate . . . physical truth; to teach . . . astronomy, or chemistry, or geology."[40]

So, what *was* the relationship between biblical language and scientific inquiry? While some deists and infidels had explicitly denigrated the reliability of the Bible, adherents to the empiricist-biblicist paradigm experienced more angst dealing with the diversity of answers emanating from within the ranks of the faithful. Over against the attempts to reconcile Genesis and geology advanced by geologists like Silliman and Hitchcock, for instance, a wave of opposition from another group of empirically minded Christians arose at midcentury. These so-called scriptural geologists opposed such reconciling schemes as dangerous concessions to infidelity. Two New York businessmen, the brothers David and Eleazar Lord, led the way in contending that all the scientific facts that anyone needed were supplied in the pages of scripture.

The writings of the Lords revealed that, despite widespread antebellum appeals to the commonsense meaning of scripture, Baconians' understanding of scripture, as it related to questions of science, was hardly uniform. The very title of David Lord's book *Geognosy; or, The Facts and Principles of Geology against Theories* (1855) was revealing.[41] The scriptural geologists dealt in Baconian "facts," Lord asserted, while the reconcilers dealt merely in "theories." Lord maintained that the plain sense of Genesis required affirming a six thousand-year-old earth, accepting a universal flood, and attributing the formation of all fossils to the waters of this deluge.[42] So sure was Lord that biblical statements were propositions of scientific truth, that if it the earth were ever proved to be ancient, he declared, then "the whole Revelation is changed at once from a heaven-descended reality into a fable."[43]

It was becoming clear that what was obvious to one biblicist was not always obvious to another. The Presbyterian Tayler Lewis, for instance, shared the scriptural geologists' vigilance for the plain sense of scripture, but Lewis and the Lord brothers hardly agreed on what that plain sense *was*. Indeed, for Lewis, professor of Greek at Union College, the ubiquitous Baconian clichés about the amicability of science and scripture merely papered over confused thinking about the subject. In order to move beyond such "verbal platitudes," Lewis contended, hermeneutics must be guided by philology, not science.[44] Both Lewis and the Lords worried that surrendering biblical interpretation to science would imperil the authority of scripture. Yet, Lewis believed David Lord was "blindly determined . . . to peril the whole veracity of the Scriptures on his narrow notion of twenty-four hours."[45] Indeed, while the scriptural geologists anathematized the notion of an ancient earth on the basis of the Genesis text, Lewis followed the *same* biblical text to the conclusion that its "days" of creation must be indefinite, ineffable periods worthy of a divine being, *not* literal twenty-four-hour days.

Similar ambiguity troubled the empiricist-biblicist paradigm among southern Presbyterians. Their conviction that God's two books could not contradict led to the establishment in 1859 of the Perkins Professorship of Natural Science in Connexion with Revealed Religion at Columbia Theological Seminary. Yet, its establishment merely institutionalized an underlying ambivalence about just what Baconianism meant for science, scripture, and their relationship. James Woodrow, the first occupant of the Perkins chair, would become known for espousing a doctrine of "noncontradiction," as opposed to harmony, between science and scripture. Indeed, he was already heading in that direction by the time of his inaugural address in 1861. While he affirmed "the absolute truth" of the biblical text, Woodrow cautioned against insisting that science base its conclusions on scripture, since its allusions to nature were of a merely "incidental character."[46] Many of Woodrow's brethren, however, still expected science to defer to scripture. For instance, Benjamin Morgan Palmer, in New Orleans, expressed outrage at the cool audacity with which geologists advanced their schemes, "just as calmly as though God had never written a book, in which was set down the age of man."[47]

Several other claimants to the mantle of science—phrenology, mesmerism, and spiritualism, for instance—garnered wide popularity, but struck others as rationalism gone rogue.[48] Charles Darwin's *On the Origin of Species*, appearing in 1859, would also elicit a mixed response, but Americans would not thoroughly engage it until the 1870s and after. Developments within mid-nineteenth-century anthropology, however, seemed to most American Protestants like unalloyed infidelity, and in assessing theories of human origins that included notions of polygenism and preadamism they spoke with a more unified voice. Despite the endorsement of such theories by Harvard's renowned scientific celebrity Louis Agassiz, believers generally identified these speculations, from the 1840s through the 1860s and beyond, as "infidel science."[49] Did scripture not affirm unequivocally that God had "made of one blood all nations of men?"[50]

Even in the slaveholding South, where the financial and polemical advantages to be gained by believing blacks and whites were different species might have incentivized its acceptance, most Christians rejected it. Southern Presbyterians, whose ranks included leading proslavery apologists like George Howe, also locked arms against polygenism. Howe, protégé of the conservative Amherst theologian Moses Stuart, was professor of biblical literature at Columbia Theological Seminary. He wrote several articles assailing polygenism, which flouted "the plain, unperverted sense of the Bible."[51] Like many other Americans, southern Presbyterians believed that genuine science presented no conflict with the "plain, unperverted sense of the Bible." In the case of polygenist ethnology, the way forward was clear enough. But with other questions of science and faith, the safe path could be shrouded with ambiguity.

Conclusion

Whatever the focus of subsequent iterations of his warfare thesis, it is noteworthy that in his 1869 speech Andrew Dickson White focused his frustrations not on Christianity per se, but on the age-old use of scriptural language as scientific data. For instance, White believed that Cosmas's sixth-century theory of the world had gone wrong because this Christian traveler and hermit had built it on texts in Job and Isaiah, turning the "splendid and precious poetry" of holy writ "into a prosaic statement" of science. Similarly, White attributed the judgment against Galileo to the Catholic Church's "applying direct literal interpretation of Scripture to science." Indeed, White explicitly disclaimed "the idea that there is a necessary antagonism between science and religion." Rather, his burden was to show "that science must be studied by means proper to itself, and in no other way."[52]

Without question, numerous adherents to the empiricist-biblicist paradigm would have taken issue with White, despite such nuanced qualifications. But this very distinction between the character of biblical language and the character of scientific data had been dawning on some who labored to hold science and faith together, even from within this paradigm, for some time. In 1861, James Woodrow uttered words that—remarkably—could have been uttered by White in 1869. Conflict between science and religion resulted, Woodrow warned, from the "disposition, that has manifested itself in every age . . . to regard [the Bible's] every mention of material objects as couched in the current scientific language of the day; and from the groundless belief that the sacred volume, besides being fitted to accomplish its chief and highest ends, is also a text-book containing the whole body of scientific truth."[53]

So, what can be made of the relationship between American science and religion in the years prior to 1869? Despite widespread affirmations of Lord Bacon, commonsense philosophy, and the inevitable accord of God's two books, the particular way forward could be unclear. Andrew Dickson White's battlefield metaphor may have overstated and oversimplified the conflict within the relationship. But if it is going too far to say there was "war," it is not going too far to say there were "rumors of war."

NOTES

1. "First of the Course of Scientific Lectures—Prof. White on 'The Battle-Fields of Science,'" *New-York Tribune*, December 18, 1869, 4, http://chroniclingamerica.loc.gov/lccn/sn83030214/1869-12-18/ed-1/seq-4.pdf.

2. Matthew 24:6 in the King James Version (KJV), the ubiquitous translation of the Bible that would rarely be out of mind as Americans related science and religion in the early national and antebellum period.

3. Numerous scholars have traced the moving targets that are "science" and "religion." See, for instance, Ronald Numbers, "Aggressors, Victims, and Peacemakers: Historical Actors in the Drama of Science and Religion," in Harold W. Attridge, ed., *The Religion and Science Debate: Why Does It Continue?* (New Haven, CT: Yale University Press, 2009), 15–16; and Peter Harrison, "'Science' and 'Religion': Constructing the Boundaries," in Thomas Dixon, Geoffrey Cantor, and Stephen Pumfrey, eds., *Science and Religion: New Historical Perspectives* (Cambridge: Cambridge University Press, 2010), 23–31.

4. Thomas Dick, *The Christian Philosopher; or, The Connection of Science and Philosophy with Religion* (New York: G. & C. Carvill, 1826), http://babel.hathitrust.org/cgi/pt?id=hvd.hw2cwb;view=1up;seq=13. On Thomas Dick's priority, see Ronald L. Numbers, "Aggressors," 16–17; and Numbers, *Science and Christianity in Pulpit and Pew* (New York: Oxford University Press, 2007), 22–23. On the Perkins Professorship and the wider context of nineteenth-century southern discourse on science and religion, see Monte Harrell Hampton, *Storm of Words: Science, Religion, and Evolution in the Civil War South* (Tuscaloosa: University of Alabama Press, 2014).

5. On the rationalization of Christianity in the wake of the Enlightenment, see Michael J. Lee, *The Erosion of Biblical Certainty: Battles over Authority and Interpretation in America* (New York: Palgrave Macmillan, 2013), esp. chaps. 1 and 2; E. Brooks Holifield, *Theology in America: Christian Thought from the Age of the Puritans to the Civil War* (New Haven, CT: Yale University Press, 2003), 5–6, 186–87, and Harrison, "'Science' and 'Religion,'" 17–31, 33–39.

6. Samuel Miller, *A Brief Retrospect of the Eighteenth Century*, 2 vols. (New York: T. and J. Swords, 1803), 2:411–12, http://books.google.com/books?id=t9PHLpfTTDoC&source=gbs_navlinks_s. Ron Numbers identifies this source as an example of "one of the earliest assessments of religion and science," in Numbers, "Aggressors," 15–16. See also Theodore Dwight Bozeman, *Protestants in an Age of Science: The Baconian Ideal and Antebellum American Religious Thought* (Chapel Hill: University of North Carolina Press, 1977), 49–51, and Holifield, *Theology in America*, 173.

7. Miller, *Brief Retrospect*, 2:432–33.

8. Miller, *Brief Retrospect*, 2:431. The biblical phrase "science, falsely so called" is from 1 Tim. 6:20. For its widespread use, see Daniel P. Thurs and Ronald L. Numbers, "Science, Pseudoscience, and Science Falsely So-Called," in Peter Harrison, Ronald L. Numbers, and Michael H. Shank, eds., *Wrestling with Nature: From Omens to Science* (Chicago: University of Chicago Press, 2011), 281–306.

9. Miller, *Brief Retrospect*, 2:433.

10. Thomas Paine, *The Age of Reason*, in William M. an Der Weyde, ed., *Life and Works of Thomas Paine*, vol. 8 (New Rochelle, NY: Thomas Paine National Historical Association,

1925), 24 (Moses), 31 (Jesus), 25 (Franklin), 41 (revelation), http://babel.hathitrust.org/cgi/pt?id=mdp.39015011315861;view=1up;seq=11.

11. On "ambient deism," see Eric Schlereth, *An Age of Infidels: The Politics of Religious Controversy in the Early United States* (Philadelphia: University of Pennsylvania Press, 2013), 6. This was brought to my attention by Rachel Spivack, "Deism in America, 1790–1810" (unpublished paper).

12. "In no century," Miller, *Brief Retrospect*, 2:432–33; practitioners as Christian clergy, ibid., 434.

13. William Buckland, *Vindiciae geologicae; or, The Connexion of Geology with Religion Explained* (Oxford, 1820).

14. Quoted in James R. Moore, "Geologists and Interpreters of Genesis in the Nineteenth Century," in David C. Lindberg and Ronald L. Numbers, eds., *God and Nature: Historical Essays on the Encounter between Christianity and Science* (Berkeley: University of California Press, 1986), 337.

15. William Buckland, *Geology and Mineralogy Considered with Reference to Natural Theology*, 2 vols. (London: William Pickering, 1836), 1:16–17. The definitive work on Buckland and this period of British geology is Nicolaas A. Rupke, *The Great Chain of History: William Buckland and the English School of Geology, 1814–1849* (Oxford: Clarendon Press, 1983).

16. Mott T. Greene, "Genesis and Geology Revisited: The Order of Nature and the Nature of Order in Nineteenth-Century Britain," in David C. Lindberg and Ronald L. Numbers, eds., *When Science and Christianity Meet* (Chicago: University of Chicago Press, 2003), 150; Martin J. S. Rudwick, *Earth's Deep History: How It Was Discovered and Why It Matters* (Chicago: University of Chicago Press, 2014), 163.

17. Benjamin Silliman, "Consistency of Geology with Sacred History," appendix to Robert Bakewell, *An Introduction to Geology*, 2nd ed. (New Haven, CT: Hezekiah Howe, 1833), 434–45. Examining Silliman's work through the lens of Bacon's "two books" metaphor, Ted Davis observes that Silliman embraced Bacon's 1605 endorsement of science's ability to help "conceive the true sense of the scriptures," but appeared to violate Bacon's warning not to "unwisely mingle or confound these learnings together"; see Davis, "The Work and the Works: Concordism in American Evangelical Thought," in Klaas van Berkel and Arjo Vanderjagt, eds., *The Book of Nature in Early Modern and Modern History* (Leuven: Peeters, 2006), 196–97, 200. For an early treatment of such harmonization schemes in the United States, see Ronald Numbers, *Creation by Natural Law: Laplace's Nebular Hypothesis in American Thought* (Seattle: University of Washington Press, 1977).

18. Herbert Hovenkamp, *Science in Religion in America, 1800–1860* (Philadelphia: University of Pennsylvania Press, 1978), 132–36. The later Buckland also found the gap theory more palatable than the day-age scheme, because it allowed all of the geologist's subject matter to be tucked neatly away, preserving the integrity of Genesis while freeing geologists to pursue most of their work outside the purview of scriptural language, Greene, "Genesis and Geology," 156–59.

19. Quoted in Holifield, *Theology in America*, 184.

20. Campbell, "Supernatural Facts," June 1839, in W. A. Morris, *The Writings of Alexander Campbell: Selections Chiefly from the Millennial Harbinger* (Austin, TX: Eugene Von Boeckmann, 1896), 147, http://babel.hathitrust.org/cgi/pt?id=mdp.39015068372849;view=1up;seq=11.

21. Joseph Le Conte, "Lectures on Coal," in *Annual Report of the Board of Regents of the Smithsonian Institution* (Washington, DC: William A. Harris, 1858), 123.

22. G. Blair Nelson, "Infidel Science! Polygenism in the Mid-Nineteenth-Century American Weekly Religious Press" (PhD diss., University of Wisconsin–Madison, 2014), 48–50.

23. These responses come from the definitive treatment of the American response to the nebular hypothesis; see Numbers, *Creation by Natural Law.*

24. Richard Gladney, "Natural Science and Revealed Religion," *Southern Presbyterian Review*, October 1859, 457.

25. Ronald Numbers, "Science without God: Natural Laws and Christian Beliefs," in Lindberg and Numbers, eds., *When Science and Christianity Meet*, 265–85; Ronald Numbers, "Simplifying Complexity: Patterns in the History of Science and Religion," in Dixon et al., eds., *Science and Religion*, 264–68.

26. [Unknown student], "Professor Woodrow's Lectures on Science and Religion," archives of Reformed Theological Seminary, Jackson, MS.

27. Edward Hitchcock, *The Religion of Geology and Its Connected Sciences* (Glasgow: Wiliam Collins, 1851), 238–40.

28. Brooks Holifield, *The Gentlemen Theologians* (Durham, NC: Duke University Press, 1978), 83.

29. Quoted in Jon Stewart, "Mediating the Center: Charles Hodge on American Science, Language, Literature, and Politics," *Studies in Reformed Theology and History* 3 (Winter 1995): 25.

30. "Lord Bacon," in Miller, *Brief Retrospect*, 1:202, quoted in Holifield, *Theology in America*, 174; "tribunal," in Miller, *Brief Retrospect*, 2:11. The latter was brought to my attention by Stewart, "Mediating the Center," 22.

31. Holifield treats dissidents from this Baconian paradigm in part 3 of Holifield, *Theology in America*. See also Walter Conser's treatment of W.G.T. Shedd, James Warley Miles, Henry Boynton Smith, and Phillip Schaff, whose romantic epistemology led them to reconcile science and religion by appealing to divinely unfolding history: Walter H. Conser Jr., *God and the Natural World: Religion and Science in Antebellum America* (Columbia: University of South Carolina Press, 1993), 37–64.

32. The abundant historiography documenting the pervasiveness of the Baconian paradigm in antebellum American Christianity includes Holifield, *Theology in America*, 159–396; Bozeman, *Protestants in an Age of Science*; Hovenkamp, *Science and Religion in America*; George M. Marsden, "Everyone One's Own Interpreter?," in Nathan O. Hatch and Mark A. Noll, eds., *The Bible in America: Essays in Cultural History* (New York: Oxford University Press, 1982), 79–100; Mark Noll, *America's God: From Jonathan Edwards to Abraham Lincoln* (New York: Oxford University Press, 2002), 93–113; 367–85.

33. Joseph Henry to Benjamin Silliman Jr., August 13, 1846, in Marc Rothenberg, ed., *The Papers of Joseph Henry*, vol. 6 (Washington, DC: Smithsonian Institution Press, 1992). (This letter was brought to my attention by Ron Numbers.) Several early-nineteenth-century commentators began to note that the epistemology of science, in actual practice, always involved the use of deduction as well as induction. These included William Whewell's *Philosophy of the Inductive Sciences* (1840), Samuel Tyler's *A Discourse on the Baconian Philosophy* (1844), and David Brewster's *Memoirs of the Life, Writings, and Discoveries of Sir Isaac Newton* (1855). But by incorporating this rehabilitated deduction into an enlarged version of induction, most Americans merely inoculated Baconian empiricism against potential critics; see Bozeman, *Protestants in an Age of Science*, 64–70, 166–68.

34. "On the Use and Abuse of Systematic Theology," *Biblical Repertory & Princeton Review* (1832); quoted in Marsden, "Everyone One's Own Interpreter?," in Hatch and Noll, eds., *Bible in America*, 84.

35. Quoted in C. Leonard Allen, "Baconianism and the Bible in the Disciples of Christ: James S. Lamar and 'The Organon of Scripture,'" in *Church History: Studies in Christianity and Culture*, 55, no. 1 (March 1986): 67.

36. [Robert Chambers], *Vestiges of the Natural History of Creation*; in James A. Secord, ed., *Vestiges of the Natural History of Creation and Other Evolutionary Writings* (Chicago: University of Chicago Press, 1994), 1–390. The phrase "nebula to man" comes from James Secord's introduction (xxiii).

37. Quoted in Bozeman, *Protestants in an Age of Science*, 152. Bozeman provides numerous examples of Christian exegetes who, exemplifying the "wide catholicity of scientific method," began to regard biblical statements as hard, empirical "facts," analogous to those studied by natural scientists, 138, 150–56.

38. Elisha Mitchell, *Remarks on Bishop Ravenscroft's Answer to the Statements Contained in Professor Mitchell's Printed Letter, of the 12th of February Last* (Raleigh, NC: J. Gales & Son, 1825), 37.

39. Elisha Mitchell, *Elements of Geology, with an Outline of the Geology of North Carolina: For the Use of the Students of the University* (N.p., 1842), 107. This work is compiled with other works in the publication, *North Carolina Geology*, in the Geological Sciences Library, University of North Carolina, Chapel Hill.

40. Mitchell, *Elements of Geology*, 104.

41. David N. Lord, *Geognosy, or the Facts and Principles of Geology against Theories* (New York: Franklin Knight, 1855).

42. Rodney L. Stiling, "Scriptural Geology in America," in David N. Livingstone, D. G. Hart, and Mark A. Noll, eds., *Evangelicals and Science in Historical Perspective* (New York: Oxford University Press, 1999), 181. For American "scriptural geology," see Stiling, and Richard Perry Tison, "Lords of Creation: American Scriptural Geology and the Lord Brothers' Assault on 'Intellectual Atheism'" (PhD diss., University of Oklahoma, 2008). These American scriptural geologists echoed many of the emphases of an earlier wave of British scriptural geologists, who had responded to similar reconciliation schemes a generation prior.

43. Lord, *Geognosy*, 45.

44. Tayler Lewis, *The Bible and Science, or the World Problem* (Schenectady, NY: G. Y. Van Debogert, 1856), 14–15. Lewis wrote this book in response to a critical review article, which Yale geologist James Dwight Dana had published in response to Lewis's book from the year prior, titled *The Six Days of Creation; or, The Scriptural Cosmology, with the Ancient Idea of Time-Worlds, in Distinction from Worlds in Space* (Schenectady, NY: G. Y. Van Debogert, 1855). Lord's insistence that philology carry hermeneutic precedence animated his bitter controversy with Dana, whose interpretation of Genesis accommodated the findings of geology and the nebular hypothesis. See Numbers, *Creation by Natural Law*, 95–100.

45. Lewis, *Bible and Science*, 143.

46. James Woodrow, "Inaugural Address," in Marion Woodrow, ed., *Dr. James Woodrow: Character Sketches and His Teachings* (Columbia, SC: R. L. Bryan, 1909), 380. For Woodrow's doctrine of noncontradiction, see Hampton, *Storm of Words*, 163–66, 226–27, 260.

47. Benjamin M. Palmer, "Baconianism and the Bible," *Southern Presbyterian Review*, October 1852, 250–52; quoted in Ernest Trice Thompson, *Presbyterians in the South* (Richmond, VA: John Knox, 1973), 1:506.

48. See Numbers, *Science and Christianity in Pulpit and Pew*, 26–31.

49. G. Blair Nelson's "Infidel Science!," a study of thirty-six American religious papers published between 1839 and 1880, concludes that the popular religious press displayed little

opposition to "science" until the appearance of this "infidel science." For Agassiz's endorsement of polygenism, see Edward Lurie, "Louis Agassiz and the Races of Men," in Nathan Reingold, ed., *Science in American since 1820* (New York: Science History Publications, 1979); William Stanton, *The Leopard's Spots: Scientific Attitudes toward Race in America, 1815–1859* (Chicago: University of Chicago Press, 1960), 100–104; Lester Stephens, *Science, Race, and Religion: John Bachman and the Charleston Circle of Naturalists, 1815–1895* (Chapel Hill: University of North Carolina Press, 2000), 94.

50. The biblical text, Acts 17:6 in KJV, was routinely cited by monogenists.

51. George Howe, "Nott's Lectures," *Southern Presbyterian Review*, January 1850, 473. For southern Presbyterian response to polygenism, see Hampton, *Storm of Words*, 63–85.

52. White, "Battle-Fields."

53. Woodrow, "Inaugural Address," 384.

CHAPTER FOUR

The Victorians: Tyndall and Draper

BERNARD LIGHTMAN

The British Association for the Advancement of Science met in Belfast, Ireland, in the summer of 1874, and John Tyndall (1820–93), professor of natural philosophy at the Royal Institution, delivered the presidential address on the evening of August 19. Presidential addresses tended to be dull reviews of the progress of science over the past year, and they typically avoided controversial topics. Tyndall chose to ignore tradition by presenting a history of materialism and how it intersected with the development of science. He began with the birth of science in ancient Greece. He discussed Democritus, Epicurus, and Lucretius, depicting them as brave men whose atomic theory was geared to eliminate the "mob of gods and demons" preventing the discovery of true knowledge.[1] He then discussed the Middle Ages, picturing it as a period ravaged by scientific drought owing to the pernicious influence of a Christianized Aristotle. Copernicus, Bruno, and Galileo later revolutionized science despite the opposition of Christian theologians. An account of Bacon, Descartes, and Gassendi stressed their contributions to atomic theory, for Tyndall an essential component of any valid scientific system. Tyndall's historical survey ended with a celebration of the achievements of Charles Darwin and the importance of the doctrine of the conservation of energy. In the last two sections of the address, Tyndall outlined his views on the philosophical implications of modern science in light of its basis in materialistic atomic theory. The region of objective knowledge belonged to science alone. Any systems that infringed "upon the domain of science" must "submit to its control." Scientists, Tyndall aggressively declared, "claim, and we shall wrest from theology, the entire domain of cosmological theory."[2] Tyndall's "Belfast Address" was energetically attacked in the pulpits of Belfast, in British periodicals, and in a series of pamphlets. The late Frank Turner has argued that "no single incident in the conflict of religion and science raised so much furoure."[3]

Tyndall was one of the most influential British scientists of the Victorian era. He was born in Leighlin Bridge, County Carlow, Ireland in 1820. His father, John Sr., was a struggling shoemaker and leather dealer. In 1839, John Jr. began work as a surveyor. In 1847, he accepted a one-year appointment as a mathematics teacher and went on to earn his doctoral degree at Marburg University, Germany. In 1853, he was appointed professor of natural philosophy of the Royal Institution, where he remained for the rest of his scientific career. Tyndall's climb up the social ladder from humble family circumstances was capped in spectacular fashion when, in 1867, he succeeded Michael Faraday as superintendent of the Royal Institution. During his career, Tyndall distinguished himself in the scientific fields of glaciers, radiant heat, spontaneous generation, and atmospheric gases. Indeed, he was among the first to recognize the role of various gases in producing the earth's natural greenhouse effect (hence The Tyndall Centre for Climate Change Research based in the United Kingdom). Owing to his flamboyant lecturing style, Tyndall also became well known as an eloquent public speaker on science to fashionable audiences, despite his noticeable Irish accent. Along with biologist T. H. Huxley, philosopher Herbert Spencer, and botanist J. D. Hooker, Tyndall argued that naturalistic, rather than theistic, explanations could (and should) account for the workings of nature.

Because of their "scientific naturalism," which brought them into opposition with Christian natural theologians, Tyndall and his allies have often been depicted as contributing to a conflict between science and religion in the Victorian period. Historian James Moore presents Tyndall's good friend Huxley as a "gladiator-general" who was one of the most "colourful contributors to the Victorian 'conflict' between science and religion."[4] Since Tyndall was Huxley's close friend, he is implicated in any statement made about Huxley, and he has also been seen as contributing to a conflict.[5] Indeed, Tyndall's links to John William Draper (1811–82) and Andrew Dickson White (1832–1918), generally seen by scholars as the founders of the "conflict thesis," would seem to offer incontestable proof that he accepted the idea that the history of science and religion was one of warfare. Tyndall's "Belfast Address" contains references to one of Draper's books, and he was on the British editorial board of the International Scientific Series, the series in which Draper's *History of the Conflict between Religion and Science* was published in 1874. Moreover, Tyndall wrote the prefatory note for White's *The Warfare of Science* (1876), a book that contained the warfare thesis later developed at length in the two-volume *A History of the Warfare between Science and Theology in Christendom* (1896). Despite the evidence that Tyndall adhered to the conflict thesis, some scholars, myself included, have argued that neither he nor Huxley actually accepted it in its entirety.[6]

Here I will analyze the relationship between two of the key figures in the dissemination of the conflict thesis—Tyndall and Draper—from 1860 up to when Tyndall delivered the "Belfast Address," and beyond. Although Draper's *History of the Conflict between Religion and Science* was published in the same year that Tyndall delivered his "Belfast Address," I will argue that this particular book had no direct impact on Tyndall's speech. So this will not be a straightforward account of how a British figure appropriated an American idea. But it will be a story about how Tyndall's experiences when he visited the United States in 1872 and 1873 affected his thinking and the role he played in promoting the conflict thesis.

An American Donkey in Oxford

It is likely that Draper and Tyndall first encountered each other at the British Association for the Advancement of Science annual meeting at Oxford in 1860. This was the famous meeting where Bishop Samuel Wilberforce and Thomas Henry Huxley sparred over Darwin's *On the Origin of Species*. Draper delivered a paper on June 30, 1860, in the life sciences section, titled "On the Intellectual Development of Europe, Considered with Reference to the Views of Mr. Darwin and Others, That the Progression of Organisms Is Determined by Law."[7] By all accounts, Draper's paper was a crashing bore.[8] Joseph Dalton Hooker, one of the Darwinians present, later told Darwin that it was "flatulent stuff" and described Draper as a "yankee donkey."[9] Hooker sat through the hour and a half paper, like other members of the audience, only because he had heard that Wilberforce planned to use the occasion to attack Darwin, and the bishop of Oxford did not disappoint. Tyndall, who was also at that session, claimed that he knew that Huxley would respond effectively to Wilberforce, and that this might have serious consequences. "So I [Tyndall] gradually edged my way through the crowd," he later recollected, "overturning in my passage a seat on which many people were standing—(laughter)—till I got close to my friend, who, I feared, incurred some risk of physical mauling."[10]

Although Hooker ridiculed Draper for being a Yankee donkey, he had actually been born in England. His father was a Wesleyan clergyman, but after entering the newly opened University of London to study chemistry and medicine, Draper slowly drifted away from the religion of his youth toward Benthamism and positivism. In 1832, at the age of twenty-one, he went to Virginia to take up a teaching position that did not materialize. After obtaining a medical degree, he was appointed in 1836 professor of chemistry and natural history at Hampden-Sydney College in Prince Edward County, Virginia. In 1839, he took up a chair in chemistry at New York University, a position that he held for the rest of his career.[11]

Whether or not Draper and Tyndall actually met at Oxford in 1860, a few years later Draper sent Tyndall a copy of his recently published book, *A History of the Intellectual Development of Europe*. The book, written in 1861 but not published until 1863 because of the Civil War, expands on the theme that Draper had dealt with in his British Association paper (which he mentions in the preface) and asserts that he had reserved the historical evidence for his physiological argument on the mental progress of Europe "... for subsequent publication. This volume contains that evidence."[12] Draper intended the book to help the American nation, as it entered a new stage in its development, to prepare for "another period of progress under new conditions" by learning from the intellectual history of Europe. An examination of this history revealed that progress was governed by natural law akin to those laws controlling bodily growth. "The life of an individual is a miniature of the life of a nation," Draper declares. This Comte-like formulation of the laws of progress was followed from the time of the ancient Greeks up to the European age of reason. Throughout this historical process "two antagonistic principles" were at work: the "ecclesiastical and intellectual." Although Draper does not encapsulate the antagonism in the phrase "conflict between religion and science," the ecclesiastical principle was embodied in religious institutions, especially Christianity after it was founded, and the intellectual was embodied in the modern period predominantly in science. However, despite his animus toward institutional religion, Draper conceives of progress in religious terms, as a law through which "one Almighty Being" ruled "the universe according to reason."[13] The *History* culminates with the vain effort of Roman Catholicism to hold back the increasing power of the scientific spirit, but Draper remains true to his Methodist roots by holding on to a concept of divine purpose in human history, though his theism was a form of rationalistic deism.[14] The proper conclusions for Americans to draw from the history of Europe, he claims, are that political institutions should work to improve and organize the national intellect while refraining from exciting theological odium against purely scientific ideas.[15] Intellectual freedom was conducive to progress.

After receiving his copy of the *History*, Tyndall wrote to Draper on July 15, 1863, that "I had the pleasure of receiving two days ago your new and important work." He told Draper that he had read the book, and that producing it required "courage as well as ability—on the part of its author." He expressed his hope that the time was coming when courage would not need to be "applied to any undertaking which consists in the earnest utterance of a man's convictions; but that each will be permitted, without fear of persecution to speak out the thought that is in him."[16]

Tyndall's comment about intellectual freedom is particularly interesting given his own situation. Tyndall's father was an ardent Orangeman who belonged to the Church of Ireland. His intense dislike of Catholicism was passed on to his son. But as a young man, Tyndall slowly discarded traditional Christianity. By 1848, he thought of himself as a freethinker, who nevertheless recognized the importance of the Bible while believing in the power of the Spirit.[17] Tyndall's hostility toward the Anglican Church resulted from the years of frustration he endured while trying to obtain a suitable scientific position after he had completed his PhD at Marburg. Since he had not attended Anglican Oxbridge, unqualified candidates with the right social connections were selected over him for a number of posts, even though he had received excellent scientific training in Germany. Through its control of the two ancient universities, the church had a virtual monopoly over potential funds for the development of a scientific professoriate. Tyndall also had principled objections to the whole notion of an established church that received funding from the British state. Tyndall's aversion to Catholicism and Anglicanism put him in opposition to traditional religious institutions and also made the notion of an antagonism between ecclesiastical and intellectual principles appealing to him.

But Tyndall was constrained in several ways from speaking out. First, he could not tackle controversial issues in his lectures at the Royal Institution, which catered to a well-to-do audience not in sympathy with Tyndall's radical religious position. As he wrote to a friend in 1877, "the clearly understood law of the Royal Institution is that neither religion nor politics shall be introduced into our lectures. I have been most scrupulous in my conformity to this law. Never in a single instance have I infringed it during the twenty four years of my connexion with the Institution."[18] But he was also constrained in what he could say outside the institution. Because of his friendship with the religiously devout Faraday, who was a father figure to him, Tyndall kept his radical religious and political views under wraps. Although he was a close friend of Huxley's, ready to defend him from the physical assault of outraged Anglicans at Oxford if need be, Tyndall did not take part in the debates over evolution in the early 1860s. Instead of publically declaring himself an adherent to evolutionary theory, he let Huxley take all the heat.

Draper and Tyndall continued their correspondence during the mid-1860s. Draper wrote to Tyndall in 1865 to thank him for sending him a copy of *Heat Considered as a Mode of Motion* (1863), praising Tyndall's researches for having "inestimable value."[19] Later that same year he expressed his gratitude for receiving a copy of Tyndall's Rede Lecture on radiation.[20] In 1866, Tyndall wrote to Draper to thank him for his letters, and to voice his regret on learning that Draper had decided to quit science for history, remarking that in his opinion "Science is a serious loser by your choice."[21] But shortly after Tyndall sent this letter to Draper, he too found that his life had changed significantly. Not only had Tyndall lost a close friend when Faraday died on August 30, 1867, he also lost an astute adviser who had helped him to keep his radical views in check. After Faraday's death, Tyndall became less and less cautious about revealing his religious radicalism and his support for Darwin, expressing such sentiments with increasing frequency in his public lectures and published articles.

While his growing outspokenness caused some to criticize Tyndall, such hostility was nothing compared with the condemnation that came when he became embroiled in a controversy over the efficacy of prayer. In 1872, Tyndall submitted an anonymous letter for publication to the *Contemporary Review*, with a brief preface that he had written. In it, he declared that on the subject of how "Providence" acted in physical affairs there were two opposing parties. One affirmed "the habitual intrusion of supernatural power, in answer to the petitions of men," while the other questioned, or even denied, any intervention. He explained that he wanted to test these two positions through scientific experiment. The experiment was designed to allow for "quantitative precision on the action of the Supernatural in Nature." Tyndall insisted that since the proposal was "fair" he could not refuse "to give it the support implied by these few lines of introduction."[22] The proposal was to compare the healing rates of two hospital wards over the course of three to five years. In one ward, prayers would be offered for the sick, while in the other there would be no prayers. This test would finally settle the question of the power of prayer. Tyndall had written about the efficacy of prayers in the past, and religious writers had generally ignored these earlier essays. But his support for this proposal was more provocative. It suggested that a spiritual matter of great delicacy could be subjected to scientific analysis. It touched off what came be known as the "prayer-gauge debate," and Tyndall found himself under attack by outraged Christians.[23] It was likely with a sense of relief that Tyndall left for a lecture tour of the United States in the middle of the prayer-gauge debate.

A British Scientist in Yankee-Land: The American Trip, 1872–1873

On January 1, 1873, Huxley wrote to Tyndall that he seemed to be making "a Royal Progress in Yankee-Land. We have been uncommonly tickled with some of the reports of your lectures which reached us, especially with that which spoke of your having 'a strong English accent.'"[24] Tyndall's American lecture tour was an unqualified success. He arrived on October 9, 1872, visiting Boston, Philadelphia, Baltimore, Washington, New York and Brooklyn, and New Haven. In four months, he delivered thirty-five lectures to large and enthusiastic audiences, which on occasion numbered as many as 1,500 listeners. Tyndall must have compared the positive reception he received in the United States with the more hostile atmosphere in England in the wake of the prayer-gauge debate. It led him to see the United States as a land of freedom, poised, as Draper had claimed, to enter into a new stage of development where science provided the engine for progress.

Though Tyndall had friends in the United States, there was some question about how hospitable the Americans would be. In an attempt to head off criticism of Tyndall, Edward Youmans, Spencer's agent in America and an adviser to the Appleton publishing house, included a piece by Tyndall titled "Science and Religion" in the November 1872 issue of the *Popular Science Monthly*. This article contains what was really Tyndall's first published systematic discussion of the relationship between science and religion.[25] It emphasizes Tyndall's belief that religion would survive the "removal of what had been long considered essential to it," such as the geocentric worldview and the reputed age of the earth. Religion would even survive the theory of evolution. "In fact," Tyndall declares, "from the earliest times to the present, religion has been undergoing a process of purification, freeing itself slowly and painfully from the physical errors which the busy and uninformed intellect mingled with the aspiration of the soul, and which ignorance sought to perpetuate." The present debate on the physical value of prayer was just another example of a physical error from which religion would free itself. The corporeal world was not prayer's legitimate domain, and "no good can come of giving it a delusive value by claiming for it a power in physical Nature."[26] By arguing that prayer belonged to a domain different from that of physical nature, Tyndall was putting forward, though in a rudimentary form, a two-spheres concept of the relationship between science and religion, or what Stephen J. Gould once referred to as the two non-overlapping magisteria. Tyndall's separation of science and religion was heavily indebted to his reading of the German romantics, and it can be found in his reverence, since at least the late 1840s, for American transcendentalists like Ralph Waldo Emerson.[27]

As it turned out, Tyndall encountered little personal animosity while he was in the United States. He wrote to his close friend Thomas Hirst from Philadelphia that people had been praying for him and writing him letters about his soul. "Some of the letters [are] so full of indescribable sweetness," he told Hirst, "that it is a pain to give such people pain by differing from them."[28] To another friend, Debus, he wrote that although the prayer question had "excited" America, the clergy "do not on the whole hold aloof from me."[29] To Bence Jones, the secretary of the Royal Institution, he exclaimed that "from beginning to end nothing has occurred to mar my visit, if I except a slight passage of arms with an ill tempered Irish Presbyterian."[30] On the whole, Tyndall was impressed by the degree of religious tolerance in the United States, even within each family. "Nothing," he wrote in his journal, "has struck me more since I came here than the widely divergent religious views among the members of the same family and still the perfect tolerance that reigns among them." The father, mother, and three children could go to "five different places of worship without the least disunion being introduced into family relations."[31]

The lectures on light that Tyndall gave at each of his stops in the United States did not deal with the prayer-gauge debate, or even with the broader issue of the relationship between science and religion. He had decided that his lectures would be more in the tradition of his Royal Institution talks, avoiding controversial issues and focusing on engaging experiments. He described, demonstrated, and explained reflection and refraction of light rays, the production of color by prisms, diffraction and interference with slits and thin films, and polarization. But he also asserted that Americans needed to be more supportive of pure science.[32] In the concluding lecture he referred to Joseph Henry and Draper as examples of true investigators inspired by the love of pure scientific research rather than being motivated by the desire to produce wealth.[33] It could be argued that Tyndall extolled the virtues of pure science to deal with a religious issue, if only indirectly. He was implying that the association between practical science and the Protestant work ethic in the United States—the notion that practical applications of science manifested humanity's God-given dominion over the earth—had undermined the American commitment to pure scientific research and become a major barrier to scientific advancement.[34]

Near the end of Tyndall's visit, his friends organized a farewell banquet in his honor. Held at Delmonico's in New York City on February 4, 1873, the attendees included Draper, White, Appleton, Youmans, the paleontologist E. D. Cope, the chemist Benjamin Silliman, and religious figures such as Henry Ward Beecher and the Reverend Dr. Bellows. Tyndall's discussion of pure research in his lectures was one of the prominent themes in the speeches heard at the banquet, but not the only one. A number of speakers focused on how science and religion properly understood were not in conflict. Andrew Dickson White praised Tyndall as the embodiment of the spirit of scientific research, a spirit sorely needed for the "political progress of our country." The Reverend Dr. Bellows insisted that there could be "no radical quarrel between . . . true science and true Christianity." He asked Tyndall to tell England that the United States was not "suspicious of true science," and to deliver "our loving compliments" to Huxley, Mill, Spencer, and the rest of his friends. The Reverend Henry Ward Beecher spoke in a similar vein. Christian ministers, he said, were "reading more of Prof. Tyndall's books, and Mr. Huxley's books, and Herbert Spencer's books, than any other profession in the United States of America." They were working together with the scientists to find truth and establish "common ground." There was not the "slightest reason to fear that religion would be overthrown by the investigations of science."[35] The sentiments expressed by those at the banquet deeply impressed Tyndall as well as his friends back home. He wrote to Joseph Henry that his English colleagues "were perfectly delighted and predict a great scientific future for a country in which such a Spirit can be shown."[36]

Undertaking the "Belfast Address"

Tyndall boarded the *Cuba* in New York on February 5, 1873 and arrived in England fourteen days later.[37] Within a few months of returning, he was wrestling with an important decision: Should he accept the presidency of the British Association? Tyndall knew as early as April 1873, that there was a possibility that he would be offered the presidency, committing him to give the presidential address in Belfast in August 1874. On April 6, he wrote to Henry, "I am hard pressed by the Council of the British Association to accept the Presidency at Belfast next year."[38] However, Tyndall may have known earlier than this and may have already decided what he was going to cover in his address when he was in the United States. In the journal that Tyndall kept during his trip there are several pages of notes on Friedrich Albert Lange's *History of Materialism and Critique of Its Present Significance* (1866).[39] Tyndall must have brought the book (not yet translated into English) with him on the trip. Either he was reading it out of personal interest and later decided that he would present a history of materialism at Belfast, drawing heavily on Lange, or he already knew that he would have to accept the offer of the presidency and had begun to prepare his speech. One of Tyndall's friends was surprised at how early he had begun to work on the address. On August 4, 1873, M. F. Egerton asked, "Are you really beginning preparations for Belfast? It seems a long time beforehand!"[40]

But by September, Tyndall was already having second thoughts about accepting the presidency. The mayor of Belfast had publically asked that someone other than Tyndall preside. Tyndall wrote to Huxley on September 24, "I wish to Heaven you had not persuaded me to accept that Belfast duty. They do not want me. Well in return I may be less tender in talking to them that I otherwise should have been. . . . So I suppose I am in for it—and so are you you know and Hooker too."[41] Tyndall's friends had to persuade him not to step down. Hirst wrote to him on September 28 that he agreed with Huxley and Spottiswoode that he should just ignore the opposition. The Council of the British Association had "deliberately selected" him, and he had accepted its invitation.[42] Tyndall decided to take this advice.

Despite his best efforts, Tyndall's progress on the address was slow throughout the fall of 1873 and the first half of 1874. His duties at the Royal Institution were all-consuming. In the spring of March 1874, he wrote to the German physicist Clausius that he would have to retire to Switzerland in order to prepare the paper. By June 24, Huxley was beginning to be worried about how slowly the address was going and about Tyndall's ability to rein in his hostility toward his critics. He reminded Tyndall about how restrained he had been when he had delivered his own presidential address in 1870 at Liverpool. Huxley wrote, "I wonder if that Address is begun, and if you are going to be as wise and prudent as I was at Liverpool. When I think of the temptation I resisted on that occasion . . . I marvel at my own forbearance! Let my example be a burning and shining light to you."[43] Tyndall wrote Hirst on July 5 that he was amused "to find Huxley expressing anxiety" and added in a letter written later that same month that, despite the slow pace, he was near the end and hoped "to make a somewhat daring but dignified wind-up."[44]

As Huxley had feared, Tyndall was unable to restrain himself. His experiences in the United States, where he had been warmly received, the opposition to his presidency, and his increasing aggressiveness since Faraday had died all combined to make Tyndall more reckless than usual. The contrast between the United States and his own country must have put him in a particularly combative frame of mind. Whereas Tyndall had been warmly received by the Americans during his lecture tour, in Britain he was attacked even before he had delivered his presidential address. While Tyndall perceived the United States as open to a diversity of religious positions, the power of the established Anglican Church in England and of the Catholic Church in Ireland combined to make Britain far less tolerant. An aggressive assault on institutionalized religion was not needed in the United States, as there was no established church and the future looked bright. A more subtle message emphasizing the importance of pure science was appropriate. But for a British audience, Tyndall decided, it was time to be outspoken.

Four days before he was to deliver the speech, he wrote to Hirst, "I do not care what the effect of the Address may be upon the audience, but I have the firmest confidence . . . that it utters a truth or two which will survive the meeting of the association."[45] He thought of the address as striking a blow for religious freedom that was similar to what Martin Luther's refusal to recant his teachings had done at the Imperial Diet of Worms in 1521. He wrote to Heinrich Debus, "I will go to Belfast as Luther did to [Worms] if necessary—and meet if requisite all the Devils in Hell there."[46]

The Belfast Address and Draper's Conflict

Draper's *History of the Intellectual Development of Europe* and Lange's *History of Materialism* must have been two of the books in the library that Tyndall had taken with him to Switzerland to prepare for his talk. The historical section of the address basically combines Lange's account of materialism with Draper's narrative on the Christian church's role in obstructing intellectual progress. Although Tyndall's historical analysis features a notion of conflict between religious institutions and science, his "dignified wind-up," where he summed up the significance of the history of the relationship between materialism, Christianity, and science, puts forward a much fuller description of his own personal beliefs, including his belief in the separation of science from religion. Knowing many would suspect him of merely advocating the crude materialism of the past, he begged his audience to be patient until the end, as the "'materialism' here professed may be vastly different from what you suppose." He then launches into a philosophical discussion, shaped by his debt to Kant's emphasis on the limits of knowledge, and to German romantics who perceived a mysterious force lying behind the natural world. Whether or not there was an external world composed entirely of matter was really impossible to prove. "Our states of consciousness are mere *symbols* of an outside entity which produces them and determines the order of their succession," he argued, "but the real nature of which we can never know." Not only was materialism philosophically suspect, it also ignored the "immovable basis of the religious sentiment in the nature of man." The religions of the world were based on this valid human sentiment, but religion became "mischievous if permitted to intrude on the region of objective *knowledge*, over which it holds no command, but [is] capable of adding, in the region of *poetry* and *emotion*, inward completeness and dignity to man." Now that Tyndall had distinguished between the realm of science, which concerned objective knowledge, and the equally important region of religion, which had to do with poetry and emotion, he could spell out the mischievous force that had led the religions of the world astray: theology. Tyndall concludes with an unapologetic demand for the intellectual autonomy of science from the restrictions of dogmatic theology. Scientists, he claims, "shall wrest from theology, the entire domain of cosmological theory. All schemes and systems which thus infringe upon the domain of science must, in so far as they do this, submit to its control, and relinquish all thought of controlling it."[47] His two-spheres approach, which retained an important role for religion, and which reconciled science and religion, was grounded in what he had learned from Kant and the German romantics. Historically, he saw conflict, but philosophically the conflict only took place between science and theology, *not* science and religion. There is nothing comparable in Draper to Tyndall's development of a philosophical reconciliation of science and religion.

Tyndall was attacked from every side. As he told Hirst a week after delivering the address, "[Y]ou can form no notion of the religious agitation. Every pulpit in Belfast thundered at me. Even the Roman Catholics who are usually wise enough to let such things alone come down upon me."[48] Despite his rejection of materialism, Tyndall was denounced in the periodical press as an immoral materialist with an unsavory connection to German forms of thought. He was condemned for abusing his position as president by speaking about religious matters outside the domain of science.[49] In October, the *Pall Mall Gazette* reported that a London merchant by the name of C. W. Stokes had sent an inquiry to the Home Secretary, asking whether Tyndall should be imprisoned for blasphemy.[50] Tyndall had realized that he would be criticized for his aggressive comments in the "Belfast Address." But he was not prepared for the virulence of the attacks or for the misreadings of his true position. The entire British religious world seemed to be against him. To one correspondent he wrote, "[R]eally on reading the religious newspapers I sometimes ask myself, are those people mad, or am I mad. Their delusions appear to me more and more grotesque and horrible."[51]

While Tyndall was dealing with the initial fallout from his talk in the summer of 1874, he received a letter from Draper dated September 15. Draper wrote, "I am at this moment finishing a book entitled 'History of the Conflict Between Religion and Science.'" Written more than a year previously, Draper told Tyndall that it would be published in six to eight weeks.[52] Draper had been asked by Youmans in 1873 to contribute a book to the International Scientific Series. *History of the Conflict between Religion and Science* (1874) was an international best seller. No other title in the more than a hundred-volume series sold as well. In the United States alone it passed through fifty printings over about fifty years. In the United Kingdom there were twenty-one editions in fifteen years, and there were numerous translations worldwide. Like the "Belfast Address," Draper's *Conflict* touched off a major controversy. The 1876 Spanish edition was entered in the *Index liborum prohibitorum*, joining works by Copernicus, Galileo, Kepler, Locke, and John Stuart Mill.[53] Since he too was under attack, Draper wrote sympathetic letters to Tyndall. On January 4, 1875, he told Tyndall that he was receiving supportive letters from friends and strangers about the beneficial impact of his *History of the Conflict*. Draper was going to repeat the central message in those letters to Tyndall, since their situations were so similar. "I say to you," Draper declared, "Stand fast. Your address is doing great good. If you need help let me know." Draper hoped that "the friends of science will stand by you in England as they are standing by me in America. Let us all fight shoulder to shoulder in our fighting—not for ourselves but for posterity."[54]

Draper's *Conflict* relies heavily on his earlier book, *History of the Intellectual Development of Europe*. It offers a more succinct historical account of the antagonism between the ecclesiastical and intellectual principles. But this time, Draper refers to the two antagonistic principles as religion and science. He also added three new chapters, two that examine the relationship of modern civilization to "Latin Christianity" and "Science," and a third, titled "The Impending Crisis," which discusses the imminent collision between the Catholic Church and modernity. In this, the concluding chapter, Draper draws attention to recent developments contributing to the crisis that had come after the publication of his *History of the Intellectual Development of Europe*. They included the Syllabus of Errors of 1864, the Vatican Council of 1869–70, the French defeat in the Franco-Prussian War of 1870–71, Bismarck's *Kultukampf*, and the evolution debates.[55]

Draper made it clear in the preface that the papacy was the real enemy. "In speaking of Christianity," he explained, "reference is generally made to the Roman Church, partly because its adherents compose the majority of Christendom, partly because its demands are the most pretentious, and partly because it has commonly sought to enforce those demands by the civil power."[56] The current battle between religion and science was merely the "continuation of a struggle" that began much further back in history when "Christianity began to attain political power."[57] The current papacy recognized that science was undermining its dogmas and its political power. Its response in the Syllabus of Errors and the Vatican Council was to declare war on scientific progress, the foundation of modernity.[58] But, significantly, while Catholicism was the true enemy, Protestantism could be redeemed. A "reconciliation of the Reformation with Science" was possible if the Protestant churches lived up to Luther's maxim of the right of private interpretation of the scriptures, which was "the foundation of intellectual liberty." When Protestants attacked scientists as infidels and atheists they did so because they were unable to emancipate themselves completely from the principles of Catholicism.[59] Draper therefore did not see all religion as being hostile toward science. As Lawrence Principe has pointed out, Draper had a religion buried underneath his critique of Catholicism.[60] It was deistical and anti-Trinitarian to be sure, but it was a religion nevertheless. Draper hoped that the conflict between science and religion would result in the purification of some form of Protestantism. He was open to some form of religion, and therefore his conception of the conflict thesis is not quite so simple or straightforward as often supposed.[61]

For Tyndall, the publication of Draper's *Conflict* provided him with a new resource to draw on in his ongoing controversies during the mid-1870s. In a note in his "Apology for the Belfast Address" (1874), he recommended "to the reader's particular attention Dr. Draper's important work entitled, 'History of the Conflict between Religion and Science.'"[62] A year later, Tyndall referred to Draper's book in a letter to the *Times* in which he protested against the Vatican's attempt to obtain control over education. Tyndall insisted that there was no reconciliation between the liberty of Rome to teach mankind, and the liberty of the human race. "This, in our day," Tyndall proclaimed, "is the 'conflict' so impressively described by Draper, in which every thoughtful man must take a part."[63]

Tyndall's positive references to Draper's *Conflict* obscure important differences in the way they handled the theme of conflict in the mid-1870s. For Draper, the main antagonist in the conflict between science and religion was Catholicism. Catholicism was the central institutional force impeding the growth of intellectual freedom in the modern world, and through its power it had poisoned the true principles of Protestantism. By contrast, Tyndall's "Belfast Address" targeted all Christendom, not just Catholicism. Although Catholicism was powerful in Belfast, Tyndall believed that the Anglican establishment was just as dangerous, perhaps even more so. In the British context, this was the religious institution opposing attempts by Tyndall and his allies to reform science. His broader critique of all Christianity was tied closely to his concerns about Christian theology. Since Tyndall adhered to a two-spheres approach, he did not see a historical conflict between science and religion, only between science and theology. Tyndall's distinction between religion and theology points to the debt that he owed to German idealism, a philosophical tradition that had little impact on Draper. Philosophy not only separated Tyndall and Draper but religion as well. Tyndall's notion of religion based on the emotions elicited by nature differed profoundly from Draper's deism. For Tyndall, a divine force was currently at work in nature. His belief that religion was intrinsically valuable if expressed through poetry and art was often lost during the controversy surrounding the "Belfast Address."

Conclusion

The "Belfast Address" and the *History of the Conflict between Science and Religion* actually contain two different conflict theses, shaped by the different contexts in which Tyndall and Draper lived. There was no single conflict thesis. It turns out that the conflict thesis was quite malleable—it could be adapted depending on the local conditions. Draper found his version useful in the American context, and Tyndall thought that his version was well suited to the British context. The story of Tyndall, Draper, and their conflict theses illustrates how we need not accept a "thesis essentialism."[64] It isn't just that the conflict thesis itself cannot provide a historiographical model for scholars—we've known this now for many years—no model can. We also need to be more sensitive to the idea that there were not a handful of fixed models—conflict, interdependence, dialogue, integration—through which historical actors viewed the relationship between science and religion.[65] That is far too simplistic. In the case of conflict theses we have a number of differing notions of the antagonisms between science and religion that melt into each other. Tyndall and Draper agree that there is conflict, but not on what is in conflict. For Draper the emphasis is on the opposition between science and the principles of Catholicism; for Tyndall it is science and theology. What would happen if we bring in a more modern figure, like Richard Dawkins, whose position is more simplistic than Draper's and Tyndall's? Where Draper holds out hope for Protestantism, Dawkins finds all Christianity, indeed, all religion, to be in opposition to science. Where Tyndall finds a positive role for a religion based on human emotion and poetic utterance evoked by nature, Dawkins rejects any positive role for religion. In some ways, Dawkins's version of the conflict thesis is the "purest" of all and conforms more closely to the version that historians have tried to explode. Perhaps, in light of a comparison of the differing conceptions of the conflict thesis held by Tyndall, Draper, and Dawkins, it is time that we rejected the notion of the fixity of theses.

ACKNOWLEDGMENTS

I would like to acknowledge the helpful suggestions I received from Ronald Binzley, Fred Gregory, Ronald Numbers, Larry Principe, and James Ungureanu. I am also grateful to Michael Barton for providing some key sources and to Julie Goriounov and Sven Pinczewski for transcribing a section from Tyndall's journals.

NOTES

1. John Tyndall, *Fragments of Science: A Series of Detached Essays, Addresses, and Reviews*, 2 vols., 8th ed. (London: Longmans, Green, 1892), 2:136–37.
2. Tyndall, *Fragments of Science*, 2:197.
3. Frank M. Turner, *Contesting Cultural Authority: Essays in Victorian Intellectual Life* (Cambridge: Cambridge University Press, 1993), 196.

4. James R. Moore, *The Post-Darwinian Controversies: A Study of the Protestant Struggle to Come to Terms with Darwin in Great Britain and America 1870–1900* (Cambridge: Cambridge University Press, 1979), 67–68.

5. Ronald L. Numbers, "Aggressors, Victims, and Peacemakers: Historical Actors in the Drama of Science and Religion," in Harold W. Attridge, ed., *The Religion and Science Debate: Why Does It Continue?* (New Haven, CT: Yale University Press, 2009), 33–34.

6. Bernard Lightman, "Does the History of Science and Religion Change Depending on the Narrator? Some Atheist and Agnostic Perspectives," *Science and Christian Belief* 24 (October 2012): 149–68; Lightman, *The Origins of Agnosticism: Victorian Unbelief and the Limits of Knowledge* (Baltimore, MD: Johns Hopkins University Press), 131–34; Ruth Barton, "Evolution: The Whitworth Gun in Huxley's War for the Liberation of Science from Theology," in D. Oldroyd and I. Langham, eds., *The Wider Domain for Evolutionary Thought* (Dordrecht: D. Reidel, 1983), 261–87.

7. For a recent revisionist account of Draper at Oxford, see James C. Ungureanu, "A Yankee at Oxford: John William Draper at the British Association for the Advancement of Science at Oxford, 30 June 1860," *Notes and Records* 70 (2015): 135–50.

8. Ian Hesketh, *Of Apes and Ancestors: Evolution, Christianity, and the Oxford Debate* (Toronto: University of Toronto Press, 2009), 78–80.

9. J. D. Hooker to C. R. Darwin, July 2, 1860, Darwin Correspondence Database, https://www.darwinproject.ac.uk/entry-2852.

10. "Professor Tyndall on Scientific Research (from Our Correspondent)," *Leeds Mercury*, December 5, 1884, 2. I am grateful to Roland Jackson for pointing out this article to me.

11. Moore, *Post-Darwinian Controversies*, 22; Donald Fleming, *John William Draper and the Religion of Science* (Philadelphia: University of Pennsylvania Press, 1950), 8–19.

12. See John William Draper, *A History of the Intellectual Development of Europe* (New York: Harper and Brothers, 1863), iii.

13. Draper, *History of the Intellectual Development of Europe*, iii-iv, 3, 421.

14. Moore, *Post-Darwinian Controversies*, 23; Fleming, *John William Draper*, 76.

15. Draper, *History of the Intellectual Development of Europe*, 616, 620.

16. Tyndall to Draper, July 15, 1863, container 7, John William Draper Family Papers, Library of Congress, Washington, DC.

17. Geoffrey Cantor, "John Tyndall's Religion: A Fragment," *Notes and Records* 69 (2015): 421.

18. Tyndall to Spottiswoode, November 11, 1877, John Tyndall Correspondence, Royal Institution, London, RI MSS T., 19/F6, 25.

19. Draper to Tyndall, July 24, 1865, container 7, John William Draper Family Papers.

20. Draper to Tyndall, November 6, 1865, container 7, John William Draper Family Papers.

21. Tyndall to Draper, January 11, 1866, container 7, John William Draper Family Papers.

22. John Tyndall and [Henry Thompson], "The 'Prayer for the Sick': Hints towards a Serious Attempt to Estimate Its Value," *Contemporary Review* 20 (1872): 205–10, on 205–6.

23. Robert Bruce Mullin, "Science, Miracles, and the Prayer-Gauge Debate," in David C. Lindberg and Ronald L. Numbers, eds., *When Science and Christianity Meet* (Chicago: University of Chicago Press, 2003), 203–24.

24. Huxley to Tyndall, January 1, 1873, John Tyndall Correspondence, RI MSS JT/1/TYP/9/2967-2969.

25. Ursula DeYoung, *A Vision of Modern Science: John Tyndall and the Role of the Scientist in Victorian Culture* (New York: Palgrave Macmillan, 2011), 98.

26. Professor Tyndall, "Science and Religion," *Popular Science Monthly* 2 (November 1872): 79–82. This piece was reprinted from the *Contemporary Review*, but, interestingly enough, with a new title. See John Tyndall, "On Prayer," *Contemporary Review* 20 (June 1, 1872): 763–66.

27. Lightman, *Origins of Agnosticism*, 99, 108, 130.

28. Tyndall to T. A. Hirst, November 23, 1872, Tyndall Correspondence, RI MS JT/1/HTYP/608-609.

29. Tyndall to H. Debus, December 26, 1872, Tyndall Correspondence, RI MS JT/1/T/269.

30. Tyndall to H. Bence Jones, January 3, 1873, Tyndall Correspondence, RI/JT/TYP 3.82.

31. Tyndall Journals, Boston, October 27, 1872, Royal Institution, London, RI MS JT/2/12, p. 13.

32. Katherine Russell Sopka, "John Tyndall: International Populariser of Science," in W. H. Brock, N. D. McMillan, and R. C. Mollan, eds., *John Tyndall: Essays on a Natural Philosopher* (Dublin: Royal Dublin Society, 1981), 196.

33. John Tyndall, *Six Lectures on Light: Delivered in America in 1872–1873* (London: Longmans, Green, and Co., 1873), 217.

34. Michael D. Barton, "The 'Efficient Defender of a Fellow-Scientific Man': John Tyndall, Darwin, and Preaching Pure Science in Nineteenth-Century America" (PhD diss., Montana State University, 2010).

35. *Proceedings at the Farewell Banquet to Professor Tyndall Given at Delmonico's, New York, February 4, 1873* (New York: Appleton, 1873), 80–81, 90, 59–61.

36. Tyndall to Joseph Henry, April 6, 1873, Joseph Henry Papers, Archives of the Smithsonian Institution, Washington, DC, 1921JE. Tyndall wrote something similar to Youmans. See Tyndall to E. L. Youmans, April 12,1873, in Eliza Ann Youmans, "Tyndall and His American Visit," *Popular Science Monthly* 44 (February 1894): 514.

37. Arthur S. Eve and Clarence H. Creasey, *Life and Work of John Tyndall* (London: Macmillan, 1945), 173.

38. Tyndall to Henry, April 6, 1873, Joseph Henry Papers, 1921JE.

39. Tyndall Journals, November 8, 1872, Royal Institution, London, RI MS JT/2/12, p. 55.

40. M. F. Egerton to Tyndall, August 4, [1873], Tyndall Correspondence, RI MS JT/1/E/43.

41. Tyndall to Huxley, September 24, 1873, Tyndall Correspondence, RI MS JT/1/TYP/9/3022.

42. Hirst to Tyndall, September 28, 1873, Tyndall Correspondence, RI MS JT/H/255.

43. Huxley to Tyndall, June 24, 1874, Tyndall Correspondence, RI MS JT/1/TYP/9/3034-3035.

44. Tyndall to Hirst, July 30, 1874, British Library, London, BL 63902-874F-10-43.

45. Tyndall to Hirst, August 15, [1874], Tyndall Correspondence, RI MS JT/1/T/917.

46. Tyndall to Heinrich Debus, 1874 (year only), Tyndall Correspondence, RI MS JT/1/TYP/7/2375.

47. Tyndall, *Fragments of Science*, 2:191–93, 196–97.

48. Tyndall to Hirst, August 26, 1874, Tyndall Correspondence, RI MS JT/1/T/715.

49. Bernard Lightman, "Scientists as Materialists in the Periodical Press: Tyndall's Belfast Address," in Geoffrey Cantor and Sally Shuttleworth, eds., *Science Serialized* (Cambridge, MA: MIT Press, 2004), 199–237.

50. "Occasional Notes," *Pall Mall Gazette* 20, no. 3026 (October 28, 1874): 1452.

51. Tyndall to Madame Novikoff, December 23, 1874, Tyndall Correspondence, RI MS JT/1/TYP/3/933.

52. Draper to Tyndall, September 15, 1874, container 7, John William Draper Family Papers.

53. Moore, *Post-Darwinian Controversies*, 28.

54. Draper to Tyndall, January 4, 1875, container 7, John William Draper Family Papers.

55. Fleming, *John William Draper*, 125–26.

56. John William Draper, *History of the Conflict between Religion and Science* (New York: D. Appleton, 1874), x.

57. Draper, *History of the Conflict between Religion and Science*, vi.

58. Draper, *History of the Conflict between Religion and Science*, 330–32.

59. Draper, *History of the Conflict between Religion and Science*, 363–64.

60. See Principe's chapter in this volume.

61. I am indebted to James Ungureanu for helping me to think this point through more carefully.

62. Tyndall, *Fragments of Science*, 2:221.

63. John Tyndall, "The Vatican and Physics: To the Editor of the *Times*," *Times* (London), December 18, 1875, 7.

64. Cantor made the point in 2010 that there were actually two versions of the conflict thesis, a weaker and a strong one. See Geoffrey Cantor, "What Shall We Do with the 'Conflict Thesis'?," in Thomas Dixon, Geoffrey Cantor, and Stephen Pumfrey, eds., *Science and Religion: New Historical Perspectives* (Cambridge: Cambridge University Press, 2010), 283–98.

65. I have taken these four models from Ian Barbour's *Issues in Science and Religion* (Englewood Cliffs, NJ: Prentice-Hall,1966), but of course models used by scholars have varied over the years.

CHAPTER FIVE

Continental Europe

FREDERICK GREGORY

An explicit assertion that science and religion represented mutually hostile endeavors made its appearance in Germany in the middle of the nineteenth century, two decades before it gained widespread attention in Britain and America. The reasons for its emergence lie deep in the rich intellectual tradition for which Germany is so famous, one that was born in the late eighteenth century when what Stephen Turner has called an ideology of *Wissenschaft* began to dominate German thought.[1] *Wissenschaft*, a word often left untranslated because there is no one word in English that captures all it entails, refers to systematic scholarly study that seeks to show how all aspects of a discipline and the first principles governing that discipline are grounded in each other.[2] It assumes that knowledge is a dynamic entity as opposed to a static body of information and it strives to establish internal consistency among all aspects of the enterprise.

Commitment to this novel intellectual endeavor involved a new role for the university professor. No longer content merely to safeguard and pass down a received canon of material, the professor now assumed responsibility to *add to* the canon through systematic research into all facets of a subject, at the same time revealing how new knowledge was bound up with the first principles of the discipline. By so doing, the goal was to make individual subjects like philosophy, theology, aesthetics, literature, history, and others *wissenschaftlich*, sometimes loosely rendered in English as "scientific." From our vantage point in the twenty-first century, it seems ironic that the systematic study of nature, *Naturwissenschaft*, did not separate itself from the more general enterprise until the middle of the nineteenth century. It can be no accident that the idea of a conflict between science and religion emerged at the same time.

The Transformation of Theology and Philosophy in the Eighteenth Century

Early in the eighteenth century, German theologians pursued natural theology with the same convictions as scholars elsewhere. The assumption was that the investigation of nature was clearly an ally of a religious worldview. Physico-theology flourished particularly between 1730 and 1760, a period that produced works on thunder theology, tulip theology, rose theology, grass theology, fire theology, water theology, snow theology, stone theology, insect theology, fish theology, plant theology, bee theology, bird theology, and no doubt more.[3] But later in the century, theology felt the impact of the emergence of the ideology of *Wissenschaft*. Many theologians became preoccupied with history and textual criticism as they grappled with the question of the scope of revelation. How trustworthy were the results of reason? Were revealed truths that were beyond reason necessarily contrary to reason? In an attempt to establish religion as reasonable, the philosopher Hermann Reimarus concluded that revelation was inconsistent with a thoroughly rational religion. Nor was he alone in his conclusion.[4]

As the result of these developments and with the appearance of two editions of Immanuel Kant's classic *Critique of Pure Reason* in the 1780s, works of physico-theology no longer seemed to address fundamental issues.[5] Kant's particular interest was to unpack the limits and proper domain of reason, which did not include what was apprehended by faith. "The separation between the life of the spirit on one side," writes Felix Flückiger, "and nature as a world of objects on the other—heritage of the critical philosophy of Kant—allowed knowledge of nature and the world to be left to the natural sciences and led to engagement at best in questions of limits."[6]

One might be tempted to assume that Kant's critique amounted to an early statement of the conflict thesis. It is true that Kant severely undermined the entire enterprise of natural theology, since it crossed the boundary beyond which reason could not operate. But Kant's radical separation of the realms of science and religion did not assert that they were mutually hostile so much as it placed limits on both religion and natural science, restricting each to its own proper sphere. Religion, now a moral concern, was denied the right to dictate what was true in nature while natural science, being restricted to the phenomenal realm, was denied the right to draw metaphysical implications from its discoveries about how nature worked. Kant's work was not a statement of the conflict thesis; in fact, it proved to be the foundation for an important German theological response to later claims of conflict, to be examined below.

Transcending Kant in the German Romantic Period

Following Kant's death in 1804, these intellectual matters were a distraction for Germans otherwise unhappy with the disruption Napoleon had unleashed. Napoleon exposed Germany's weakness as a power in Europe, generating calls for the German states to unify. Yet nowhere on the horizon was there prospect of the meaningful change needed to bring about a more modern Germany. Germans could take heart from the words of the king of Prussia, uttered during the dark days of French domination, that "the state must replace intellectually what it has lost physically."[7] They could look with considerable pride to the remarkable cultural and intellectual achievements that had been made in the late eighteenth and early nineteenth centuries. And they could indeed detect signs of unification, if not of the political kind, at least on the religious horizon. In 1817, the king supported the union in Prussia of the Lutheran and Reformed congregations into a new unified *Evangelische Kirche*.

In addition, filtering down from the halls of academe was a new outlook, the occasional subject of a popular lecture, that was also capturing public attention in the proliferation of new journals intended, as the phrase went, "for the educated of all classes [*für gebildete aller Stände*]." At the precise time when public opinion was becoming a new and undeniable force in German social and intellectual life, expressions of what would later become known as the romantic age in articles and lectures about culture and philosophy promised a holistic, unified conception of life and experience.[8] Emerging against the background of Kant's critical achievements of the late eighteenth century, the new perspective promised to move beyond the limits Kant's philosophy had imposed when he restricted our knowledge of the world to the phenomenal realm treated by mathematically expressed mechanical explanations of natural science.

To romantic minds, Kant's claim that there could be no scientific explanation of living things—indeed, that there was no hope of uncovering the ultimate truth of nature—was simply unacceptable. The romantic mood was everywhere, exemplified in reviews of the new *Naturphilosophie* of the young genius Friedrich Schelling or discussions of the latest product from the pen of the revered poet, dramatist, and essayist Johann Wolfgang von Goethe. Both were critical of Kant's restriction of natural philosophy to the mechanical interactions of phenomena. Goethe's fame meant that his poetic depiction of nature and his biting critique of Newton received wide attention. Schelling in particular insisted that the fundamental metaphor for nature was organism, not the machine. Since we too are organisms, we can transcend the limits Kant imposed and know nature as we know ourselves. At a time when calls for German unification sounded loud and clear among German youth, the rejection of what many regarded as Kant's dualistic outlook held great appeal. Dualism became the enemy, while a unified conception of life proved more and more irresistible. Because of what this anti-dualistic stance inspired later in the century, I have elsewhere referred to it as "proto-monism."[9] For those attracted to a romantic outlook like Schelling's or Goethe's there was a price to pay. For all their spiritual tone, there was nothing in their works remotely resembling traditional religion. Embracing the new outlook ran the risk of being denounced as a pantheist by the religiously orthodox, as Schelling frequently was. But traditional religion had been under attack for some time in Germany. The theologian Karl Barth assures us that the spirit of the rationalist critique of such matters as miracles, original sin, and the special status of the biblical narrative, as articulated by professors of theology like Julius Wegscheider in his 1816 *Institutes*, "reached the masses of citizens and peasants through the pulpits of countless villages and small towns."[10] When David Friedrich Strauß published his notoriously scandalous *Life of Jesus* in 1835, followed six years later by Ludwig Feuerbach's exposure of the human origin of Christian doctrine in his *Essence of Christianity*, it was difficult to imagine that there was anyone one left in Germany who was unaware that traditional religious doctrine was undergoing a severe critique from the side of German *Wissenschaft*. What was required to give rise to a bona fide conflict thesis was to link the conclusions of *Wissenschaft* to *Naturwissenschaft*.

Free Religion and the Transition to a Conflict Thesis

The beginnings of a transition to a full-fledged conflict thesis occurred in the 1840s, a time when, according to historian Todd Weir, "philosophical controversies became caught up in social, political, and religious conflicts in new ways."[11] Weir, historian of science Peter Ramberg, and others have focused on the Free Religion movement of the 1840s and 1850s as a key locus for the growth of secularization in German life. Weir documents how in its early stages devotees of Free Religion embraced a form of Christian rationalism, in which the assumption was that the use of reason led to the inner truth of biblical revelation. But Christian rationalism became an oxymoron as Free Religion was caught up in the dynamics of dissent.[12] Now dogma of all sorts became the target of dissatisfaction. One vehicle for supporting the emerging secularized worldview was a newly minted form of popular natural science.

The popularization of natural science in Germany in the nineteenth century has been dealt with in Andreas Daum's magisterial study of 1998, followed a year later by Angela Schwartz's comparative work on popularization at the end of the century in both Germany and Great Britain.[13] Key foci for Daum in particular are the Free Religion movement of the 1840s and the materialists of the 1850s and beyond. What becomes clear from the work of Daum, Weir, and Ramberg is that in Germany on the eve of the century's midpoint the concern evident in the changing relationship between religion and science was not so much to depict warfare between science and religion, although there was some of that, but to celebrate natural science as a replacement for religion. The devotees of natural science understood it variously as merely a romanticized image of nature, traditional *Wissenschaft*, or the increasingly emerging modern *Naturwissenschaft*. Warfare is thus more implicit than explicit, but the mutual hostility between traditional religion and natural science was hard to miss.

Daum points out that participants in the Friends of Light and German Catholic movements refused to give churches and state institutions the last word in questions of knowledge and worldview. They valued the independence of the individual researcher over the church and referred to "rationalism," "the use of reason," and "science" (*Wissenschaft*) as foundations for the congregation. The appeal was to the autonomy of the mind over holy writ, one's own experience over tradition as the basis for knowledge. They thus opened themselves to the claims that were being made by some natural scientists at the very same time, in particular, natural scientists who retained the pantheistic outlook of the tradition of Schelling, with its rejection of the dualism of God and the world, but who at the same time discarded the speculative trappings of his *Naturphilosophie*.[14] As we have seen above, Schelling's *Naturphilosophie* had captured the attention of many earlier in the century and inspired them. By the 1840s, the enthusiasm for Schelling's nature philosophy was giving way to an emphasis on empirical observation of the natural world and a prescribed process of dispassionate, methodical analysis.

The most visible representative of the new authority of the natural scientist was Alexander von Humboldt, whose five-volume *Kosmos* began appearing in 1845. *Kosmos* was, according to historian William Langer, the most widely read book in Europe outside the Bible.[15] Humboldt delighted in reporting on recent discoveries, but his goal of recognizing unity in diversity and grasping nature's essence resonated with virtually all readers. Natural science, the reader understood, could supply truth unbiased by the prejudice of dogma. And it could be the foundation for a new worldview. For many in the Free Religion movement this worldview was their new religion. The "Catechism of the Christian Religion of Reason" asked what it was that preserved the wonderful harmony in the immense expansion of the universe? It was preserved, wrote the author of the catechism, Herbert Rau, in 1848, "by the eternal laws of nature."[16]

Todd Weir characterizes the shift in outlook here as a shift from idealistic monism to naturalistic monism, from a unified system that privileged spirit, whose disciplinary foundation was philosophy, to a unified system that privileged matter, whose disciplinary foundation was natural science.[17] The term "monism" would not become widely used until Ernst Haeckel made it so later in the century, but the resistance to dualism, either the epistemological dualism of Kant's thought or the classical matter-thought dualism of René Descartes, was already under way in the context of dissent and revolution in the 1840s.[18] Although Humboldt had not exploited the lessons of natural science to express dissent, others, today known as scientific materialists, would begin to do so in the late 1840s and 1850s.

Scientific Materialism, Popular Science, and the Emergence of the Conflict Thesis

There was little doubt among the readers of the scientific materialists that they believed the progress of natural science in history had undermined belief in the traditional religious doctrines of creation, the existence of the soul, and even the existence of God. But such an explicit declaration of the conflict thesis only occasionally surfaced in their works. Heinrich Czolbe, for example, held that the course of history was proof that systems relying on immaterial entities like the soul or the will naturally succumbed to the onslaught of the natural sciences.[19] But to believe that science and the church stood in conflict does not require that this conviction has to be expressed in terms of the *history* of science and religion.

Among the earliest scientific figures who did wage a hostile campaign against tenets of traditional religion in Germany was the physiologist and paleontologist Karl Vogt.[20] Son of a physician, he came from a family of liberal political activists. Vogt himself was forced to follow in his father's footsteps in escape from Germany to Switzerland in the tumult of the *Vormärz* period. His critical stance against established religion, then, must be seen in conjunction with his general antiestablishment demeanor.

In 1846, as a young natural scientist without a permanent position, Vogt did not shy away even in one of his first publications from declaring that natural science stood against the idea of a Creator.[21] He saw natural science as the antithesis of the religious and political status quo. Every advance in natural science, because of the materialism on which it was based, contributed to the erosion of the Christian state. Every new truth uncovered in research tore down propositions of catechism and bourgeois law. He urged scientists not to adopt Humboldt's stance of erecting limits beyond which faith reigned.[22] Hermann Misteli sums up Vogt's preface to *Ocean und Mittelmeer* in the simple declaration that "to do science is to make revolution." That coincided, according to Misteli, "with the conclusion of the materialistic Bible: unrestrained war against church and religion."[23] As an elected delegate to the Frankfurt Parliament, Vogt's speech of August 22, 1848, made clear his stance: "Every church, without exception, is as such a restriction of the free development of the human spirit." Because he wanted the free development of the human spirit, Vogt left no doubt about what followed: "I want no church."[24]

No other scientific materialist was as stridently anticlerical and antireligious in the name of natural science as Vogt. The writings of Ludwig Büchner and Jakob Moleschott broke onto the scene in the early 1850s with their loud proclamations of a materialistic worldview. In his *Theory of Nutrition: For the People*, Moleschott acknowledged that his religion arose from humankind's dependence on matter in motion. Genuine piety, he said, was to cherish this relationship with the universe, a sentiment close to the similar view of Büchner.[25] In his next book, *The Cycle of Life* (*Der Kreislauf des Lebens*) of 1852, he criticized Justus Liebig soundly for asserting in his *Chemische Briefe* that the laws of nature pointed to a Creator. He marveled that in the land of Ludwig Feuerbach one would still try to defend "the insoluble contradiction between the omnipotence of a Creator and natural law."[26]

Midcentury also marked the appearance of new popular journals of natural science that helped to spread the new gospel of scientific materialism. Among them stands *Nature: Journal for the Spread of Natural Scientific Knowledge and the Naturalistic Outlook for Readers of All Classes (Die Natur: Zeitschrift zur Verbreitung naturwissenschaftlicher Kenntniß und Naturan-*

schauung für Leser aller Stände), edited by the botanist Karl Müller, the popular science writer Emil Roßmäßler, and Moleschott's brother-in-law Otto Ule. Calling it a classic and the unrivaled model of the journal market, Andreas Daum declares that this journal can rightfully be seen as having founded the movement to publicize popular natural science in Germany.[27]

Indicative of the scientific materialists and the editors of and contributors to popular natural scientific journals was a conviction that made subscribing to a conflict thesis easier: that natural science described the world as it really is.[28] These writers were all convinced that natural science was unlike theology because it was based in the here and now and could deliver truth where theology could not; further, the conclusions of natural science were objective truth, uncontaminated by personal preference of religious belief or political party. Emil Roßmäßler expressed it well as he looked back over his career as a writer who celebrated the wonders of natural science: "Natural science is as it must be according to eternal laws. No one—not even Urban VIII—can say it should be so and so. It is only one thing—and therefore it is what it is by having factual content and is thus a genuine science. Only a genuine science is one that . . . provides true knowledge, which dogmatic theology cannot do and will never be able to do."[29] On his three-year lecture tour between 1849 and 1852, Roßmäßler recalled that he had delivered natural "scientific travel sermons [*naturwissenschaftliche Reisepredigten*]" because the people were total novices in this "battle [*Kampf*]."[30]

Monism and the Congealing of the German Conflict Thesis

The initial impact of scientific materialism and the rapid increase in popular journals of natural science all happened prior to Darwin's *On the Origin of Species*. It would take time, as we know, for nineteenth-century minds to appreciate how natural selection focused the question of the relationship between science and religion. By explaining how purposeful forms of organization could arise from purposelessness Darwin had solved a problem that had long haunted German philosophy. It was left to a new voice, that of Ernst Haeckel, to capture the attention of Germany where the issue of evolution and religion was concerned.

Writing against the background of Haeckel's achievement, historian of science Robert Richards has eloquently defended a claim close to the following: Because of particularities in the biography of Haeckel, who more than anyone else is responsible for the popular perception that evolutionary science was materialistic and a-theistic (if not atheistic), the emergence of the conflict thesis in the nineteenth century was contingent, not inevitable.[31] It was, after all, with Haeckel that monism, the heritage of Goethe and Schelling, became a watchword in the German public sphere after Haeckel borrowed the term from his friend August Schleicher for use in his *General Morphology* of 1866. Given the widespread presence of anti-dualist sentiments within the Free Religion movement and among the materialistic writers we have examined, the ground had been well prepared to receive a term that expressed directly what so many felt. Haeckel's stature as a natural scientist and his unique gifts as an artist and spokesman brought him into the limelight of German public life like no other in the latter decades of the nineteenth century.

Haeckel did not want himself to be seen as an enemy of religion as such, even choosing categories of religion to describe his goals. Readers were asked to let the new understanding of human origins motivate them to penetrate deeper into the "sanctuary [*Heiligtum*]" of nature, and from this source of "natural revelation [*natürliche Offenbarung*)]" they could obtain the purest enjoyment of the mind and moral purification of reason that could not be attained in any other way than through "simple natural religion [*einfache Naturreligion*]."[32] While thoroughly rejecting all notions of the traditional God, Haeckel made clear that he did believe in God, the monistic idea of God to which the future belonged. It was the God of Goethe, the most noble and lofty idea of which humans were capable—the unity of God and Nature.[33]

Haeckel may have believed that his views were not hostile to religion, at least as he understood it, but he made perfectly clear that faith, which in his mind was not involved in his religion, had nothing to offer the scientist. He claimed to have nothing against those who insisted that a supernatural Creator had formed matter. But that was irrelevant to the scientist, who assumes that matter has always existed and will not disappear. One gained nothing for a scientific knowledge of nature from a belief in supernatural creation. In a sentence emphasized in the text he proclaimed: "Where faith begins, science stops."[34]

Haeckel's words seemed to require acceptance of a pantheistic God as the only one compatible with modern science. It was the joining of his embrace of monistic materialism, which involved the elimination of a personal God, that would prove offensive. Eventually it would blossom into a full blown religion of monism, especially as seen in the widely read pamphlet of 1892, *Monism as the Bond between Religion and Science*.[35] Of the period after 1880 to Haeckel's death in 1919, Richards writes: "The antagonism between conservative religion and evolutionary theory, brought to incandescence at the turn of the century and burning still brightly in our own time, can be attributed, in large part, to Haeckel's fierce broadsides launched against orthodoxy in his popular books and lectures."[36]

More support for this claim comes from an episode just prior to the explicit announcement by Haeckel that monism was the bond between religion and science. It sheds light on the way in which Haeckel, while claiming to be religious, was regarded as a central player in the warfare between science and religion. And it occurred around the time that the two most widely influential promoters of the conflict model, the scientist-historian John William Draper and the educator Andrew Dickson White, were sounding their attacks in the United States. The episode in question was the confrontation between Haeckel and the famed zoologist Rudolf Virchow that occurred in the wake of the fiftieth meeting of the Gesellschaft Deutscher Naturforscher und Ärzte (GDNA), held in the fall of 1877 in Munich. Haeckel's address, "The Theory of Development of the Present Day in Relation to Science in General," came on the first morning of the conference. He used the opportunity, as everyone expected, to praise Darwin's achievement and went on to claim that evolutionary theory was a historical science, though one in which the causal explanations of the natural sciences could be applied. But it was his aggressive entry into the political realm of public education that provided a new cause célèbre. Haeckel called for "a thoroughgoing reform of instruction." He called evolutionary theory the most significant "lever [*Hebel*]" of progressive knowledge and ennobled education. The education of the youth was weak. "Therefore evolutionary theory must lay claim to its legitimate influence in the schools as the most important means of education; it will not be merely be tolerated, but will become standard and leading."[37]

Haeckel had to leave the conference the next day and did not hear Virchow's rejoinder. Among his many objections to Haeckel's lecture one was especially effective. If unbridled theorizing is done within the context of natural science, imagine, he said, what a socialist thinks of the theory of descent. Socialism has, after all, exhibited sympathy for the theory.[38] In an unambiguous reference to Haeckel's monism, Virchow opposed the attempt "to push our purely theoretical and speculative constructions so much into the foreground that we want to construct an entire worldview from there."[39] As for the call to include Darwinian evolution in the schools, Virchow declared, "We must say to the schoolteachers, do not teach it."[40] Of course,

Haeckel had to respond, and he did so, denying that science was limited to certainty, as Virchow had implied. He also vigorously asserted that evolutionary theory did not support socialism; on the contrary, the drive for competition made it incompatible with socialism.[41] But he had been stung badly, and letters of support and denunciation came thick and fast. Opposing sides within the German scientific establishment had congealed into a major conflict for all to see.

The debate spilled over into the Reichstag during deliberations over the antisocialist laws of 1878. Taking the offensive, the social democrat August Bebel aggressively declared that it was obvious science was on the side of social democracy, though he regretted that Haeckel had no idea that Darwinism was necessarily useful to socialism. What started out as a debate about Darwinism in the schools now appeared to have ballooned into something even more dangerous, as the conservative deputy Wilhelm Freiherr von Hammerstein made clear. If "Haeckel-Darwinism" was sanctioned and students could have materialism implanted in them, he cried, the administration would have to take the responsibility for a generation "whose confession of faith is atheism and nihilism and whose political view is communism."[42] And this was just the beginning of a prolonged public exchange.[43]

What made it all the more visible was the earlier parallel opposition between Haeckel and the celebrated Berlin physiologist Emil Du Bois-Reymond over the question of the limits of knowledge. Du Bois-Reymond's lecture at the 1872 meeting of the GDNA was what Daum has called "the decisive signature" of the debates about the limits of knowledge that had been going on in Germany since the 1850s.[44] With his famous proclamation that there were things in natural science that "we shall not know" Du Bois-Reymond drew a line in the sand separating those who resented any limits being placed before science and those whose convictions forced them to acknowledge that we are not equipped to answer all the questions we might ask about the natural world. Haeckel weighed in on this issue in his *Anthropogenie* of 1874. Referring to "the universally known lecture by the famous physiologist Du Bois-Reymond," Haeckel said there was no better example of how even highly visible biologists have missed the deeper meaning of evolutionary history. While the church militant gave thanks for this "we will not know," said Haeckel, he had to protest most decisively. Du Bois-Reymond's lecture had produced vivid regret among all friends of intellectual progress.[45]

Andreas Daum has underscored the enormous publicity these debates received in the daily and monthly press, in the cultural and popular science journals, in sessions of scientific societies, in clubs, and even in the Reichstag. Virchow, Du Bois-Reymond, and Haeckel stood out as the clear foci of what a headline in the journal *Die Gartenlaube* styled as a "Duel in the Realm of the Mind."[46] Daum points out that the combatants opened the semantic field of the debate by linking the scientific position of their opponents to loaded extrascientific terms. Haeckel's opponents strung together the sequence Darwinism-atheism-materialism-socialism, all of which were dangerous-unchristian-demoralizing-pernicious; From Haeckel's side came the combination evolutionary theory-freedom-reason-fact, which were causal-natural-objective.[47]

All this points to the conclusion that at the same time warfare between science and religion was announced in England, it was already in full bloom in Germany. Daum argues that the public debates over Darwinism, with their intellectual sound bites and with the political overtones they acquired in Germany, reduced the question of the church's acceptance of modern natural science to the question of human origins as defined by the radical materialism of Haeckel's monism. Daum, who is fully aware that British and American historians of science have found the conflict paradigm severely wanting, argues not only that it has been preserved in German historiography and in the German cultural consciousness, but that it remains the best way to describe this situation. "What experienced rapid dissemination was the imagery of a battle between natural science and religion and the largely unquestioned assumption that, with the appearance of Darwinism, a fundamentally unbridgeable contradiction had arisen between both areas."[48]

Otto Zöckler's *History of the Relations between Natural Science and Religion*

While many German theologians did not react publicly to the appearance of a conflict thesis in their homeland, a few refused to sit idly by. Among the most prolific was the conservative theologian at Greifswald, Otto Zöckler.[49] It was in the historical work for which he is remembered that Zöckler explicitly acknowledged the existence of the conflict thesis. It came at the very opening of his massive two-volume book, *History of the Relations between Theology and Natural Science, with Special Attention to the History of Creation* (1877, 1879).[50] His approach would not be like that identified by Zöckler as "a certain school of English and North American historians": "Under the relations of theology to natural science we do not understand merely the hostile points of contact between the two. It would perhaps be more timely and our work would raise more interest in the eyes of many had we wished to write a history of just the *conflicts* between theology and natural science. . . . Such an approach would probably be able to win the approval of many in Germany too, especially given the Kulturkampf that is still going on. . . . A conflict-history [*Conflicts-Geschichte*] would be popular, but it would not be true."[51]

Zöckler went on to say that he would leave it to Draper, White, and others to regale their public with the glories of the "warfare" ("*Kriegeszug*") and the "unstoppable triumphant advance" ("*Siegeszug*") of science (*Wissenschaft*).[52] He singled out White's *History of the Warfare of Science with Theology in Christendom* as a typical work from this genre, with its thesis that throughout history interference in science by religion had invariably hurt both religion and science while all untrammeled scientific investigation had resulted in good for both. Had White cited examples for both parts of this thesis, one might have approved his approach, Zöckler observed. But White only cited examples of how religion had interfered with science and never could free himself from his hostility to Christianity. He confined himself to listing a register of sins as confirmation of the thesis he assumed.

According to Zöckler, White's selection of scientific material was full of gaps. Light fell only on science, leaving the church always in the shadows. Zöckler questioned White's use of sources, listing several questionable references provided to support baldly stated claims and catching errors of several particulars concerning Galileo's trial. It was hard for Zöckler to grasp how a scholar like John Tyndall, who had endorsed White's work, could recommend such a weak work.[53] It was just not a work of good scholarship.

Zöckler classified Draper's *History of the Intellectual Development of Europe* and his *History of the Conflict between Religion and Science* to be "of similar caliber" to White's. Zöckler relied heavily on his appeal to the shallowness of his enemies. His refutation of materialism in his work was based largely on his characterization of it as a "model disease of the scientifically half-educated."[54] But he recognized that however unjustified the conflict thesis and the materialism from which it had grown might be, the reasons for their popularity in his day were not superficial. Whatever they were, they were connected to developments from earlier in the century.

Zöckler in fact identified three general arenas that in his mind had given rise to fundamental *wissenschaftliche* challenges to religion. Zöckler's use of

Wissenschaft here is especially interesting because of what Denise Phillips has taught us about how slowly the use of "*Naturwissenschaft*" congealed into the meaning we all too quickly assume it must have had throughout the century. Her conclusion that it was not until midcentury that *Naturwissenschaft* began to narrow into something akin to what we mean by science today is borne out by my contention that a true conflict thesis only emerges at just that same time. Even among writers like Zöckler in the 1870s, who uses *Naturwissenschaft* in the title of his major work, it is still the more general term, *Wissenschaft*, that he chooses when he looks back over the century to understand how the conflicts with religion have come about.[55]

Accordingly, Zöckler was ready to enter the lists first with regard to negative critics of biblical documents, second with pantheistic philosophy and the materialistic natural science closely allied with it, and third with the ultramontanism of the Catholic Church.[56] It is the second arena in particular, which is tied to and somewhat naturally entails the third, that has provided us with a rich repository of material for treating the genesis of the conflict thesis in Germany.

Wilhelm Herrmann's Refusal of the Conflict Thesis

Among the neo-Kantian voices who opposed materialism was an important theologian of the late nineteenth century, one who objected on quite different grounds from Otto Zöckler. During the heyday of the debates surrounding Haeckel, Wilhelm Herrmann characterized his concern as the problem of "religion in relation to knowledge of the world."[57] Du Bois-Reymond's 1872 lecture on the limits of knowledge was especially important to Herrmann.[58] Herrmann refused the notion of a warfare between science and religion because he believed them to be so radically separate from each other that they could not even intersect, let alone do battle.

As he viewed the theology and natural science of the early 1870s, the budding theologian was struck by how many of his colleagues in theology bought into what he saw as a flawed goal in the popular science of the time: the construction of a worldview based on scientific knowledge. It was easy enough, Herrmann wrote, to see where this came from. Scientists forget the limitations of knowledge in the face of the impressive mastery over phenomena that they have achieved. And theologians, happy to acknowledge limits of knowledge, assume that theology's task is to supply what human wisdom cannot achieve.[59] For Herrmann the neo-Kantian, it was a fundamental mistake to ignore the limits of knowledge by treating a metaphysical claim as a matter of cognition, whether it came from theology or natural science. The message of his work was that we must remove metaphysics from both theology and natural science. Doing so makes clear what the tasks of theology and natural science were.

What drove the natural scientist was mastery of the world, something accomplished by adopting "the hypothesis of the comprehensibility of the world." Using this hypothesis, the scientist discovered new information about the material world and established causal relationships among phenomena. The scientist also responded to the drive for knowledge, which is not always linked to practical ends. Here the goal was to achieve a final theory, so to speak, but with the understanding that it would never be attained.[60] The scientist must experience complete freedom in the pursuit of knowledge of the world, acknowledging no restrictions from religion.

As for theology, its task was to represent systematically a religious view of life, which is not based on our knowledge of the world or on human cognitive powers. It subjects the world, however it is in itself, to the highest purpose humans can have: to declare their supreme value by forming the idea of a power dominating everything in their behalf. So the world is to be judged in theology, but not by virtue of our knowledge of how it works, not according to the fruits of the hypothesis of the comprehensibility of the world. It is to be judged in accordance with the assumption that there is a highest good and that how we behave matters. "Every religious worldview is an answer to the question: how is the world to be judged if there really is a highest good?"[61] Religion becomes ethics and holds its focus on personal authentication through the transformative experience of personal redemption.

Although Herrmann was not a participant in the battle between science and religion being waged in the German public sphere at the time his work appeared, he is important because he rejected the stance of protagonists from *both* sides of that battle. But his rejection could not be heard in his day nor since. He was unlike Zöckler, who rejected the idea of conflict in deference to his conservative religious beliefs. He rejected Haeckel, who passed off his monistic worldview as the results of cognition. For similar reasons he rejected those liberal thinkers who argued that science and religion could be harmonized because they too held out hope that somehow the worldviews of naturalism and supernaturalism could be reconciled. Nor was his position the old doctrine of double truth, since in science truth rests on the shifting sands of coherence, and religion has been subsumed by the personal declarations of ethics.

There have been, not surprisingly, echoes of Herrmann's thought in the theology of two of his students at Marburg, Karl Barth and Rudolf Bultmann. Barth's vehement "Nein!" to the possibility of natural theology and Bultmann's respect for natural science while emphasizing kerygma were reminiscent of their teacher's views. Ueli Hasler has written that "Herrmann's recommendation that one [can] exclude all possibility of conflict between theology and natural science by clarifying the epistemological jurisdiction of science has acquired almost canonical significance," adding that this is evident not only in Bultmann and Barth, but even in Tillich.[62] Nevertheless, existential theology in the twentieth century, in spite of its considerable influence, has not been any more successful in quashing the notion of conflict than Herrmann was in the preceding one.

Conclusion

The question remains why Herrmann's rejection of conflict has largely been a failure. The difficulty, in a word, might be that Herrmann simply draws too sharp a dividing line between science and religion. It is apparently just too difficult to rest content with a methodological naturalism and pretend that it lends no credence to the metaphysical naturalism lurking in its shadow. And, as Richard Olson has written about Bultmann, the boundaries between science and religion cannot be made so sharp because both science and religion must use language, including the language of ordinary people with the meanings and shared cultural experience in which it is embedded. "There is, therefore, bound to be a crossover—whether intended or not—between the concepts associated with science, or natural knowledge, and those associated with religious beliefs at any particular time and place."[63] Perhaps this is one reason why the conflict between science and religion will not die.

NOTES

1. R. Stephen Turner, "The Growth of Professional Research in Prussia, 1818–1848—Causes and Context," *Historical Studies in the Physical Sciences* 3 (1971): 142, 147, 153, 156, 172.

2. See Johann Christoph Adelung, *Grammatisch-kritisches Wörterbuch der Hochdeutschen Mundart*, 4 vols. (Vienna: Pichler, 1808), 4:1582–83.

3. William Clark, "The Death of Metaphysics in Enlightened Prussia," in William Clark, Jan Golinski, and Simon Schaffer, eds., *The Sciences in Enlightened Europe* (Chicago: University of Chicago Press, 1999), 434.

4. For a discussion of rational theology in the eighteenth century, see Frederick Gregory, *Nature Lost? Natural Science and the German Theological Traditions of the Nineteenth Century* (Cambridge, MA: Harvard University Press, 1992), 24ff.

5. That is not to say that a concern for natural theology disappeared completely. Cf. Denise Phillips, *Acolytes of Nature: Defining Natural Science in Germany, 1770–1850* (Chicago: University of Chicago Press, 2012), 34.

6. Quoted from Flückiger's "Die protestantische Theologie des 19. Jahrhunderts" in Svenja Meindl, *Otto Zöckler: Ein Theologe des 19. Jahrhunderts im Dialog mit den Naturwissenschaften* (Frankfurt am Main: Peter Lang, 2008), 112.

7. Quoted by Notker Hammerstein, "Universitäten und gelehrte Institutionen von der Aufklärung zum Neuhumanismus und Idealismus," in Gunther Mann and Franz Dumont, eds., *Samuel Thomas Soemmerring und die Gelehrten der Goethezeit* (New York: Fischer Verlag, 1985), 327.

8. For discussion of early-nineteenth-century German science in the context of the public sphere, see Frederick Gregory, "Entstehung und Voraussetzungen alternativer Wissenschaften," in Klaus Vondung und K. Ludwig Pfeiffer, eds., *Jenseits der entzauberten Welt* (Munich: Wilhelm Fink Verlag, 2006), 83–97, esp. 92ff.; Phillips, *Acolytes of Nature*, chap. 5.

9. Frederick Gregory, "Proto-Monism in German Philosophy, Theology, and Science, 1800–1845," in Todd Weir, ed., *Monism: Science, Philosophy, Religion, and the History of a Worldview* (New York: Palgrave Macmillan, 2012), 45–69.

10. Karl Barth, *Protestant Theology in the Nineteenth Century: Its Background and History* (London: SCM Press, 1972), 474. On miracles, original sin, and the status of the biblical narrative as "burning questions of rational religion," see Claude Welch, *Protestant Thought in the Nineteenth Century*, 2 vols. (New Haven, CT: Yale University Press), 1:39–41.

11. Todd H. Weir, *Secularism and Religion in Nineteenth-Century Germany: The Rise of the Fourth Confession* (Cambridge: Cambridge University Press, 2014), 68.

12. Weir, *Secularism and Religion*, 70.

13. Andreas Daum, *Wissenschaftspopularisierung im 19. Jahrhundert: Bürgerliche Kultur, naturwissenschaftliche Bildung und die deutsche Öffentlichkeit, 1848–1914*, 2nd ed. (Munich: R. Oldenbourg Verlag, 2002); Angela Schwartz, *Der Schlüssel zur modernen Welt: Wissenschaftspopularisierung im Großbritannien und Deutschland im Übergang zur Moderne* (Stuttgart: Franz Steiner Verlag, 1999).

14. Daum, *Wissenschaftspopularisierung*, 198–99.

15. William Langer, *Political and Social Upheaval, 1832–1852* (New York: Harper and Row, 1969), 535.

16. Quoted in Daum, *Wissenschaftspopularisierung*, 199. Rau later penned *Das Evangelium der Nature: Ein Buch für jedes Haus* in 1853.

17. Weir, *Monism*, 16.

18. See esp. Weir's discussion in "Free Religious Worldview: From Christian Rationalism to Naturalistic Monism," chap. 2 in *Secularism and Religion*, 66–104.

19. Heinrich Czolbe, *Neue Darstellung des Sensualismus* (Leipzig: Hermann Costenoble, 1855), 232.

20. On Vogt, see Frederick Gregory, *Scientific Materialism in Nineteenth Century Germany* (Dordrecht: Reidel, 1977), chap. 3. See also Timan Matthias Schröder, *Naturwissenschaften und Protestanismus im Deutschen Kaiserrreich* (Stuttgart: Franz Steiner Verlag, 2008), 66ff.

21. Karl Vogt, *Lehrbuch der Geologie und Petrafactenkunde.* 2 vols. (Braunschweig: Vieweg, 1846), 2:370.

22. Karl Vogt, *Ocean und Mittelmeer*, 2 vols. (Frankfurt am Main: Literarische Anstalt, 1848), 1:19–21.

23. Hermann Misteli, *Carl Vogt: Seine Entwicklung vom angehenden naturwissenschaftlichen Materialisten zum idealen Politiker der Paulskirche, 1817–1849* (Zurich: Leemann, 1938), 89–90.

24. Quoted from the stenographic report by Wilhelm Vogt, *La vie d'un homme: Carl Vogt* (Paris: Schleicher, 1896), 66n1; and Schröder, *Naturwissenschaften und Protestantismus*, 58ff.

25. Jakob Moleschott, *Die Lehre der Nahrungsmittel: Für das Volk* (Erlangen: Enke, 1850), 256.

26. Jakob Moleschott, *Der Kreislauf des Lebens: Physiologische Antworten auf Liebig's "Chemische Briefe,"* 2nd ed. (Mainz: Zabern, 1855), 18–19.

27. Daum, *Wissenschaftspopularisierung*, 346.

28. For a discussion of the naive realism of the scientific materialists, see Gregory, *Scientific Materialism*, 145ff.

29. Emil Roßmäßler, *Mein Leben und Streben: Im Verkehr mit dem Natur und dem Volk*, ed. Karl Russ (Hanover: Ruempler, 1874), 24.

30. Roßmäßler, *Mein Leben und Streben*, 141.

31. Robert Richards, *The Tragic Sense of Life: Ernst Haeckel and the Struggle over Evolutionary Thought* (Chicago: University of Chicago Press, 2008), 15.

32. Ernst Haeckel, *Natürliche Schöpfungsgeschichte* (Berlin: Georg Reimer, 1868), vi.

33. Haeckel, *Natürliche Schöpfungsgeschichte*, 58.

34. Haeckel, *Natürliche Schöpfungsgeschichte*, 7.

35. Given originally as remarks in response to comments on science and religion at a professional meeting, the monograph went to seventeen editions and served as the basis for his even more popular *Welträtzel* of 1899. See Richards, *Tragic Sense of Life*, 353.

36. Richards, *Tragic Sense of Life*, 244.

37. Quoted in Daum, *Wissenschaftspopularisierung*, 66–67.

38. Rudolf Virchow, *Die Freiheit der Wissenschaft im modernen Staatsleben* (Berlin: Wiegandt, Hempel und Parey, 1877), 12.

39. Virchow, *Freiheit der Wissenschaft*, 22.

40. Virchow, *Freiheit der Wissenschaft*, 15.

41. For an insightful discussion of Haeckel's response, see Richards, *Tragic Sense of Life*, 324ff.

42. Quoted in Daum, *Wissenschaftspopularisierung*, 71–72.

43. Daum, *Wissenschaftspopularisierung*, 71–75.

44. Daum, *Wissenschaftspopularisierung*, 68.

45. Ernst Haeckel, *Anthropogenie, oder Entwicklungsgeschichte des Menschen* (Leipzig: Verlag von Wilhelm Engelmann, 1877), xiv. These comments come from the foreword to the first edition of 1874, included in this third edition.

46. "Zweikampf auf dem Gebiete des Geistes," *Die Gartenlaube* 25 (1877): 823–25. Cf. Daum, *Wissenschaftspopularisierung*, 76.

47. Daum, *Wissenschaftspopularisierung*, 77–78.

48. Daum, *Wissenschaftspopularisierung*, 79.

49. For a treatment of Zöckler's life and works, see Gregory, *Nature Lost?*, chap. 4.

50. Otto Zöckler, *Geschichte der Beziehungen zwischen Theologie und Naturwissenschaft, mit besonderer Rücksicht auf der Schöpfungsgeschichte*, 2 vols. (Gütersloh: Bertelsmann, 1877, 1879).

51. Zöckler, *Geschichte der Beziehungen zwischen Theologie und Naturwissenschaft*, 1:1–2.

52. Zöckler, *Geschichte der Beziehungen zwischen Theologie und Naturwissenschaft*, 1:2. Zöckler put both of these military words in quotes. He translates *Kriegeszug* as "warfare" in his reference to A. D. White's *Warfare of Science*.

53. Zöckler, *Geschichte der Beziehungen zwischen Theologie und Naturwissenschaft*, 1:12–13n1.

54. Zöckler, *Geschichte der Beziehungen zwischen Theologie und Naturwissenschaft*, 2:397.

55. On the lingering appreciation of *Wissenschaft* as the linchpin of *Bildung*, as opposed to *Naturwissenschaft* with its empirical focus and restricted methodology, see Phillips, *Acolytes of Nature*, 218, 232ff.

56. Meindl, *Otto Zöckler*, 48.

57. On Herrmann, see Gregory, *Nature Lost?*, chaps. 6–7.

58. Gregory, *Nature Lost?*, 320n5.

59. Wilhelm Herrmann, *Die Metaphysik in der Theologie* (Halle: Niemeyer, 1876), 4–5.

60. Herrmann, *Metaphysik in der Theologie*, 5–6.

61. Herrmann, *Metaphysik in der Theologie*, 8.

62. Ueli Hasler, *Beherrschte Natur: Die Anpassung der Theologie an die bürgerliche Naturauffassung im 19. Jahrhundert* (Frankfurt am Main: Peter Lang, 1982), 294–95.

63. Richard G. Olson, *Science and Religion, 1450–1900: From Copernicus to Darwin* (Santa Barbara, CA: Greenwood Publishing, 2004), 2.

CHAPTER SIX

Roman Catholics

DAVID MISLIN

While proponents of the warfare thesis cast their claims in the broad categories of religion and science, in reality their attacks frequently focused on Roman Catholicism. For these critics, the Catholic Church exemplified the ways in which a religious institution impeded scientific progress. Barely a page into his *History of the Conflict between Religion and Science*, John William Draper, a nineteenth-century English-American scientist and self-styled historian, denounced the papacy's refusal to allow "reconciliation" between itself and "modern civilization." Indeed, Draper explicitly declared that his discussion of Christianity almost entirely focused on Roman Catholicism, which he found more worthy of criticism than Protestantism or the Greek Orthodox tradition. The hands of the Vatican, according to Draper, "have been steeped in blood" for its persecution of scientists. As he traced its history, Draper blamed the Catholic Church for both stifling "intellectual advancement" and fostering an "illiterate condition" during the Middle Ages. Closer to his own time, the church had experienced "the endangering of her position" of political, social, and intellectual authority by the progress of science, and the papacy had responded by attempting "to define its boundaries and prescribe limits to its authority."[1]

Draper was not alone in arguing that Roman Catholicism was militant in its opposition to science. Though Draper's contemporary Andrew Dickson White, the first president of Cornell University, found much to criticize in Protestantism's history, he nevertheless emphasized the "long and bitter" war against Galileo as a prime example of the Vatican's tendency to respond forcefully to scientific advancements. For White, the case provided the definitive instance of the church's unwillingness to tolerate ideas that led to consequences that were "destructive to Christian truth." So, too, the English scientist Thomas Huxley, one of the leading proponents of evolutionary theory, observed that one of the "greatest merits" of the theory was its "complete and irreconcilable antagonism to that vigorous and consistent enemy of the highest intellectual, moral, and social life of mankind—the Catholic Church." In other words, he wrote, "for those who hold the doctrine of evolution, all the Catholic verities about the creation of living beings must be no less false."[2]

Indeed, as Draper suggested, several developments during the late nineteenth century provided powerful new evidence for these critics of the church. First, the papal publication of the *Syllabus of Errors* in 1864 seemed to confirm that the Holy See had little interest in embracing modern thought. Such critics found additional fuel for their rhetoric in the doctrinal declaration of papal infallibility at the First Vatican Council in 1870. Though infallibility was in reality limited to matters of faith and morals, observers like White delighted in emphasizing the "infallible" character of the pontiff as they pointed out the now-outmoded aspects of knowledge that popes had historically supported. Moreover, Draper was not persuaded by the proscribed limits on infallibility to "moral or religious" areas. "Where shall the line of separation be drawn?" he asked. "Omniscience cannot be limited to a restricted group of questions; in its very nature it implies the knowledge of all, and infallibility means omniscience."[3]

Such accusations of Catholicism's hostility to science and of the oppressive behavior of ignorant pontiffs proved commonplace during the late nineteenth and early twentieth centuries. The Catholic intellectuals who reflected on the relationship between their religious tradition and scientific progress were well aware of these allegations. The initial reviews of Draper in the Catholic press excoriated his claims. The American periodical *Catholic World*'s review of the *History of the Conflict between Religion and Science* blasted the volume as "a farrago of falsehoods" in which "a cynical and sneering spirit betrays him into errors which a Catholic Sunday-school child would blush to commit." Meanwhile, the London-based *Dublin Review* lamented that "the misconception of Catholic dogmas" in Draper's work was "pitiable in the extreme." As Draper, White, Huxley, and others advanced the conflict thesis, Catholics on both sides of the Atlantic worked strenuously to refute such assertions. They noted that historically the church was Europe's major patron of scientific advancement and insisted that the doctrine of papal infallibility did not make the pontiff the arbiter of all new scientific research.[4]

And yet, given the frequency with which Catholics found themselves defending their tradition against the proponents of the conflict thesis, it is clear that the metaphor of warfare proved compelling to outside observers of the church. Indeed, the anxious response of Catholics suggests that perhaps they, too, recognized such critiques as having at least an air of plausibility. Thus it was not sufficient for Catholic authors simply to attack works like Draper's. Rather, they needed to muster evidence showing that Roman Catholicism had historically supported science and continued to do so even in light of new developments such as the theory of organic evolution.

Critics who argued for an ongoing conflict between Catholicism and science adopted a three-pronged argument. First, they resorted to stereotypes of the ignorance of the Middle Ages and suggested that it was Catholic institutions that had impeded the progress of knowledge. Second, they invoked the Galileo case as a prime example of church authority run amok. Finally, they pointed to contemporary scientific developments, most notably the theory of evolution, and embraced the claims of Huxley and others that evolution was inimical to Catholic doctrine.

In this chapter, I examine the responses Roman Catholic intellectuals offered to these three arguments. The first two they easily dismissed with evidence of the church's support for science in preceding centuries and a frank acknowledgment that the Galileo case represented an unfortunate anomaly. The issue of contemporary scientific theories required greater finesse. Nevertheless, Catholic intellectuals crafted a response to evolutionary theory that allowed them to express conditional support for its science without challenging their church's doctrine.

But even as intellectuals sought to muster support for these arguments, proponents of the conflict thesis found new ammunition for their case in a series of events at the close of the nineteenth century and the dawn of the twentieth. These included the Vatican's prohibition of several books that came close to arguing for the development of human life through evolutionary processes, its censure of leading British intellectual St. George Jackson Mivart, and finally its condemnation of modernism, a transnational school of thought within the church that made the acceptance of evolution a core aspect of its intellectual platform. As I will show, these new censures of particular voices and ideas by church officials required Catholic intellectuals to shift their arguments for the compatibility of religion and science. They increasingly placed greater emphasis on the longer history of Catholicism and scientific progress and significantly less on contemporary developments. While some Catholic thinkers did continue to advocate the reconcilability of their faith with contentious theories like evolution, most instead focused their efforts on criticizing the way in which the scientific theory of evolution had been carried into the realm of political policy and social thought. In so doing, they effectively argued that apparent conflict stemmed not from science but from the misuse of science.

In considering the Catholic response to the conflict thesis from different countries and regions and over the course of a century, it is important to acknowledge the significance of geographical distinctions and developments over time. Reactions in places like France and Ireland, where Catholicism was the dominant Christian tradition, differed from those in the United States and England, where Catholic intellectuals sought to prove that their minority tradition was not at odds with the values of dominant culture, and from those in Latin America, where traditionally Catholic nations worked to prove their intellectual equality with Europe. So, too, understandings of Catholicism and science changed constantly amid rapidly shifting social and political realities.

Notwithstanding these differences of time and place, however, it remains possible to perceive recurring characteristics in the relationship between Catholicism and science. On the one hand, Catholic intellectuals could consistently muster ample evidence to counter claims of centuries of warfare between the church and science. On the other hand, the mere fact that these intellectuals felt the need to devote such effort to refuting these claims, combined with the events of the 1890s and early 1900s, suggests that the relationship was not purely harmonious either. Catholics consistently faced accusations that their tradition was uniquely inimical to scientific progress. Ultimately, these divergent opinions reflected a wider disagreement between Catholics and non-Catholics about the role of the Catholic Church's authority. For people within the institution, it seemed entirely reasonable that the church should seek to protect its members from ideas that challenged fundamentals of the faith. For critics, such actions were not benevolent protection but rather constituted intolerable interference with the scientific enterprise. Disputes over the conflict thesis ultimately had less to do with the relationship of religion and science and more to do with a disagreement between Catholics and non-Catholic critics about the ability of the church's institutions to serve as intellectual guardians of the faithful.

The Nineteenth Century: Proclaiming Harmony amid Scientific Innovation

The first argument proponents of the conflict thesis commonly raised was that the Catholic Church had historically impeded the progress of scientific advancement. This was, in many respects, the easiest point for Catholic intellectuals to refute, as they too had historical evidence to support their position. Bertram Windle, a child of English parents who was raised in Ireland, converted to Catholicism, and ultimately migrated to Canada, defended his adopted faith's historical record. "During the Middle Ages," he wrote, a great deal of scientific investigation, "often of fundamental importance, especially perhaps in the subjects of anatomy and physiology, emanated from learned men attached to seats of learning in Rome." Moreover, the scientists responsible for this work "quite frequently held official positions in the Papal Court." In the United States, James Joseph Walsh, a faculty member of the Fordham University school of medicine and a prolific writer on the history of science, agreed. He emphasized the church's role as the major driver of knowledge during the medieval period. Walsh blasted the "ridiculous intolerance of writers who knew practically nothing of the real history of science in the Middle Ages" who "wrote their own prejudices large into the story of the times." In reality, Walsh observed, medieval Italy boasted "the best medical schools in the world, to which the most ardent students from all over the continent and even England went." Indeed, "the nearer to Rome the university the better its medical school."[5]

But for all the history that these intellectuals could invoke to support their argument of Catholicism's long-standing support of science, there remained an element of history that seemed to work against them: the Vatican's condemnation of Galileo. In some respects anticipating the even more forceful arguments of seventy-five years later, Catholic writers in the late nineteenth and early twentieth centuries saw little point in debating or rationalizing away the events of the early seventeenth century. Windle conceded that the condemnation of Galileo was "a mistake" but one that had not been repeated. Walsh likewise admitted that Galileo was indeed prosecuted by the church and that the events had been "a deplorable mistake" of church officials trying to evaluate "scientific truth." The Galileo case represented "an unfortunate incident, but not a policy," and, according to Walsh, its significance had been overstated by "those who know nothing about the history of medicine and of science." Moreover, although not releasing the church from culpability in the matter, Catholic writers nevertheless placed some blame on Galileo. First, they correctly pointed to Galileo's combative personality and his refusal to abide by the church's demands that he cease discussing the heliocentric theory. Moreover, using an argument similar to one that was deployed in their discussions of contemporary scientific debates, Catholic intellectuals such as American Augustine Hewit observed that Galileo had cast his unproved theory as established science. Hewit considered this a major source of Galileo's difficulty, and observed that once the science became proven, the church embraced Galileo's teaching.[6]

But while Catholic intellectuals were able to deftly parry arguments about the historical relationship of the church and science, there remained the issue of that relationship in the present, amid new scientific knowledge such as the theory of evolution. As is evident from the extent to which proponents of the conflict thesis emphasized evolution in their discussions of Catholicism, the theory did pose some difficulties for Christian doctrine.[7]

At the most basic level, the theory of biological evolution challenged the biblical narrative of creation, thereby threatening all biblically based religions, not just Catholicism. Yet, because of the oversight of church officials over biblical interpretation, Catholics had a means of reconciling new science with the scriptural narrative that their Protestant counterparts lacked. From the perspective of intellectuals within the church, the Vatican hierarchy—far from impeding science as critics accused—actually served as a crucial mediator in situations where science and scripture seemed at odds. Moreover, Catholic writers observed that moments of tension between science and the Bible were unavoidable, given that the Bible was not meant to serve as a scientific textbook. John Gmeiner an American priest, offered a straightforward summary of the commonly held view that "the Bible does not intend to teach any purely . . . scientific truths." In making such claims, Catholic intellectuals drew directly on a position articulated by Pope Leo XIII in his 1893 encyclical *Providentissimus Deus*, which, among other things, affirmed that the purpose of scripture was not to educate Catholics in matters of natural science.[8]

In some respects, this encyclical had given fuel to the arguments of supporters of the conflict thesis, as the major thrust of the pontiff's position was a denunciation of the historical-critical study of the Bible separate from the guidance of Catholic doctrine. But other elements of the encyclical's argument worked in favor of those Catholics who thought their faith was reconcilable with emerging evolutionary science. *Providentissimus* essentially confirmed that church teaching trumped literal interpretation of the Bible in cases where such a reading appeared inconsistent with science. According to the encyclical, the authors of the biblical text, "did not seek to penetrate the secrets of nature but rather described and dealt with things in more or less figurative language, or in terms which were commonly used at the time." Thus, as one American bishop noted, Catholics had no obligation to imbue biblical "descriptions of natural processes" with "scientific meaning" if "the Church has pronounced no definitive judgment in the matter." Before the end of the nineteenth century, Catholic officials had made no official pronouncement on the theory of evolution, and therefore some Catholics insisted that speculation among Catholics about the theory of evolution was entirely legitimate.[9]

Catholic intellectuals had a precedent in their response to the new geology, another apparent conflict between religion and science during the nineteenth century, for this argument against a literal interpretation of the Bible in scientific matters. In the decades before evolution became a central issue, developments in the science of geology had likewise raised the specter of conflict between Catholicism and science. The Irish priest and scientist Gerald Molloy summarized the commonly held view that Genesis and geology stood at odds by observing that many believed "we cannot pursue the study of Geology, if we are not prepared to renounce our belief in the doctrines of Revelation." At issue was the increasingly incontrovertible evidence of the vastly older age of the earth than the biblical narrative described, the much slower development of life, and especially the large gap before the emergence of humans.[10]

Molloy's response typified the Catholic approach to new science. He first affirmed the theoretical nature of geologists' assertions, and he then insisted that scripture could be interpreted to accommodate those theoretical claims with little difficulty. From the standpoint of biblical interpretation, "an interval of countless ages may have elapsed" between the first act of creation and the beginning of the six days described in Genesis. Moreover, there was no reason to deny the possibility that the earth existed "for millions of years before man was introduced upon the scene." For Molloy, the key issue was that church authorities had made "no definite judgment" about the science of geology. Thus there was no danger of "irreverence to the Holy Scripture" in considering such views—views, he noted, that were shared by many Catholic thinkers, including members of the conservative Jesuit order. Other intellectuals within the church went even further. Augustine Hewit insisted that it was not "necessary to propose and prove any definite and positive theory" that made "the brief texts in which the account of creation which we have received through Moses is contained" consistent "with the scientific data and conclusions of geologists." In his view, the goal of scripture was so different from the aims of science that it was unnecessary to reconcile the two.[11]

Ultimately, the responses honed in discussions of geology, combined with their invocation of *Providentissimus,* allowed Catholics to refute claims of inherent conflict between evolutionary theory and their faith on scriptural grounds. Indeed, Catholics were adamant—at times even boastful—that their reliance on doctrines and institutions other than the Bible left them in a far better hermeneutical position than the Protestant churches that Draper and others looked on more favorably. Catholics found themselves on "higher and safer ground" because the church could offer nuanced theological interpretations in situations where scientific knowledge seemed to contradict the plain reading of the Bible.[12]

At least in the case of evolutionary theory, however, the interpretation of scripture was not the only issue at stake. There was a more substantive problem with evolution, which explained why that theory posed greater difficulties for Catholic intellectuals than had the earlier developments in geology. Thomas Huxley identified the crux of the issue when he noted the incongruence of the theory of evolution with Scholastic philosophy, a vital part of the Catholic intellectual tradition. According to its tenets, every living thing consisted of both generic matter and a "substantial form," which accounted for its uniqueness. For humans, part of their substantial form was the *anima rationalis*, which was understood to be a unique, original creation for each human being. For Huxley and others, this notion of human beings becoming human only with the special implantation of a soul was "absolutely inconsistent with the doctrine of the natural evolution of any vital manifestation of the human body." Drawing heavily on the writings of the sixteenth-century Jesuit theologian Francisco Suarez, who was a major figure in later Scholasticism, Huxley insisted that any inclusion of humans in the scheme of natural evolution flew in the face of Catholic theology. Among other problems, Huxley found that such a view necessitated the belief that women had developed in the same way. Such a belief, he wrote, proved irreconcilable with Suarez's insistence that the church taught that "woman was, in the strictest and most literal sense of the words, made out of the rib of man."[13]

During the late nineteenth century, Scholastic philosophy, which had waned in importance during the two previous centuries, experienced a resurgence within Catholicism, ultimately becoming the foundation of nearly all theological scholarship within the church by century's end. Scholasticism was a philosophical tradition that originated in the medieval universities and received its best-known and influential expression in the work of the thirteenth-century Dominican friar Thomas Aquinas, who built his system on the philosophy of Aristotle. So great was Scholasticism's hold on Catholic intellectual life by the end of the nineteenth century—Pope Leo XIII's 1879 encyclical, *Aeterni Patris*, made it the sole basis for all study at Catholic institutions—that no scientific theory could possibly win the Vatican's approval without proving its compatibility with Scholastic thought.[14]

Most significant for the Catholic debate over evolution, the Scholastic outlook insisted on the intimate relationship of the body and soul, in contrast to the philosophy of Descartes, which had argued for the dual nature of humans. The belief in the unity of body and soul presented an obvious problem when it came to including humans in the evolutionary process. In the Aristotelian worldview, the human body represented the outward manifestation of the soul. In other words, the two were intrinsically connected. As a result, many proponents of Scholasticism found it impossible to posit simply that the human body had evolved through evolutionary processes while the soul was a direct creation of God.[15]

Yet, even on this thorny issue, a number of Catholic intellectuals rejected the idea of an inherent conflict between Catholicism and the theory of evolution. Writers on both sides of the Atlantic vehemently insisted that it was entirely possible to reconcile Scholastic theology with evolution—even human evolution—and, moreover, that a strong precedent already existed for doing so. One of the first intellectuals to advance such an argument was the English Catholic St. George Mivart. In his 1871 treatise *On the Genesis of Species*, Mivart noted that the fourth-century theologian Augustine had argued for a "derivative" process of creation akin to modern understandings of evolution. In other words, God made "organic forms . . . by conferring on the material world the power to evolve them under suitable conditions." Even more important, Mivart argued that Thomas Aquinas, the most revered exemplar of Scholastic theology, had endorsed Augustine's views.[16]

In the years that followed, numerous Catholic intellectuals echoed Mivart's claims. The American bishop John Spalding concurred that "the doctrine of derivative creation has been familiar to theologians for centuries" and had been "advanced by St. Augustine and approved of by St. Thomas [Aquinas]." Spalding's fellow American John Gmeiner was even bolder, noting that following a comparison between Augustine and Darwin "one might be tempted to look upon St. Augustin as the venerable teacher who advanced some grand comprehensive ideas, which his disciple Darwin has explained more in detail." In France, the priest Maurice Dalmas Leroy repeated the same sentiments. While affirming Aquinas's insistence that "the human body could not arise from the transformation of an animal," he nevertheless argued that it was not at odds with Catholic theology that the "matter" that combined with the "substantial form" to make the human might have developed through a system of "limited evolution."[17]

What is especially notable is that—at least initially—even some Catholic intellectuals who opposed the theory of evolution on scientific grounds nevertheless conceded the existence of theological precedent for supporting it. The conservative American bishop Francis Chatard pointed to the same passages in Augustine as had Mivart, Gmeiner, and Spalding, and he affirmed that the fourth-century theologian has offered a "theory of the potentiality of matter" by which in an initial act of creation by God "gave to matter a power to develop the germs of everything that is material." While Chatard refrained from endorsing evolution, he did not consider the theory to be at odds with his beliefs.[18]

Conflict after All?

Difficulties soon emerged, however, for Catholics who insisted that no conflict existed between their faith and science, particularly with regard to the theory of evolution. Many of the arguments against the conflict thesis rested on the premise that the church had not pronounced judgment on contentious scientific topics and that Catholics were thus free to hold a variety of views. During the 1890s and early 1900s, however, the Vatican did issue a series of decisions that appeared to be negative judgments of scientific theories. Moreover, some of these pronouncements were targeted at the very intellectuals who so forcefully opposed the notion of a conflict between religion and science.

The first sign of trouble appeared in the early 1890s, when Leroy came under attack from French Jesuits for his suggestion in *L'évolution restreinte aux espèces organiques* that prehuman organic life had evolved through evolutionary processes. Though he had insisted on the "incessant action of divine Providence in the universe," Leroy declared that he found it easier to believe that the hundreds of thousands of known species had developed as "the result of evolution from previous types with the exception of the first ones" rather than to accept that each had been individually made through "a special act of creation." Although Leroy had stressed that it was not until God specially and directly imbued existing organic matter with a soul that it became human, his critics nevertheless objected to this position as coming too close to an endorsement of human evolution. After protracted conflict and a negative (although unpublished) judgment from the Congregation of the Index, the Vatican congregation tasked with examining and censoring books suspected of content contrary to faith or morals, Leroy publicly retracted his publication in 1895.[19]

The next few years witnessed a series of similar events. In 1896, the American priest and scientist John Augustine Zahm published *Evolution and Dogma*. Over the previous decade, Zahm had inched toward ever-bolder declarations that support for the modern theory of evolution might be found in the writings of the church fathers, including Augustine and Thomas Aquinas. According to Zahm, Aquinas "laid down principles concerning derivative or secondary creations"—two phrases synonymous with the development of life through evolutionary processes—"which scientists and theologians now recognize to be of inestimable value." In this same book, Zahm raised the possibility of including humans in the evolutionary process in an argument akin to Leroy's. He suggested that earlier forms of organic matter had evolved into an object that, with the imputation of the soul, became human. Thus, in one book, the American intellectual had not only suggested his endorsement of evolution of the human body but he, like others, linked his support to the teachings of Aquinas, whose work formed the core of the nineteenth-century Scholastic revival. As sales of *Evolution and Dogma* exploded, and as preparation began for French and Italian translations, Zahm's book was denounced to the Congregation of the Index. As in the case of Leroy, the Congregation chose not to publish a formal decree against Zahm, and as a result his book was never placed on the Index of Prohibited Books. Nevertheless, at the insistence of the Congregation, Zahm agreed to remove his volume from publication in 1899.[20]

That same year, St. George Mivart, who had been the first to invoke Augustine and Aquinas as intellectual progenitors of the idea of evolution, entered into a conflict with church leaders in England. Mivart's case was exceptional. Whereas Zahm and Leroy had remained insistent on the absence of conflict between church doctrine and scientific theories, Mivart himself came to believe that the two were "fatally at variance." The fundamental issue for Mivart, however, was not evolutionary theory, but rather the nature of scripture. The origins of his difficulties lay in a series of articles he published during the early 1890s titled "Happiness in Hell," in which he challenged church teaching about the unending alienation of humans in hell from God. While this argument itself proved controversial, the larger issue was the bolder implication that underlay Mivart's claim: that doctrine might evolve to reflect new knowledge and understanding. This assertion went too far for Vatican officials, and the "Happiness in Hell" series was placed on the Index, a decision that Mivart blamed for the loss of his faculty appointment at the University of Louvain, one of the centers of Catholic intellectual life.[21]

The Vatican's formal condemnation of his views managed to anger Mivart but failed to alter his thinking. In the winter of 1900, he published a new set of articles. Church officials interpreted these as an assault on a host of doctrines, but the key point of the debate centered on the interpretation of the Bible. In response to Mivart's publication, Herbert Vaughan, the archbishop of Westminster, demanded that he sign a profession of faith. In language that closely paraphrased the encyclical *Providentissimus*, the proposed profession affirmed that scriptures "are sacred and canonical—not because . . . they were afterwards approved by the Church's authority" but rather because "they have God for their author." Moreover, the profession required Mivart to accept "as false and heretical the assertion that it is possible at some time, according to the progress of science, to give to doctrines propounded by the church a sense different" from that traditionally understood.[22]

Mivart refused to sign the profession of faith and accused the church of expecting a greater adherence to traditional doctrine and scripture than it had in previous years. Vaughan denied Mivart's claim, but the aging British scientist was adamant that Catholicism now demanded adherence to doctrines and scriptural passages at odds with modern science. Although neither man touched on specific scientific issues in their exchange, the clear implication was that Mivart believed doctrine or biblical text could easily be used to challenge modern scientific theories, and the church would prove unwilling to set aside long-standing doctrines in the face of new theories. Mivart thus concluded that "a vast and impassable abyss yawns between Catholic dogma and science." After refusing to sign Vaughan's profession of faith, he was, on the archbishop's orders, prevented "from approaching the sacraments . . . until he shall have proved his orthodoxy." Mivart died shortly thereafter without becoming reconciled to the church.[23]

Whether instigated by critics of scientists or by scientists themselves, these various instances had the cumulative effect of silencing some of the most prominent voices in the church who had argued not only for the compatibility of Catholicism and science in general terms but also for the harmony of their faith with one of the most significant scientific ideas of the time: the theory of evolution. Critics inclined to see a conflict between Catholicism and science received seemingly more confirmation in 1907, when Pope Pius X issued a condemnation of modernism in the encyclical *Pascendi dominici gregis*. Modernists were a loosely defined group of intellectuals in both the United States and Europe, but their common characteristics included a belief in the critical scholarly study of scripture and an acceptance of the theory of evolution. Although *Pascendi* did not address the question of organic evolution, it did offer a strong condemnation of the belief that church doctrine might progressively develop (a line of argument similar to that which Mivart had embraced). The encyclical drove a number of prominent intellectual leaders from the church and further confirmed the growing impression that Catholicism favored the preservation of doctrine over the acceptance of new findings in science.[24]

Toward a Nuanced Interpretation of Catholicism and Science

Without question, the events of the 1890s and early 1900s signaled a turning point in Roman Catholic discussions of the relationship between religion and science. The optimistic insistence that church doctrine and new scientific theories would always be easily reconcilable, which characterized so much of the writing of Mivart, Leroy, Zahm, and others, diminished after 1910. Yet Catholic intellectuals remained adamant in their opposition to the warfare hypothesis and insisted that apparent conflict stemmed from issues unrelated to science. Writing in 1920, for example, Bertram Windle recalled the "old, far-off, unhappy events" of the Mivart case and explained that "the troubles" of the late scientist "did not originate in, nor were they connected with, any of the scientific books and papers" Mivart had written "but with those theological essays which all his friends must regret that he should ever have written." Even after the upheaval of the 1890s and 1900s, Catholic intellectuals sustained their view that conflict emerged only when scientists impinged on the realm of theology.[25]

But, as Windle and others seemed to recognize, the cases of Zahm, Mivart, and the modernists had given new ammunition to those inclined to argue for conflict. Moreover, in light of the recent Vatican pronouncements that seemed to foreclose consideration of ideas that had seemed acceptable, resting too much of the argument for harmony on contemporary scientific theories represented a risky proposition. Thus, discussions of Catholicism and science moved away from efforts to reconcile contemporary theories with doctrine and instead toward broader considerations: first, of the church's history of supporting the advancement of human knowledge, and second, of the perilous political and social implications of modern scientific theories and the church's role in guarding Catholics from them.

In the United States, James Joseph Walsh emerged as a leading defender of Catholicism's record of support for science and education with books that included the provocatively titled *The Thirteenth, Greatest of Centuries*. In several histories of science of the late Middle Ages, Walsh credited Catholic institutions with considerable advancement of medical science. Citing his assessment that "anatomy and surgery and every branch of medicine was cultivated much more assiduously and with much better opportunities provided for students down in Italy, than anywhere else in the world," the American Catholic physician concluded that it proved "the utter absurdity" of claims that his tradition "was opposed to medical progress."[26]

Nor did Walsh deal in platitudes and generalities. He challenged the specific criticisms leveled against the church, particularly by Andrew Dickson White. One of his frequent targets was the erroneous claim that the Vatican had prohibited dissection, which, according to proponents of the warfare thesis, had impeded the development of knowledge about anatomy. Beyond the evidence he mustered that dissections had, in fact, occurred frequently under the auspices of Catholic authorities, Walsh also found a potent retort in his suggestion that proponents of the conflict thesis failed to distinguish between the church's teaching and wider social values of the Middle Ages. It was medieval "popular prejudice" that opposed dissection of human bodies, not the Vatican, and when impediments rose to the practice, it was not because of church policy, but rather because Catholic leaders could not overcome existing prejudice.[27]

Yet just because Catholics believed that their faith had historically been in front of popular opinion did not mean that they always viewed that to be the case. On the contrary, in the early decades of the twentieth century, Catholics who defended the reconcilable nature of their faith with science increasingly bristled at many of the claims of scientists. This was particularly true as eugenics movements grew in popularity worldwide, and the theory of natural selection was more commonly used to advocate an active policy of social engineering. In the late nineteenth century, proponents of evolutionary theory—including Darwin himself—had expressed deep concern about the prospect of humans attempting to exert some control over the evolutionary process. By the dawn of the twentieth century, however, growing anxieties about expanding social ills such as poverty and divorce combined with advances in the emerging field of genetics provided an impetus to the eugenics movement.[28]

Around the globe, Catholics emerged as outspoken critics of eugenics movements, and of the way in which natural selection had come to undergird it. In part, this reflected a very natural response to the reality that supporters of eugenics often wedded that commitment to a staunch anti-Catholicism. In the United States, the most ardent champions of eugenics held fast to a racial and cultural hierarchy that situated Anglo-Saxon civilization above the Catholic countries of eastern and southern Europe. So, too, increasingly popular eugenics campaigns in Latin American nations such as Chile stressed the importance of protecting the nation from southern European immigrants who might too readily support the church.[29]

Yet Catholic intellectuals remained adamant that their objection was not to Darwin's original theory, but rather to the claims of eugenicists who carried natural selection into the larger realm of social science. "The great and gentle-hearted originator of that theory" had never argued for its being viewed as a general "rule of life." Darwin, Bertram Windle noted, "during his life had to protest" against "the ignorant and exaggerated ideas" that emerged from his theories, and "were he now alive, would certainly be shocked at the teachings which are supposed to follow from his theory and the dire results which they have produced." The problem was not the original biological theory. Rather, it was the way in which the theory had influenced social policy and ultimately inspired people to view conflict as a natural part of existence rather than "the curse and disaster which all reasonable people, not to say all Christians, feel it to be." For Windle and others, science was not the issue. Rather, the issue was a political vision of social progress through force and competition, which took the form of "political nostrums of some who wrest science to a purpose for which it was not intended."[30]

This growing skepticism concerning the manipulation of science also reflected deeper anxieties following World War I. Not only did natural selection lead to eugenics when it expanded into the realm of policy, but it also bred military conflict—and bloody, destructive conflict at that. "Natural Selection means the Survival of the Fittest in the Struggle for Life," Windle observed, suggesting this had been "the proposition by which we were confronted" during the Great War. He laid bare what he believed to be the inevitable conclusions that followed from adopting such a philosophy as the basis of statecraft: "if Natural Selection be our only guide, let us sink hospital ships, destroy innocent villages and towns, exterminate our weaker opponents in any way that seems best to us." It was incumbent upon the church to protect its members from just such an outlook.[31]

The crucial point here is that while Catholics might have seemed less favorably disposed to science in the first decades of the twentieth century, their reservations were in fact limited to a small subset of scientific theories that seemed to raise genuine concerns regarding the ease with which they could influence social policy. These Catholics did not reject science as a whole, nor did they view it as inimical to their faith.

Yet even in the early decades of the twentieth century, as Catholic apologists shifted to emphasize the church's historical support for science while denouncing the political and social implications of modern scientific theories, a few continued to advocate a compatibility between the theory of evolution and church teaching. These advocates, however, initially retreated from efforts to include humans in the evolutionary process. In 1907, the Jesuit priest and scientist Erich Wasmann offered a series of lectures in Berlin on the relationship between evolution and the Catholic faith. Wasmann's choice of city was critical: Darwin's theory was extremely popular in Berlin, and the city was home to many who subscribed to the view that evolution could serve "as a kind of battering-ram against Christianity" because it eliminated the need of a divine creator. The Jesuit firmly rejected this position, emphasizing, as Catholic intellectuals had done for decades, the difference between evolution as a biological theory that was "*absolutely unconcerned with any theory of the universe*" and as a philosophical system. The former, Wasmann argued, was entirely compatible with church teaching, provided it acknowledged the need of an original creator and accepted the existence of a directly created soul that "forms one complete being and substance" with "the human body." He was thus careful to separate humans from the process of organic evolution. Although Wasmann issued these lectures in the same year that modernism was condemned, their publication was approved by the Jesuit order, and the work received favorable reviews in the Catholic press.[32]

Other European Catholics continued to argue for the compatibility of the faith and the theory of evolution in the decades that followed. Henry de Dorlodot, a geologist from Belgium, affirmed in a 1918 book that the evolution of the human body was a conclusive determination of science, a finding that he insisted the church would have to accept. Like others who had written on the subject, he found support for the theory of evolution in the works of Augustine. The fourth-century theologian, according to Dorlodot, "affirms with emphasis that living beings were only virtually created at the beginning, and that they subsequently appeared by an absolute natural evolution." While Dorlodot received widespread criticism, he did not face the official condemnation that his intellectual predecessors had endured two decades earlier. Finally, in his 1950 encyclical *Humani generis*, Pope Pius XII affirmed that Catholics had freedom to explore the creation of the human body as an unproved conclusion, though he urged "the greatest moderation and caution" by those scientists who addressed the issue. This, in turn, paved the way for the greater liberty allowed to Catholic proponents of evolutionary theory following the Second Vatican Council, an event widely interpreted as an expression of the church's greater openness to scientific endeavor.[33]

Conclusion

By tracing discussions within the church about the relationship between Roman Catholicism and science from the middle of the nineteenth century to the middle of the twentieth, some larger conclusions become clear. From the perspective of Catholic intellectuals, no conflict existed between their faith and scientific knowledge. The church boasted a long history of support for scientific endeavor, the frequently invoked Galileo case was anomalous, and Catholic doctrine and hierarchy served as intermediaries to help the faithful reconcile seeming disagreements between science and scripture. Indeed, one can easily view the rich discussion of evolutionary theory—including the theory of human evolution—in the final decades of the nineteenth century as evidence of Catholicism's openness to scientific innovation. Catholic intellectuals were thus quite convinced that apparent conflicts emerged only under specific circumstances: when scientific theories were carried into the social and political realms (as with natural selection and eugenics), when scientists began to speculate on theological matters (as Mivart had done), or when they overconfidently portrayed disputed theories as settled fact, especially when those theories related to human origins (as Zahm had done).

Yet, as demonstrated by the cases of Leroy, Zahm, and especially Mivart (who, after all, himself declared a conflict to exist in the aftermath of *Providentissimus*), critics had evidence to invoke when making the assertion of warfare between the Catholic Church and science. Therein lay the difficulty for Catholic apologists. Those within Catholicism found it entirely reasonable for the Vatican to exert some control over the publication of scientific theories for the protection of the faith of, as Windle described them, "weaker members" who lacked the ability to judge science for themselves.[34]

Ultimately, the perceived conflict between Catholicism and science might more accurately be understood as a conflict over what constituted reasonable church authority. Catholic apologists could reasonably argue that no disharmony existed between contemporary theories and church teaching. Yet there were some implicit assumptions in this position. Many Catholics presumed that science and religion occupied different spheres and that it was possible, especially for scientists, to resist venturing "beyond the scope of their own proper branches." But this assertion obscured the way in which some questions of science—the origins of human life, for example—could not easily be disentangled from questions of Catholic theology. So, too, many apologists assumed that it was entirely reasonable for the church to maintain some control over the intellectual life of Catholics. The belief in such control stemmed not out of any desire to stifle scientific innovation but rather to serve as a "perpetual and unerring tribunal," protecting its members just as a government would seek to protect its citizens. From the perspective of those inclined toward the conflict thesis, such views confirmed their suspicion that the Catholic Church was hostile to the development of science.[35]

This divergence helps to explain why the warfare thesis has persisted, especially in discussions of Roman Catholicism. Catholics were adamant that no such phenomenon existed. Yet the strong hand of the Vatican hierarchy and its insistence on oversight of science that intersected with church doctrine—the very thing that comforted many Catholic intellectuals—provided sufficient evidence to convince skeptical critics that their assumptions of conflict were indeed correct.

ACKNOWLEDGMENTS

I would like to thank Ron Binzley, Jeff Hardin, Ron Numbers, and Jon Roberts for their insightful comments on this chapter.

NOTES

1. John William Draper, *History of the Conflict between Religion and Science* (1874; New York: D. Appleton, 1898), vi, x–xi, 268, 332.

2. Andrew Dickson White, *A History of the Warfare of Science with Theology in Christendom*, 2 vols. (New York: D. Appleton and Company, 1896), 1:130–31; see 130–48 for White's full discussion of the Galileo case; Thomas Henry Huxley, *Darwiniana: Essays* (New York: D. Appleton, 1897), 146–47.

3. White, *History of the Warfare*, 342; Draper, *History of the Conflict*, 225.

4. "Draper's Conflict between Religion and Science," *Catholic World* 21 (May 1875): 179; review of *Harmony between Science and Revelation* by J. de Concilio, *Dublin Review* 26 (October 1891): 488–89.

5. Bertram Windle, *Science and Morals and Other Essays* (New York: Kenedy & Sons, 1920), 75; James Joseph Walsh, *The Popes and Science* (New York: Fordham University Press, 1915), 7–8, 16; on Windle, see Patrick Allitt, *Catholic Converts: British and American Intellectuals Turn to Rome* (Ithaca, NY: Cornell University Press 1997), 171–75.

6. Windle, *Science and Morals*, 90; Walsh, *Popes and Science*, 17, 19; Augustine F. Hewit, "The Warfare of Science III" *Catholic World* 53 (August 1891): 682–84.

7. For studies of the church's response to evolution, see R. Scott Appleby, *Church and Age Unite! The Modernist Impulse in American Catholicism* (Notre Dame, IN: University of Notre Dame Press, 1992), chap. 1; Appleby, "Exposing Darwin's 'Hidden Agenda': Roman Catholic Responses to Evolution, 1875–1925," in Ronald L. Numbers and John Stenhouse, eds., *Disseminating Darwinism: The Role of Place, Race, Religion, and Gender* (New York: Cambridge University Press, 1999); Mariano Artigas, Thomas F. Glick, and Rafael A. Martinez, *Negotiating Darwin: The Vatican Confronts Evolution, 1877–1902* (Baltimore, MD: Johns Hopkins University Press, 2006); Stefaan Blancke, "Catholic Responses to Evolution, 1859–2009: Local Influences and Mid-scale Patterns," *Journal of Religious History* 37 (September 2013): 353–69; Barry Brundell, "Catholic Church Politics and Evolutionary Theory, 1894–1902," *British Journal for the History of Science* 34 (March 2001): 81–95; David Mislin, "'According to His Own Judgment': The American Catholic Encounter with Organic Evolution, 1875–1896," *Religion and American Culture* 22 (Summer 2012): 133–62.

8. John Gmeiner, *Modern Scientific Views and Catholic Doctrines Compared* (Milwaukee, WI: J. H. Yewdale and Sons, 1884), 13; on *Providentissimus*, see Don O'Leary, *Roman Catholicism and Modern Science* (New York: Continuum, 2006), 68–72.

9. John Lancaster Spalding, "Religious Faith and Physical Science," in *Lectures and Discourses* (New York: Catholic Publication Society, 1882), 59; *Providentissimus Deus*, reprinted in Dean P. Béchard, ed., *The Scripture Documents: An Anthology of Official Catholic Teachings* (Collegeville, MN: Liturgical Press, 2002), 54.

10. Gerald Molloy, *Geology and Revelation, or, the Ancient History of the Earth, Considered in the Light of Geological Facts and Revealed Religion* (London: Longmans, Green, Reader, and Dyer, 1870), 26. On Catholic discussions of geology more broadly, see William J. Astore, "Gentle Skeptics? American Catholic Encounters with Polygenism, Geology, and Evolutionary Theories from 1845 to 1875," *Catholic Historical Review* 82 (January 1996): 40–76.

11. Molloy, *Geology and Revelation*, 29, 288, 296; Augustine F. Hewit, "Scriptural Questions: No. 1," *Catholic World* 40 (November 1884): 153.

12. Spalding, "Religious Faith and Physical Science," 59.

13. Huxley, *Darwiniana*, 142–43, 145.

14. On the Scholastic revival, see Gerald A. McCool, *Catholic Theology in the Nineteenth Century: The Quest for a Unitary Method* (New York: Seabury Press, 1977).

15. See, for example, Appleby, "Exposing Darwin's 'Hidden Agenda,'" 193–94.

16. St. George Jackson Mivart, *On the Genesis of Species* (London: Macmillan, 1871), 264–65.

17. Spalding, "Religious Faith and Physical Science," 56; Gmeiner, *Modern Scientific Views*, 158; Maurice Dalmas Leroy, *L'évolution restreinte aux espèces organiques*, 259–60, 266–67, 283, trans. and quoted in Artigas et al., *Negotiating Darwin*, 58–60.

18. Francis Silas Chatard, "The Brute-Soul," *Catholic World* 57 (July 1893): 448.

19. Leroy, *L'évolution restreinte*, 31, trans. and quoted in Artigas et al., *Negotiating Darwin*, 56–57; O'Leary, *Roman Catholicism and Modern Science*, 96–97, 100–123.

20. John Augustine Zahm, *Evolution and Dogma* (Chicago: D. H. McBride, 1896), 29–30; on Zahm, see Appleby, *Church and Age Unite!*, 48–52; Artigas et al., *Negotiating Darwin*, chap. 4.

21. Mivart to Vaughan, January 23, 1900, in *Under the Ban: A Correspondence between Dr. St. George Mivart and Herbert Cardinal Vaughan* (New York: Tucker, 1900), 22; on the Mivart case, see Artigas et al., *Negotiating Darwin*, chap. 7

22. See the profession of faith, printed in *Under the Ban*, 8.

23. Mivart to Vaughan, January 23, 1900, and Vaughn to the Clergy of the Diocese of Westminster, in *Under the Ban*, 22, 31.

24. On *Pascendi* and its consequences, see Appleby, *Church and Age Unite!*, chap. 2; Thomas E. Woods, Jr., *The Church Confronts Modernity: Catholic Intellectuals in the Progressive Era* (New York: Columbia University Press, 2004).

25. Windle, *Science and Morals*, 96.

26. Walsh, *Popes and Science*, 78.

27. Walsh, *Popes and Science*, 52; Walsh, "The Supposed Warfare between Medical Science and Theology," *Medical Library and Historical Journal* 4 (September 1906): 263–86.

28. For a particularly useful discussion of the global embrace of eugenics in the early twentieth century, see Nancy Leys Stepan, *"The Hour of Eugenics": Race, Gender, and Nation in Latin America* (Ithaca, NY: Cornell University Press, 1991), 23–35.

29. Appleby, "Exposing Darwin's 'Hidden Agenda,'" 194–96; Sarah Walsh, "'One of the Most Uniform Races of the Entire World': Creole Eugenics and the Myth of Chilean Racial Homogeneity," *Journal of the History of Biology* 48 (2015): 622.

30. Windle, *Science and Morals*, 19–20, 27.

31. Windle, *Science and Morals*, 122–23.

32. Erich Wasmann, *The Berlin Discussion of the Problem of Evolution* (1907; London: Kegan Paul, Trench, Trübner, 1912), 22, 25, 32; see the review in *Dublin Review* 145 (July 1909): 422–25.

33. Henry de Dorlodot, *Darwinism and Catholic Thought*, trans. Ernest C. Messenger (London: Burns, Oates and Washbourne, 1922), 1:82; see also Raf De Bont, "Rome and Theistic Evolutionism: The Hidden Strategies behind the 'Dorlodot Affair,' 1920–1926," *Annals of Science* 62 (October 2005): 457–78; O'Leary, *Roman Catholicism and Modern Science*, 126–40; Pius XII, *Humani generis*, quoted in Blancke, "Catholic Responses to Evolution," 366.

34. Windle, *Science and Morals*, 83.

35. Hewit, "Scriptural Questions: No. 1," 147, 149.

CHAPTER SEVEN

Eastern Orthodox Christians

EFTHYMIOS NICOLAIDIS

The two most influential nineteenth-century books positing the existence of "warfare" between science and religion, John William Draper's *History of the Conflict between Religion and Science* (1874) and Andrew Dickson White's *A History of the Warfare of Science with Theology in Christendom* (1896), are almost completely unknown to Greek intellectuals. They are certainly unknown to those without expertise in the history of science and largely ignored even by those with such expertise before the 1980s. Therefore, it is hardly surprising that in his recent book, *From the Pyre to the Pulpit* (2009), the Greek left-wing intellectual and philosopher of science Efthychis Bitsakis fails to mention either Draper or White even as he promotes his own version of the conflict thesis by attacking the Orthodox Church for promoting mysticism and nonrationality in order to support the ruling class. Other contemporary Greek intellectuals who share Bitsakis's belief in a conflict between religion and science also take no notice of Draper's and White's books.[1]

To a certain extent, Greek intellectuals' lack of interest in these works seems puzzling because Draper and White themselves were not ignorant of Eastern Orthodoxy, nor did they fail to treat the Orthodox tradition in their histories. Draper, an English-American chemist who later wrote works developing his own theories of history, presents in detail the ecclesiastical debates of the Early Byzantine period (but with almost no reference about science-religion relations), and he mentions the little-known fifth-century Greek language writer Cosmas Indicopleustes in order to upbraid him for advocating a flat earth. Yet despite his criticism of Cosmas, he largely treated Eastern Orthodoxy with respect and even denied that any kind of conflict exists between Orthodoxy and modern European science.[2] Indeed, Draper claimed that the Greek churches, like the Protestant ones, had adopted moderate opinions regarding science and, unlike the "extremist" Catholics, had always welcomed the discipline:

> In thus treating the subject [of the conflict between religion and science] it has not been necessary to pay much regard to more moderate or intermediate opinions, for, though they may be intrinsically of great value, in conflicts of this kind it is not with the moderates but with the extremists that the impartial reader is mainly concerned. Their movements determine the issue.
>
> For this reason I have had little to say respecting the two great Christian confessions, the Protestant and Greek Churches. As to the latter, it has never, since the restoration of science, arrayed itself in opposition to the advancement of knowledge. On the contrary, it has always met it with welcome. It has observed a reverential attitude to truth, from whatever quarter it might come. Recognizing the apparent discrepancies between its interpretations of revealed truth and the discoveries of science, it has always expected that satisfactory explanations and reconciliations would ensue, and in this it has not been disappointed. It would have been well for modern civilization if the Roman Church had done the same.[3]

As we will shortly see, the Eastern Orthodox Church did not always welcome "new discoveries" with the enthusiasm Draper claimed. Nonetheless, it is important to note that for Draper the real religious enemy of science was the Roman Catholic Church and not Christianity as a whole. In his view, the Protestants and especially the Orthodox were constructive partners with science.

White, the cofounder of Cornell University and a vigorous supporter of secular education, also discussed Eastern Orthodoxy in his history and with more sophistication than Draper, no doubt owing in part to the fact that he knew the Orthodox tradition firsthand. For several years in the early 1890s, White served in St. Petersburg as the United States minister to Russia. Compared with other scholars of his day (and even compared with many in our day), White devoted unusual attention to Eastern Orthodoxy. His book included discussions of the fathers of the church on the Creation, of Byzantine theologians, and of the seventeenth-century Russian Orthodox patriarch Nikon. In contrast to Draper, who never mentioned the Eastern Church after the Middle Ages, White referred to its attitude toward biblical interpretation in the eighteenth century and to the "Greek church" in Russia using scripture to forbid peasants from raising and eating potatoes. Furthermore, in his autobiography, White included a lengthy tale of the Russian Church's credulity about miracles.[4]

Despite these mentions of Orthodoxy, Draper and White, like most other well-known historians who endorsed the warfare thesis, were interested in the mainstream history of science (a history focused chiefly on western Europe), and thus their polemics concerned mainly the Catholic Church. In the Greek world of the late nineteenth and early twentieth centuries, where Eastern Orthodoxy prevailed and many people often dismissively referred to Catholicism as "papism," it is not strange that Draper's and White's books went mostly unnoticed by intellectuals: Orthodox thinkers in particular did not feel the need to respond to attacks on Catholicism, which they saw as distinct from and mostly alien to their own church.

Although the works of Draper and White or any other historians of science played little role in its development, a version of the warfare thesis did emerge in the Greek world in the final years of the nineteenth century. The thesis took hold as a consequence of a conflict between traditional forces of society, such as the Orthodox Church, that were suspicious of western Europe, and modernizing forces that looked to integrate the eastern and western halves of the continent. These latter forces were found most prominently in the milieu of engineers who had studied in western Europe or were influenced by the idea that technological progress is social progress. The roots of this conflict between traditionalists and modernizers go back a century from this period and are to be found in the debate about scientific knowledge during the Greek Enlightenment, an intellectual movement of the second part of the eighteenth century. The scholars involved in this earlier debate were typically clerics of the Orthodox Church, and the point at issue among them was not any purported warfare between science and religion but rather the proper attitude that the Orthodox Church ought to adopt toward science.

This eighteenth-century Greek debate over the church's posture toward science reflects an important difference between Greek scientists of that period and those practicing in western Europe at the same time: the former to a great extent were affiliated with church institutions, and the latter were to a great extent secular. The connection of Greek scientists to the church was the result of the organizational system of the Ottoman Empire, which controlled the great majority of Orthodox adherents outside of Russian lands. The Ottoman *millet* system gave ecclesiastical leaders power over their believers in almost all matters excepting taxes. Therefore, the Patriarch of Constantinople controlled education for members of his church, and

Orthodox people received nearly the whole of their education, including any scientific education, from schools that were under the supervision of the Orthodox Church. Because the Ottoman Empire did not have any universities, Orthodox students went abroad for higher education, at first attending the University of Padua and later other European universities as well. Except for those training to be physicians, most of these university students were priests or monks who were destined to become the teachers in Orthodox communities.

During the mid-eighteenth century a controversy arose about the content of science teaching in Greek schools. Because of their innovative teaching of Newtonian physics, important scholars and theologians, such as Eugenios Voulgaris (1716–1808) and Nikiforos Theotokis (1731–1800), both future archbishops of Cherson in Russia under Catherine the Great, were obliged to leave the Ottoman Empire, where they had taught at important Orthodox schools, and to go to Russia. Indeed, during the second half of the eighteenth century, conservative forces within the Orthodox Church began to fear that innovative scientific teaching would alienate Orthodox believers. One reason for their fear was that some Orthodox regarded the new scientific ideas from the West to be pro-Catholic or pro-Protestant and therefore necessarily incompatible with the traditional Orthodox teaching of ancient Greek science, which had its roots in the Byzantine period. The debate about the place of the sciences in the curriculum of Orthodox schools culminated in March 1829, when the Patriarch of Constantinople edited an encyclical that dealt with education:

> Everywhere there reigns a disdain for matters of grammar, and the arts of logic and rhetoric and the teaching of the elevated theology are completely ignored. This disdain and ignorance come from the exclusive love of students and professors for mathematics and science, and turning cold toward our faith. . . . For the Nation, the teaching of grammatical classes is more beneficial and necessary than mathematical or scientific classes. . . . [F]or what is the advantage for the students who follow these courses to learn figures and algebra, cubes and cubocubes, and triangles and trigonosquares, and logarithms, and symbolic calculations, and projected ellipses, and atoms and voids and whirlpools, and forces and attractions and weight, and qualities of light, and polar auroras, and optics, and acoustics, and thousands of similar and monstrous things, in order to count the sand on the shore and the drops of dew, and to move the earth—if support is offered via Archimedes[?] Yet they are

> barbarous in their speech and poor in their writing, ignorant in their religion, perverse and corrupt, and noxious to politics, these obscure patriots who are unworthy of the hereditary vocation.[5]

The patriarch's emphasis on the teaching of grammar was animated not only by his disdain for science but also by his commitment to the tradition of the Greek Orthodox Church (still practiced today) of not translating the sacred ancient Greek texts into modern Greek. Common believers had difficulties understanding the language of these texts, but at least, according to the church, pupils educated in the schools should be able to understand them.

The ideas against scientific teaching expressed by this patriarchal encyclical are deeply rooted in Orthodox history and from time to time have constituted its main trend. These ideas stretch back at least to the fourteenth-century Hesychast movement, which denied that scientific knowledge is essential to understanding the created world or that it mediates between humans and the Creation. The Hesychasts, following the theologian Gregory Palamas (1296–1359), sought *Theosis*, that is, the union with the spiritual through prayer and exercise. They believed all needed knowledge to be revealed this way. While the Hesychasts did not directly condemn science, they neglected the importance of it, believing the teaching of science unnecessary for society.

Throughout the centuries, the Hesychast movement waxed and waned, and the attitude of Orthodoxy toward science (whether or not it was a gift of God) changed many times. But even during those periods when the Hesychast movement was dominant, there remained within the Orthodox Church supporters of the idea that science mediates between nature and God. Thus in the case of the Orthodox controversy over the teaching of science it is more accurate to speak of a theological debate rather than of warfare between science and religion. Byzantine society and, later, Greek society of the Ottoman period encompassed both reverence for the mystical prayer cultivated in the Hesychast tradition and admiration for ancient Greek knowledge, which many believers considered to be their exclusive heritage. Throughout the centuries, the position of the church served to balance these two contradictory sets of convictions, with one usually having more sway than the other, depending on politics, individuals, and historical developments.[6] Until the end of the nineteenth century, Orthodox intellectuals considered this debate to be essentially a theological one. As the century came to a close, many of them began to understand the stakes in the debate differently, as a result of their apprehension over the efforts of some secular scholars to modernize the new states that had emerged from the dismantlement of the Ottoman Empire.

The Warfare Thesis at the End of the Nineteenth Century

Owing largely to the influence of the Greek Enlightenment, interactions between science and religion in nineteenth-century Greece differed markedly from such interactions in other Orthodox countries, especially those of the Slavic world. The Greek national revolution, which culminated in the formation of the independent Greek state in 1830, was unique from national revolutions undertaken by other Orthodox peoples in that the Greek revolutionaries shared not only an Orthodox identity but also certain Enlightenment convictions. Most important among these convictions was the notion that Western civilization found its origin in ancient Greece and that the modern Greeks were the heirs of the Hellenes. Orthodox Greeks embraced the Hellenic ideal of reviving ancient Greece in its homeland and of course believed that this revival should include the revival of ancient Greek science. In Slavic countries, the Hellenic ideal had no influence on Orthodoxy, and as a consequence, believers there were torn between Western modernity and the Orthodox tradition without the intermediary of a "glorious past," such as the Greeks had. The absence of such an intermediary helps explain why the Hesychast tradition and Greek theologians important to that tradition, such as Maximus the Confessor (c. 580–662) and Gregory Palamas, influenced the Orthodox of the nineteenth-century Slavic world far more than they did the Greek Orthodox of the same period.[7]

Besides the ideal of reviving ancient learning, the relationship of the Orthodox Church to the Greek state also greatly affected religious-scientific interactions in the new Greek nation. The Greek revolutionaries had understood well the importance of Orthodoxy to the national cause; indeed, they took the cross as the symbol of their struggle for independence. Greek leaders regarded the Greek state as a Christian one that rested on two chief pillars, the Orthodox Church and Hellenism. In 1833, the Greek government proclaimed the Greek Orthodox Church, which was now the church of the Greek state, to be autonomous from the Patriarchate of Constantinople, which remained under Ottoman rule.

The autonomy of the Greek Orthodox Church, together with the growing influence of the Protestant world—where voluntary religious organizations flourished—led to the emergence of a number of Orthodox religious societies and groups throughout Greece. Although not directly controlled by the Greek Church, these societies had strong ties with its hierarchy. As they did not formally belong to the church, the societies could promote and propagandize in behalf of ideas that representatives of the official church were not willing to fight for openly, because of the church's relations with the state. The most powerful of these societies was Anaplasis (Regeneration), which began publishing a journal of the same name in 1887. Orthodox Greeks who felt threatened by modernity relied on the same tool as their modernists opponents to advance their cause, namely, the press. Activist believers, such as Ioannis Skaltsounis (1821–1905), a merchant of the Greek diaspora who lived in Trieste, wrote for several decades on every subject they thought threatened Orthodoxy. These included Darwinism, materialism, and sometimes cosmology. Although not official church teaching, the polemics of Skaltsounis and others enjoyed the tacit and sometimes explicit approval of high-ranking clergy.

Prior to the revolution, secular scientists had been the exception in Greece. In the newly founded Greek state, they made up the overwhelming majority of all scientists. Generally Greek scientists worked in state-funded institutions (University of Athens, Polytechnic School, Military School, state ministries), and many had contacts with western Europe and directly assisted with efforts to modernize the country. They were also very active in writing popularizing books and articles and also in publishing journals. As a rule, Greek scientists praised Christianity and did not instigate religious confrontations. For instance, the cosmological or naturalistic articles of professors of the University of Athens, such as Theodor Orfanidis (1817–86), a botanist, or Demetrios Eginites (1862–1934), an astronomer and also the director of the Observatory of Athens and minister of education in 1917 and 1926, are full of references to the creator of the world, whose wisdom can be gleaned in nature itself. Other scientists, such as Constantine Mitsopoulos (1844–1911), a geologist, were in fact devout believers who defended Orthodox Greece against anything seen as a threat.

Among the Greeks, the first references to the historical warfare thesis are to be found in the writings of three modernists at the University of Athens: Dimitrios Stroumpos (1806–90), the first professor to hold the physics chair; Anastasios Christomanos (1841–1906), a leading Greek chemist as well as the organizer of modern chemistry education at the university; and Nikolaos Apostolides (1856–1919), a professor of zoology and for a time the Greek government's minister of finance. All three men were prominent scientists and held important positions in the newborn Greek state. They were also all advocates for modernizing the nation by means of scientific and technological development and through the integration of Greece into the culture of western Europe.[8]

Over the course of their careers at the University of Athens, Stroumpos, Christomanos, and Apostolides each found occasion to give speeches about the dark Middle Ages and their catastrophic influence on science. Despite being favorable to Western ideas as well as proponents of modernization, these scientists accused neither the Greek Orthodox Church nor the Orthodox religion of being historically responsible for a war against science and reason or for the decay of science and the nation.

Thus, for example, when Stroumpos was elected dean of the University of Athens in 1858, he choose to give a lecture on the history of science titled "About the Knowledge and Beliefs of Ancients and Moderns on the Natural Phenomena and the Methods of Their Investigation."[9] His decision to focus on history rather than contemporary developments in physics (his own area of expertise) reflected the tradition for such speeches—one rooted in the fact that most professors at the university studied the humanities and not science. When giving major addresses, professors had generally sought to promulgate the ideology that the new Greek state was the re-foundation of ancient Greece and had usually looked to history to make their arguments.

In his speech, Stroumpos drew on French translations of Henri Ritter's *History of Philosophy* and Sir Humphry Davy's *Elements of Chemical Philosophy* to present an account of ancient Greek science.[10] To exalt Greece, he found in his foreign sources legitimation of the notion that the modern sciences originated in ancient Greece. For instance, he quoted from Ritter's preface that "we owe to the Greeks almost all the discoveries and the principles of almost all sciences and most of the arts that today are fulfilling our lives" and repeated from Davy's work equally fulsome praise for Hellenic science. Stroumpos recited to his listeners Davy's opinion that the Greek philosophers "had inherent feeling of the Good, the Brilliant, the Important." The ancient Greeks were, according to Davy, "wrong not because they lacked of geniality and persistence but only because they followed a wrong path, and founded their claims about physics on a system based on imagination rather than on its aspect and its perception by the sense of touch."

After finishing his discussion of ancient science, Stroumpos made only a brief mention of the first Byzantine period, which he viewed as a time of political and religious instability unfavorable to science, before turning to the history of alchemy. He insisted that Muslims had helped to develop science through their work in alchemy and registered his surprise that they had done so since the principles of their religion, according to him, should have caused Muslims to betray science. In contrast to the Islamic world, the Latin West had persecuted alchemists as sorcerers, and the Catholic Church had defended its dogmas by punishing anyone working in the sciences or the arts who displayed a "courageous spirit." Turning to the Galileo affair, he claimed that "it was said that in Rome some people proposed to drag his dead body on the streets of the city and burn his writings" and that the Jesuits regarded the idea that the earth moves as far worse than that of the mortality of the soul.[11] He ended his oration by contrasting the relationship between science and religion in the ancient world with its relationship in the contemporary one, observing that while science was related to religion in Egypt, Greece, and Rome, in present times it has become entirely independent.

In 1864, six years after Stroumpos offered his blistering assessment of the medieval Catholic Church's impact on science, Christomanos offered a very similar one when giving his inaugural lecture after becoming assistant professor of chemistry. Like Stroumpos, he took as his subject the history of science and argued that science had reached its highest development in contemporary western Europe. He also insisted on the importance of promoting the development of the modern sciences in Greece. In his historical review of scientific progress, Christomanos both underlined the accomplishment of ancient Greek science and the decay it suffered with the rise of Christianity.[12] To say the least, much of his audience was unlikely to have received his antichurch polemic with equanimity. His listeners included the four faculties of the university: those of law, medicine, philosophy (which included the sciences), and theology. Despite the undoubted hostility of many in the room, he declared:

> During the Middle Ages Christianity did not at all help the progress of science. On the contrary, fanatic and intolerant views stopped all spiritual development of humankind by explaining wrongly natural phenomena for self-interested aims. Christianity, this best and divine teaching, departed from its truths and filled fanatically the hearts of its followers who rushed for the spread of this new religion . . . . And our religion was founded on rivers of blood because for long centuries it prevailed in an endless fight for the introduction of Christianity instead of paganism and Judaism. During

> this period, dark for all history of science, the nations could not of course take care of sciences and especially of physical sciences, because many miracles and misunderstandings of nature were used by the active establishers of the religion to prove the truth of their sermons. Today, the great and liberal religion that we are supporting contributes to the most precise and better diagnosis of nature, offering instead of obstacles the means of research. The period of the introduction of Christianity was erosive for the natural sciences, and not only obstructed every kind of research but guided the human mind to contradictory and improbable theories."[13]

Two decades later, in 1884, Nikolaos Apostolides in his own inaugural lecture as a professor of zoology offered a similarly bleak assessment of the scientific record of the medieval church. But some believers may have found Apostolides's supreme confidence in the authority of science and his high notions of its place in society almost as troubling as his harsh words about the church. For instance, in a lecture delivered in the 1890s, he not only urged the importance of founding modern zoological laboratories but also maintained that "[h]uman progress is so assimilated to science that both seem to be necessary conditions of civilization" and that "scientific truths are like atoms of matter; they can never be destroyed."[14]

The prominent modernists—Stroumpos, Christomanos, and Apostolides—promoted the conflict thesis, but only for the specific case of Western Christianity. They claimed that modern Europe developed science against the dogmas of the Catholic Church and thereby rediscovered the philosophical freedom of ancient Greece. They avoided making any negative assessments of Orthodox Christianity, probably for at least two reasons: the importance of Orthodoxy to the political ideologies prevailing in nineteenth-century Greece and their fear of provoking the powerful faculty of theology at the university where they taught.

From the foundation of the Greek state until the end of the nineteenth century, only a small minority of Greek scientists directly confronted scholars from religious circles. The most prominent confrontation of this type resulted from the debate between the modernist journal *Prometheus* and the Orthodox journal *Anaplasis* (see above) over the matter of Darwinism and materialistic natural philosophy. Constantine Mitsopoulos, a modernist who was at the same time devoutly Orthodox, founded *Prometheus* in 1890 and intended the journal to be a "periodical of the physical and applied sciences" that would promote an interest in the sciences among the Greek public. Writing in the very first issue of *Prometheus*, Christomanos declared that, like the Prometheus of myth, the journal aimed to spread knowledge about nature, which until then had been hidden in a fog. The debate over Darwinism began when *Prometheus* published an article by Stamatios Valvis in 1890 on human creation that defended Darwinian ideas. Vivid and impassioned rejoinders soon appeared in *Anaplasis*, and the back and forth between the two journals would last until November 24, 1892, when the publication of *Prometheus* was interrupted.[15] Despite this hiatus in the debate over Darwinism in print, it continued in others way—as, for example, when Alexandros Valvis gave lectures on Darwinian evolution in the Medical School of Athens on the eve of the twentieth century. A slightly different case was that of Spiridon Miliarakis, a professor of botany who translated and defended Darwin in the 1890s but who was milder in his tone and attracted less criticism from para-ecclesiastical circles.

Generally, high-ranking officials of the state church avoided directly confronting prominent scholars who might be their colleagues at the university or state officials, but many laymen or members of the lower clergy showed no such restraint and expressed extreme positions. For instance, during the debate on Darwinism in the early 1890s, Philopoimin Stephanides (1873–1961), a young student in medicine and philosophy, exhorted the Holy Synod to intervene: "The enemy is among us and we must take measures against him. First the Church must show that it lives. Second the State must dismiss from the University and the Colleges every teacher teaching materialist theories . . . . And if our politicians do not understand this obligation, so the Holy Synod must remind them. It is its duty and its right."[16] Regardless of such impassioned calls for action, the Holy Synod did not make any official response, and no high-ranking church officials exhorted the state to dismiss the professors.

Despite the reluctance of state church officials to confront scholars, some prominent churchmen did enter into public controversies, at times fiercely, by writing articles on cultural and political matters, which appeared in various journals and periodicals. Nevertheless, it was rare even for the most outspoken church official to write on matters concerning science, with the sole exception of materialism. But even in the case of materialism, the condemnations were directed more at social trends than at science, reflecting the church's fear of a materialistic way of life rather than of unwelcome scientific theories. The only scientific question on which the church took a public and aggressive stand was the reformation of the calendar. The state church was eager to block the introduction of the Gregorian calendar, and indeed Greece was the last European country to abandon its opposition, only introducing it in 1923. But again in this case, as with that of materialism, the church's opposition was chiefly driven by something other than scientific concerns per se. The church resisted calendar reform in large part because of its animosity against Western Christianity: the Catholic Church had initiated the new calendar, and therefore the Orthodox clergy viewed it with suspicion.

The visceral public debate over Darwinism and materialism in the last decade of the nineteenth century was not simply an argument among intellectuals over arcane scientific theories. Although most of the authors published in *Anaplasis* and *Prometheus* were intellectuals, the readership of these journals included a much broader cross section of the population, and it would be accurate to say that the Darwinism debate took place in the public sphere and not only among intellectuals. While the debate centered on the question of the role of God in the creation of the universe and humanity, partisans on both sides realized important social questions also lurked in the background. The debate was not only an argument over the mechanisms of creation but also a contest between two very different conceptions of society: who would intellectually lead and direct the new Greek state—the church or the engineers and scientists? The fierce attacks of *Anaplasis* on the journal *Prometheus* engendered by the Darwinism controversy caused the circulation of *Prometheus* to contract sharply and ultimately led to the journal's closing in November 1892.

Conflicts between science and faith in the new Greek state were focused on materialism and concerned only indirectly the history of science and arguments such as Draper and White were making. Very few on either side of the Greek materialism debate made historical arguments to bolster their positions. The most fanatical supporters of the Orthodox Church accused the scientists with modernist leanings and their supporters of spreading false materialism that weakened the church and society. They evoked scientific arguments rather than theological ones to fight evolution and the other materialist theories they despised.

Although the historical warfare thesis played little role in the debates over materialism and Darwinism, a version of the thesis would become more prominent in the twentieth century. Already during the heat of the Darwinism controversy, proponents of evolutionary theory began to accuse their Orthodox critics of opposing science and reason. Eventually, the ideas of the scientific milieu around *Prometheus* and later around the "progressives" encouraged many Greek intellectuals to embrace the notion that during periods when the Orthodox Church dominated culture, the sciences had failed to develop. However, Michael Stephanides (1868–1957), the first professional Greek historian of science, would challenge this belief in a conflict between Orthodoxy and science.

The Warfare Thesis during the Twentieth Century

Stephanides had an ideal background to tackle the *terra incognita* that at the turn of the century was the history of science. In 1888, he entered the faculty of philosophy of the University of Athens, where he attended courses on philology and also science. His studies prepared him for his subsequent career as scientist, philologist, and historian. In 1896, he published the *Mineralogy of Theophrastus*, which was the first Greek book on the history of science to use the tools of modern historical research and to be noticed by the international community of historians of science. Prior to the appearance of Stephanides's book, Greek works on the history of science had nearly all been encyclopedic volumes written by scientists who wished to present the development of their scientific field.

After the publication of *Mineralogy*, Stephanides successfully pursued a career as a researcher, and later as a teacher of the history of science. Indeed, he institutionalized the history of science in Greece when in 1924 he managed to create a chair for the history of science in the faculty of science of the University of Athens. He himself would occupy this chair until his retirement in 1939. The best known of all his books, *The Psammurgic Art and Chemistry*, compares Greek, Arabic, and Western alchemy and argues that the *Chrysopea* of the Greeks, Alexandrians, and Byzantines constituted a distinct school from Arabic and Western alchemy. George Sarton, leading historian of science at that period, esteemed Stephanides's work because it highlighted Byzantine science, ignored by contemporary scholars.

Despite the groundbreaking character of his scholarship, Stephanides was still deeply imbued with the historical vision of the new Greek state, which stressed the continuity of the Greek people from antiquity to the nineteenth century. Thus his works divided that history into three distinct stages, those of ancient Greece, Byzantium, and modern Greece (starting in 1453).[17] In 1938, he published his *History of Natural Sciences*, which reflected his historical vision of continuity and paid great tribute to ancient Greek science. In fact, 240 out of the 290 pages of the book were devoted to ancient Greek and

Byzantine authors. What is more, Stephanides argued in this history that ancient Greek science did not differ fundamentally from modern European science: the latter was the development of ancient Greek knowledge. When presenting the development of the new science in Europe, Stephanides never mentioned the warfare thesis, not even in discussing the Galileo affair. In the sole paragraph of his book devoted to that matter, he wrote: "Of course the new approaches and the new ideas provoked the reaction of a large number of conservative scientists, who considered Galileo's support of Copernicus as scandalous. For that reason, Newton's England inherited the scientific ideas from Galileo's Italy, where the war against heretical ideas was fierce."[18] This terse summary of opposition to Galileo's heliocentrism is revealing of Stephanides's entire approach; a vague reference is made to "conservative scientists" without mentioning Galileo's clerical opponents or his trial at the hands of the Roman Inquisition.

Stephanides's tendency to ignore conflicts between Orthodoxy and those pursuing the study of nature is on full display in his small book *The Physical Sciences in Greece before the Revolution* (that is, the national revolution against the Ottoman Empire in 1821), which was published in 1926. The work contains not even the slightest suggestion of a conflict between religion and science. Nor does the text mention that a debate took place during the Greek Enlightenment among Orthodox clergymen concerning the kind of science that should be taught. In the preface of the book, Stephanides wrote of Greek scholars, "The teaching of the new scholars (having studied at the Universities of Paris, Trieste, Livorno, Leipzig, Vienna, and Munich) about Hellenism was enthusiastic. Citizens or clerics came to hear them from everywhere where Light was needed, in order to diffuse the new knowledge. . . . All the European movement in a loyal miniature."[19]

This portrayal of harmony between the Orthodox Church and science completely ignored all contrary evidence, including the few cases in which the Orthodox Church excommunicated scholars for promoting the new European science and philosophy.[20] Stephanides's sanguine portrait of Orthodoxy's relationship with science was an indirect political reply to those supporting the conflict thesis and would ultimately be challenged, but chiefly by intellectuals from the left-wing parties rather than historians of science.

The most prominent of the left-wing intellectuals to promote the conflict thesis was Nikolaos Kitsikis (1887–1978), who also became one of the top Greek engineers of the twentieth century. Kitsikis studied engineering in

Berlin, mathematics in Paris under Henri Poincaré, and philosophy at the College de France under Henri Bergson. He joined the Communist Party after World War II and was made president of the Greek-Soviet Association in 1947. The same year, he published *Philosophy of Modern Science*, which proved highly influential. In this book, he presented the warfare thesis in connection with the struggle between the social classes. Konstantinos Sotiriou, a left-wing intellectual teacher and friend of Kitsikis, presenting the book writes:

> In philosophy we have two major and noncompatible currents, the idealistic and the materialistic. The idealistic philosophy is by its nature antiprogressive and conservative because it promotes the antiscientific explanation of the world. It erects barriers to science; it strangles it. A classic example is the Middle Ages. The materialistic philosophy is by its nature progressive and revolutionary; it knocks down the idols. It expresses the desires and the interests, and it is closely related to the progress of the social class that it upraises. The capitalist class, which today promotes idealism as its support against the revolutionary movement, was in the past, as I said, the vector of science and cultivated in parallel the materialistic philosophy. This materialistic philosophy was its main weapon, as today the weapon of the working people in its fight is dialectical materialism.[21]

Kitsikis's somewhat indirect articulation of the conflict thesis was typical of the Greek left, which at the time was heavily influenced by Soviet Communism. For Greek left-wing thinkers, the chief enemy of progress was not religion per se but the capitalists and the idealist philosophies the capitalist class promoted for its own self-interested reasons. Kitsikis and other left-wing intellectuals attacked the church and in particular the Orthodox Church for its preaching of idealism and thus for being a "bastion of reaction," but the conflict thesis between Christianity and science did not strongly appear in their narrative. In fact, when attacking the church, they made very few arguments based on the history of science, and this inattentiveness to history and the conflict thesis was also true of intellectual discourse in the Soviet Union. For example, in the Soviet Union's ten-volume *World Encyclopedia of Youth*, although an important interest is expressed in the history of science, one finds only a few passages concerning the historical warfare thesis. The few conflicts discussed are famous episodes from the history of Western Christianity, such as the Catholic Church's persecutions of Bruno and Galileo.

As in the nineteenth century, the debate in Greece during the twentieth century concerned mainly contemporary scientific issues rather than historical ones. The relations between the Orthodox Church and science were approached from a political point of view. During the nineteenth century, the debate was between parareligious associations and pro-European modernists; during the twentieth, it was between parareligious associations and left-wing intellectuals. The actors in the debate changed for a while after World War II, with the creation of the Christian Union of Scientists, which in 1946 issued a Declaration on Science and Religion attacking materialism.[22] The formation of the union was a response to the preeminence of left-wing scholars and their Marxist views about materialism in Greek intellectual life of this period. The positions promoted by the Christian Union of Scientists contained little that was original and consisted largely of quoting well-worn platitudes from various European scientists, such as "science explores the beauty of the creation," "the new scientific discoveries are compatible with faith," and "science has limits, and beyond these is the world of God."

Some years later, during the 1950s, the revival of the Christian association Zoe (founded in 1907) renewed Christian popular literature about science. Zoe promoted science and scientific education in its journal *Aktines* (Rays) and the journal's openness to scientific knowledge complicated the religion and science debate in Greece and made it more difficult for left-wing intellectuals to defend the warfare thesis. Views similar to those expressed in the Declaration or in *Aktines* were sometimes expressed by Orthodox scientists of other nations, such as the Serbian Michael Idvorsky Pupin, who compared Faraday's lines of force with the omnipresence of God.[23] However, a number of Orthodox thinkers, especially numerous in the Slavic nations, rejected all attempts to find God in the material world as a Western way of perceiving the divine that was alien to their own tradition. Instead, thinkers of this stripe looked to the Hesychast tradition that denied the importance of scientific knowledge and embraced the theology of Maximus the Confessor or of the Hesychast leader Gregory Palamas.[24]

The military dictatorship in Greece (1967–74) promoted the ideal of a Christian Orthodox state that would have as its main pillars the church and the army; as a consequence liberal scholars came to identify the Orthodox Church with the extreme political right. In the late 1970s, after the fall of the dictatorship, an innovative wind blew through the Greek scientific community with the arrival of many scholars who, because they had been political exiles, had made their careers abroad, especially in France. At the same time, a group of left-wing Orthodox believers, who called themselves the Neo-Orthodox, sought a renewal of the Greek Orthodox Church, discredited in their eyes through its association with the dictatorship, by reviving the "true Orthodox values" of fraternity and devotion.

It was in this context that Yannis Karas arrived in Greece. Born in 1934, the son of a family of political exiles of the civil war (1946–49), he grew up and studied philosophy in socialist Romania. From 1972 through 1974, he continued his studies in Paris and defended a PhD thesis on the history of science in the Greek world during the Ottoman domination. Returning to Greece after the restoration of parliamentary government, he was appointed researcher at the National Hellenic Research Foundation. Yannis Karas included in his work new approaches that had been evident for several decades in the international community of historians of science: research on published and manuscript sources; analyses of science within its political, economic, and cultural context; and an appreciation for how the modern Greek world had interacted with Europe in the development of the sciences. Despite his communist political background, Karas adopted a moderate version of the warfare thesis when speaking about the debates between the hierarchy of the Orthodox Church and scholars during the period after Byzantium. In his many books on history of science during the period from 1453 (fall of Constantinople) to 1821 (Greek national revolution) he presented the relationship of Orthodoxy with science more as an internal debate within the Orthodox Church rather than as warfare between the church and science.

As had been the case earlier in the century, during the twentieth century's final decades, the warfare thesis was again not to be found in the works of historians of science but rather in the writings of left-wing intellectuals, such as the historians of modern Greece Alkis Aggelou (1917–2001) or Philippos Iliou (1931–2004). Both men insisted on the backwardness of the Greek Orthodox Church during modern history, especially in areas concerning science and reason. The title of Iliou's small book *God, Blind Your People: Prerevolutionary Crisis and Nicolas Piccolo*, which analyzes the ideas of the church in Constantinople during the first decades of the nineteenth century, is revealing of the author's approach.[25] In his own book *The Lights*, Aggelou also defends, albeit in a more moderate way, the warfare thesis.[26]

In a poll among students of the department of education of the University of Athens and the Hellenic Open University during the academic year

2014–15, a majority of respondents supported the warfare thesis (specifically that the church opposes science). When asked to provide evidence for the thesis, students tended to reply vaguely, and a large majority referred to Copernicus and Galileo. Despite the apparent widespread belief in the warfare thesis among Greek students, there are very few Greek works of the nineteenth and twentieth centuries presenting the thesis as a historical fact. The few works that do so tend to speak chiefly about Western Christianity. The main historiography of the warfare thesis in Greece is to be found among the writings of scientists of the end nineteenth and beginning of the twentieth centuries who participated in the debates about materialism or evolution, and among socialist intellectuals who presented the warfare thesis as an aspect of the Marxist notion of class struggle. However, these theses tend not to be linked to the Greek Orthodox tradition but rather to international intellectual trends and are therefore quite similar to theses one can find at the same periods in other countries.

Conclusion

During the nineteenth and twentieth centuries, the relationship between religion and science in the Greek case was unique compared with other Orthodox nations. Like other Orthodox peoples, the Greeks inherited a strong antiscientific tradition rooted in Byzantine Hesychasm. But unlike the rest of the Orthodox world, the Greeks also inherited the Hellenism of the Greek Enlightenment, which insisted that ancient Greece had given birth to science and rationality. In Greece, the more traditional minded among the Orthodox shared with their modernist opponents the conviction that the modern Greek state was the heir of ancient Greece and its intellectual accomplishments. Thus, the church avoided intervening in scientific matters or attacking secular knowledge; similarly scientists avoided attacking Orthodoxy, which they viewed as a supporter and pillar of the national revolution.

In other Orthodox nations, the relationship between religion and science was usually quite different. For example, in Russia, antimodernist currents inspired by Hesychasm strongly influenced the public sphere. In Greece, a latent antimodernism only came to the surface in those periods when society faced cultural crises that touched on the eternal question of whether Greece belongs to the East or the West. In such periods, the anti-Western sentiments of certain Orthodox circles included the rejection of modern science as part of the Western way of life. However, such a position toward science always remained marginal, and it did not seriously affect the relationship of Greek Orthodoxy to science during the twentieth century.

ACKNOWLEDGMENTS

This chapter was written as part of the NARSES (Nature and Religion in Southeastern European Space) Project, funded by the Aristeia—National Strategic Research Framework and implemented in the Institute of Historical Research, National Hellenic Research Foundation. I would especially like to thank Paul Erickson for his commentary and assistance.

NOTES

1. Efthychis Bitsakis, *Από την πυρά στον Άμβωνα* [From the pyre to the pulpit] (Athens: Topos, 2009). Unless otherwise noted, all translations of passages quoted in the text and titles of works in the notes are my own.

2. John William Draper, *History of the Conflict between Religion and Science* (New York: D. Appleton, 1874), 68–75, 91–95 (Eastern Christianity), 154 (Cosmas Indicopleustes).

3. Draper, *History of the Conflict*, x.

4. Andrew Dickson White, *A History of the Warfare of Science with Theology in Christendom*, 2 vols. (New York: D. Appleton, 1896), 1:6, 53 (fathers), 1:182 (Nikon), 2:285 (potatoes); 2:311 (eighteenth century); White, *Autobiography of Andrew Dickson White*, 2 vols. (New York: Century, 1905), 2:67–69 (miracles).

5. Philippos Iliou, *Τύφλωσον κύριε τον λαόν σου: Οι προεπαναστατικές κρίσεις και ο Νικόλαος Πίκκολος* . . . [God, blind your people: Prerevolutionary crises and Nicolas Piccolo] (Athens: Poreia, 1988), 47–48.

6. On Hesychasm and the sciences, see E. Nicolaidis, *Science and Eastern Orthodoxy: From the Fathers of the Church to the Era of Globalization* (Baltimore, MD: Johns Hopkins University Press, 2011), 93–105.

7. Y. M. Rabkin and S. Rajapopajan, "Les sciences en Russie: Entre ciel et terre," in M. Blay and E. Nicolaidis, eds., *L'Europe des sciences: Constitution d'un espace scientifique* (Paris: Seuil, 2001), 215–62.

8. For the relations between science and religion in the Greek state during the nineteenth century, see Kostas Tampakis, "Sciences and Religion: Their Interaction in the Borders of Europe (1832–1915)," international colloquium, "Europe et sciences modernes: Histoire d'un engendrement mutuel," Nantes, February 21–22, 2013, http://narses.hpdst.gr/sites/narses.hpdst.gr/files/1302-tampakis-narses-project-sciences-and-religion-interaction.pdf.

9. Dimitrios Stroumpos, *Περί των γνώσεων και των δοξασιών των τε αρχαίων και νεωτέρων ως προς τα φυσικά φαινόμενα εν γένει και των μεθόδων που ερευνά αυτά* [On the knowledge and beliefs of the ancients and moderns about the physical phenomena and the methods that explore them] (Athens: Kopida and Gavriel Printing House, 1858).

10. Henri Ritter, *Histoire de la philosophie*, trans. C. J. Tissot (from German) (Paris: Ladrange, 1835–61). The first six volumes (history before Descartes) were not published until 1844. Chevalier Homfrede Davy, *Elements de philosophie chimique*, trans. J. B. Van Mons (Paris, 1813).

11. Stroumpos, [On the knowledge and beliefs of the ancients and moderns], 23 (Middle Ages), 30 (Galileo).

12. Anastasios Christomanos, *Εναρκτήριος λόγος εκφωνηθείς τη 2 Μαΐου 1864* [Inaugural speech given on May 2, 1864] (Athens: Mavromattis Printing House, 1864).

13. Christomanos, [Inaugural speech given on May 2, 1864], 15–16.

14. Nikolaos Apostolidis, *Εισιτήριος λόγος εις το μάθημα της Ζωολογίας, υπό Νικολάου Χ: Αποστολίδη, εκφωνηθείς εν τω Εθνικώ Πανεπιστημίω τη 10 Ιανουαρίου 1884* [Inaugural lecture for the lesson of zoology, given on January 10, 1884] (Athens: Asmodaios Printing House, 1884); Apostolidis, *Τα θαλάσσια ζώα και τα επιθαλάσσια εργαστήρια* [Marine animals and marine laboratories] (Athens: Estia Printing House, 1894), 20.

15. George Vlahakis, "Προμηθεύς, Σύγγραμμα περιοδικόν των φυσικών και εφηρμοσμένων επιστημών" [*Prometheus*, periodical of natural and applied sciences], in Loukia Droulia and Gioula Koutsopanagou, eds., *Encyclopedia of the Greek Press, 1784–1974* (Athens: National Hellenic Research Foundation, 2008), 3:540–41.

16. Philopoimin Stephanides, *Αι υποθέσεις του υλισμού και του δαρβινισμού και το ελληνικόν Πανεπιστήμιον* [The cases of materialism and Darwinism and the Greek university] (Athens, 1895), 68.

17. On Stephanides, see Efthymios Nicolaidis, "Historiography of Science in Greece," *Osmanli Bilimi Arastirmalari* [Studies in Ottoman science] 6, no. 2 (2005): 201–12.

18. Michael Stephanides, *Εισαγωγή εις την ιστορίαν των φυσικών επιστημών* [Introduction to the history of natural sciences] (Athens, 1938), 264.

19. Michael Stephanides, *Αι φυσικαί επιστήμαι εν Ελλάδι προ της επαναστάσεως* [The natural sciences in Greece before the revolution] (Athens: Sakellarios, 1926), 10–11.

20. The most important case was that of Methodios Anthrakites (c. 1660–c. 1736), excommunicated by the Patriarchate of Constantinople as a heretic, but in fact for his teaching of liberal ideas of Malebranche and Descartes; the excommunication was abolished after he confessed his Orthodox faith (see Efthymios Nicolaidis, "Les condamnations d'idées scientifiques par l'Église orthodoxe" (The condemnations of scientific ideas by the Orthodox Church), in Michela Malpangotto, Vincent Jullien, and Efthymios Nicolaidis, eds., *L'homme au risque de l'infini: Mélanges d'histoire et de philosophie des sciences offerts à Michel Blay* (Turnhout: Brepols, 2013), 323–34.

21. Konstantinos Sotiriou, "Ν. Κιτσίκης, Φιλοσοφία της νέας φυσικής. Συμπεράσματα και διδάγματα" [N. Kitsikis, Philosohy of modern physics: Conclusions and lessons], *Νέα Οικονομία* (December 1947–January 1948), http://arxeiokdsotiriou.blogspot.gr/2014/01/blog-post.html.

22. *Διακήρυξις της Χριστιανικής Ενώσεως Επιστημόνων* [Declaration of the Christian Union of Scientists] (Athens: N.p., 1946).

23. Edward Davis, "Michael Idvorsky Pupin: Cosmic Beauty, Created Order, and the Divine Word," in Nicolaas Rupke, ed., *Eminent Lives in Twentieth-Century Science and Religion* (Frankfurt: Peter Lang, 2007), 201.

24. For positions of the Russian Orthodox tradition, see, e.g., Alexey V. Nesteruk, *Light from the East: Theology, Science, and the Eastern Orthodox Tradition* (Minneapolis, MN: Fortress Press, 2003); Nesteruk, *The Universe as Communion: Towards a Neo-patristic Synthesis of Theology and Science* (London: T. & T. Clark, 2008).

25. Iliou, [God, blind your people].

26. Alkis Aggelou, *Των Φώτων* [The lights] (Athens: Ermis, 1989).

CHAPTER EIGHT

Liberal Protestants

JON H. ROBERTS

Few American Protestants who reviewed John William Draper's *History of the Conflict between Religion and Science* (1874) regarded that work as a disinterested examination of the past. Although they acknowledged that it was possible to point to instances in which religious believers had wrongly attacked scientific ideas—the heliocentric universe espoused by Galileo was a case in point—they insisted that Draper's narrative of western thought as an ongoing struggle between courageous investigators of the natural world and bigoted religious authoritarians presented readers with a "series of caricatures" rather than sober historical analysis.[1] Reviewers commonly ascribed Draper's erroneous interpretation to his refusal to devote much attention to the "moderate or intermediate opinions" expressed by exponents of Protestant "confessions" and to his equation of Christian orthodoxy with the Roman Catholic Church.[2]

Still, if Protestant reviewers were quick to reject Draper's interpretation of the past, many acknowledged that the relationship between religion and science was, as one commentator put it, "*the* question which is now agitating the world of thought."[3] Many recognized, too, that Draper was hardly alone in perceiving significant tension between religious thought and the scientific worldview. In 1860, Thomas Huxley, Darwin's "bulldog" and an ardent scientific naturalist, had declared that "whenever science and orthodoxy have been fairly opposed, the latter has been forced to retire from the lists, bleeding and crushed, if not annihilated; scotched, if not slain."[4] In 1875, E. L. Youmans, an American editor who had chosen to include Draper's book in Appleton's International Scientific Series, observed that the conflict between science and religion was "as much a reality of human experience as the collisions of nations."[5] For their part, many mainstream Protestant thinkers also called attention to the acrimonious character of the relationship between science and religion in their day. Charles Hodge, Princeton Seminary's professor of exegetical, didactic, and polemic theology, declared that "the fact is painfully notorious that there is an antagonism between scientific men as a class, and religious men as a class."[6] At the other end of the theological spectrum, the Unitarian clergyman James Bixby concluded that "the conflict now going on between the physical discoveries and theories of these latter days, and the forms of faith which have hitherto ruled the mind of Christendom, is one of the most noticeable phenomena of the intellectual movement of the times."[7]

Some Protestants charged that the rancor was primarily due to irreligious attitudes on the part of many practitioners of science. Hodge, for example, complained that scientists as a group refused to consider evidence not derived from the senses, dismissed ideas put forward by nonscientists, and accorded biblical revelation little or no respect.[8] Not all Protestant thinkers, however, shared that perspective. During the last quarter of the nineteenth century, a growing number of them, in fact, joined Draper in affirming that "opinions on every subject" were subject to alteration in the face of "the irresistible advance of human knowledge."[9] Those Protestants maintained that the progress of scientific investigation required Christians to make significant revisions in their apologetics, doctrine, and biblical interpretation.

The willingness to make such revisions in an effort to harmonize the Christian world view with the tenor of modern thought has rightly been viewed as one of the defining features of the American liberal Protestant theological tradition. Although the roots of that tradition could be found well before the 1870s, the explosion of knowledge concerning the natural world and the increasing awareness of the value of historicism in understanding human affairs had the effect of swelling the ranks of Protestant thinkers in the United States who became committed to finding a middle ground between unbelief and allegiance to biblical inerrancy by modifying theological formulations in response to advances in knowledge. Many of the early proponents of what came to be known as the "New Theology" or "Progressive Orthodoxy" were associated with the Congregational tradition, but during the 1880s and 1890s many members of the other "mainline" Protestant denominations—Episcopalians, Presbyterians, northern Baptists, Methodists, Disciples of Christ, and the Lutheran Church in America—became proponents of liberal theology.[10] Practitioners of the natural and human sciences also often played important roles in advancing the cause of liberal

Protestantism. The breadth of support for the liberal perspective prompted one well-known clergyman to suggest that the ranks of liberalism included "the names of those who are eminent in every form of thought."[11]

Convinced that God was the ultimate source of all truth, liberal Protestant opinion leaders blurred the distinction between sacred and secular knowledge and insisted that the full range of human experience be brought to bear in interpreting and formulating the Christian worldview. Although those thinkers acknowledged that every individual was capable of acquiring knowledge, they denied that all people were equally competent to evaluate truth claims. Instead, they insisted on relying on the judgment of specialized "experts" to determine what counted as trustworthy knowledge.[12]

The conviction that it was necessary to defer to the views of experts on a whole range of subjects played a fundamental role in differentiating American liberal Protestants from more conservative evangelical thinkers. Liberals emphasized that opposition to advances in knowledge in such areas as science and history would go a long way toward confirming the charges of critics who associated Christianity with obscurantism. From that perspective, they regarded the accommodation of the precepts of Christian theology to ideas and theories resulting from the ongoing investigations of credentialed specialists as an essential aspect of defending the faith.

That position led liberal Protestants to maintain that the interpretation of natural phenomena produced by practitioners of science and accepted by a consensus of those specialists provided humankind with the "most certain and valuable knowledge" that was available concerning the nature of the created order.[13] Liberals acknowledged that the judgments of scientists were not infallible. Nevertheless, they held that "science, imperfect though it is, is the only proper witness for interpretation of the facts of nature" and that humanity "incurred greater danger of doing violence to the truth by rejecting the general verdict of science than by devoutly accepting it."[14]

Liberal Protestants also assumed that their respect for the knowledge of experts was widely shared.[15] That assumption lent a certain sense of urgency to their desire "to think in harmony with other departments of thought and study."[16] Convinced that they were living in an age in which the credibility of the Christian worldview was no longer taken for granted, they reasoned that "the future of Religion would be vastly more sure and prosperous if she could make science an ally instead of a rival."[17] Failure to harmonize Christian doctrine with the tenor of modern thought, they warned, would convince the literate masses to abandon their commitment to Christianity.[18]

Liberal Protestants were convinced that in order to preserve the credibility of the Christian worldview, it would be necessary to come to grips with important changes that had taken place in scientific thought during the course of the nineteenth century. Owing to the explosion of scientific knowledge that had occurred in that period, scientists found it increasingly possible to ascribe the origin and behavior of phenomena to natural processes rather than supernatural intervention. The growing explanatory power of naturalism was exemplified by the theory of organic evolution, which provided natural historians with a credible alternative to the idea that the origin of species had been the result of "half a million distinct miracles."[19] What is more, by the end of the nineteenth century most practicing scientists—even those sympathetic to theism—had concluded that the attempt to "narrow the domain of the supernatural, by bringing all phenomena within the scope of natural laws and secondary causes" was central to the very nature of the scientific enterprise.[20] Authentic science had no place for appeals to the supernatural.[21]

The increasing commitment of scientists to the precepts of what came to be called "methodological naturalism" convinced liberal Protestants that in order to resolve tensions with science, it would be necessary to engage in a fundamental "reconstruction" of Christian theology. They insisted that such work was a periodic necessity, for the essential truths of Christian theology were characteristically expressed in idioms and categories drawn from the cultural conceptions that happened to prevail at any given time.[22] When changes took place in the larger culture, inherited doctrinal formulations ceased to express theological truth, and it became necessary to engage in the revision of those formulations.

During the last quarter of the nineteenth century, the increasingly prevalent conviction within the scientific community that divine fiat was simply "not the way in which Nature does business" played an important role in impelling liberal Protestants in the United States to alter their apologetic strategies.[23] In the face of that conclusion, liberal Protestants in America and elsewhere developed a new arsenal of arguments for God's existence.[24] Most eschewed efforts to call attention to gaps in scientific explanation and chose instead to emphasize that the order and intelligible patterns in the natural world disclosed by scientific investigation could most plausibly be explained as the work of a divine Mind. "That which is intelligible," they maintained, "has intelligence in it."[25] Proponents of that argument welcomed the efforts of practitioners of science to find additional instances of intelligible behavior in the created order, for they interpreted such behavior as "an expression or revelation of the Eternal Reason, which is another name for God."[26]

Some liberal apologists, finding "external arguments for the being of God" less than compelling, chose to emphasize that the "foundation of religion" was to be found "in human nature, in the great primal instincts of the soul."[27] Those thinkers typically located an "inner manifestation" of God in the testimony of feelings, which they regarded as "facts" that possessed "evidentiary value."[28] The most sustained exposition of the apologetic value of feeling written in the United States was *The Religious Feeling* (1877), a work by the Congregational clergyman Newman Smyth. Drawing on the work of the German theologian Friedrich Schleiermacher, Smyth held that the "perennial source of religion, opened afresh in every new-born soul" was the feeling of "absolute dependence" that human beings experienced in contemplating their existence.[29] That feeling, he maintained, attested to the existence of an "Unconditioned" Absolute, which he identified as God.[30] Other liberals held that the sense of moral obligation constituted strong evidence for God's existence. Lewis French Stearns, the occupant of the chair in systematic divinity at the Congregationalists' Bangor Theological Seminary, described the "moral intuition" that humanity was ruled by an "absolute Right" as the "argument *of* arguments" for belief in the existence of God.[31] Proponents of arguments for the existence of God based on humanity's inner convictions maintained that just as the source of the sense data that served as the foundation for scientific claims lay outside the mind, in material objects, the source of the feeling of dependence and the sense of moral obligation also existed outside the mind, in the person of a divine Creator and Moral Governor.[32] That reasoning led Smyth to conclude that "scientific and religious knowledge stand or fall together, with the trustworthiness of our mental faculties."[33]

Seeking an understanding of the relationship between God and the created order that was consistent with the spirit of methodological naturalism, liberal Protestants concluded that the notion that God's immediate action in the created order was limited to areas in which there were gaps in scientific explanation must be replaced by a more lively awareness of God's pervasive activity in nature.[34] That conviction prompted liberal Protestants to make the doctrine of divine immanence, which affirmed the ubiquity of divine activity within nature, history, and human consciousness while maintaining a distinction between God and the created order, the "guiding conception of philosophical theology."[35] In explicating the meaning and implications of that doctrine, liberal Protestant thinkers held that the forces present in the natural world should be regarded not as "independent, efficient, self-acting agents" but as "the ever-present, all-pervading, ever-acting energy of Deity."[36] Those thinkers maintained that a robust conception of divine immanence enabled Christians to accept with alacrity efforts to extend the scientific mode of analysis while requiring no alteration in the way that scientists and others apprehended natural phenomena.[37] The "laws" that scientists used in discussing natural phenomena, they asserted, were simply descriptions of the processes by which God had chosen to act.[38] Although immanentists granted that practitioners of science played an important role in describing the behavior of the natural world, they assigned the task of actually providing the ultimate *explanation* of that behavior to theologians.[39]

Mainline liberal Protestants were aware that much of the tension that existed between Christian theology and the disclosures of science centered on the issue of revelation. Accordingly, their desire to accommodate the Christian worldview to the results of scientific inquiry led them to espouse important departures from the ways Protestants had previously conceived of revelation. Most of them were disinclined to reinterpret the words of the Bible for the purpose of harmonizing them with the latest disclosures of science. That practice, they warned, would "excite a general distrust of Scripture."[40] Instead, they sought to develop a "new canon of interpretation" predicated on the notion that the biblical "messengers" were "limited by their conditions, and bound under the burdens of their own generation" and hence possessed no greater knowledge of the natural world than any other of their contemporaries.[41] Scientific investigation and the Bible, they insisted, were fundamentally different sources of insight. The conviction that the scriptures addressed spiritual concerns rather than scientific issues prompted Washington Gladden to flatly assert that "the Bible is not an infallible Book."[42]

Many mainline liberal Protestants also took the position that changes in humanity's understanding of religious truth could be discerned within the pages of the scriptures themselves. The Bible, they asserted, constituted a "serial history" of humanity's changing views of God and the scheme of redemption.[43] Some liberal thinkers concluded that this suggested that the Deity had employed the same principle, "the same law of evolution" in nature and the Bible alike.[44]

Even as liberal Protestants were moving toward a somewhat more limited conception of the scope of biblical authority, they also began placing greater emphasis on the idea that divine revelation extended well beyond the confines of the Bible. Indeed, they asserted, revelation should be interpreted to include "everything that makes God known to men, and everything that is made known of him."[45] Far from being an "abnormal thing," one liberal clergyman declared, revelation was actually "involved in the very constitution of man and nature."[46] Proponents of that perspective paid homage to the scriptures as an important record of humanity's religious experience.[47] In addition, they emphasized that the Bible contained the teachings of Jesus, as well as a number of other components of the Christian message that could not be found elsewhere.[48] At the same time, however, liberal Protestants insisted that all of human history and the natural world should be valued as important vehicles of divine revelation.[49] From that perspective, liberal Protestants praised the scientific enterprise as "a tracing of the presence of God in the universe."[50]

Finally, their desire to create a "living theology" moved liberal mainline Protestants in the United States to emphasize that human beings in every generation possessed a faculty of consciousness that could serve as a vehicle of divine self-disclosure.[51] Religious experience, they asserted, constituted not only a source of spiritual discernment but also an instrument for verifying the truth of many of the most essential tenets of Christianity. Liberals most commonly employed the terms "religious," "spiritual," or "Christian" consciousness in describing the faculty to be used in what they regarded as an "inductive process" of validation.[52] By carefully examining the religious experience mediated through that faculty, believers could distinguish "that which is not and that which is Christianity," thereby liberating themselves from "servile dependence on tradition and blind idolatry of the past."[53] In attempting to minimize the possibility that the religious experience of an "illuminated spirit" would wrongly legitimize subjective "idiosyncrasies" or lend credence to the charge that "every Christian makes his own Bible," liberals emphasized the importance of comparing the testimony of the individual's religious consciousness with both the experience of other Christians, past and present, and with the testimony of the Bible.[54] Such comparisons, they asserted, would go a long way toward making the verification of the legitimacy of an individual's religious experience analogous to the practice of verification associated with the sciences.[55]

Although liberal Protestants were intent on reconciling their doctrinal formulations with truths gleaned through scientific investigation, they remained committed to what they regarded as the essential tenets of Christian theology. They were rarely explicit in describing just what all of those tenets were, but they characteristically agreed that the belief that human beings were unique in having been created in the image of God was a precept central to the Christian conception of the scheme of redemption.[56] Their commitment to that belief prompted them to reject interpretations of human nature that seemed to be irreconcilable with its privileged status. Mainline liberals were characteristically content to ascribe the origin of human species to an evolutionary process, for they were convinced that such an origin did not entirely determine either the nature of human beings or their status within the created order.[57] At the same time, virtually all of them inferred from their belief in the "inherent sanctity" of human nature that during the course of evolution, an immanent Deity had endowed human beings with the attributes associated with personality. Personality, they held, constituted the "crowning attribute of man" and the site of humanity's "immortal spirit," different in kind from those of other organisms.[58] Indeed, liberals held that the salient characteristics of human nature—reason, the moral sense, free will, religiosity, and self-consciousness—should be seen as evidence that "man is the chief object of the creative energy and care."[59]

In taking the position that the mental attributes of human beings were fundamentally different from those of other organisms, liberal Protestants were affirming an anthropocentric view of reality that was implicit in the Christian worldview. If there was one theme that dominated the discussion of liberal Protestants in the United States in the late nineteenth and much of the twentieth centuries, it was that the concept of personality provided the key to understanding the nature of God and humanity alike, as well as the dynamics of the divine-human encounter. Indeed, belief in the privileged status of personality convinced liberal Protestants that it was necessary to begin with mind rather than matter in attempting to understand the nature of reality.

The commitment of mainline liberal Protestants to the exalted status of mental attributes prompted many of them to conclude that the most credible metaphysical framework available to Christians who wished to combine a commitment to the disclosures of scientific investigation with an affirmation of the primacy of a divine Mind within reality was philosophical idealism. The period between 1870 and 1920 witnessed the emergence of a number of different versions of idealism. They were all predicated on the idea that "the only thing absolutely real is mind; that all material and all temporal existences take their being from mind, from consciousness that thinks and experiences."[60] Not surprisingly, given their belief in the sanctity of personality, liberal Protestants typically regarded personal idealism as particularly compelling. Borden Parker Bowne, a professor of philosophy at Boston University who became well known in liberal Protestant circles as the founder of "personalism," presented a succinct statement of the central claim of personal idealism when he took the position that reality should be understood not as a system of matter in motion but as "a world of persons with a Supreme Person at the head."[61] Bowne and other proponents of personal idealism made concerted efforts to show that it was possible to endorse the revelations of modern science concerning the structure and behavior of natural phenomena while affirming that the ultimate ground of those phenomena was a personal deity. They praised scientists for demonstrating that the universe was "an orderly totality" by describing the continuity and uniformity of the natural world.[62] They also emphasized, however, that the tasks of accounting for the pervasiveness of order within the universe and ascertaining the ultimate nature of the agencies and entities that accounted for the behavior of natural phenomena lay beyond the purview of science. Noting that "active will endowed with intelligence" constituted the only known source of causation within nature, they concluded that "causal explanation must always be in terms of personality."[63] No less important, they concluded that the ubiquity of order in the cosmos attested to the fact that it was the product of "an absolute Will and Intelligence—an intelligent Will, a willing Mind."[64]

Proponents of personal idealism in the United States emphasized the harmony of their position with the precepts of liberal Protestantism. Bowne, for example, maintained that "religiously, there is no difference between idealistic theism and immanent theism."[65] On the other side of the coin, many liberal Protestants invoked ideas and rhetoric associated with personal idealism in describing their understanding of the nature of reality. William Newton Clarke, whose *An Outline of Christian Theology* (1898) provided readers with the first systematic exposition of the "New Theology," thus asserted that "the universe is the work and expression of a personal Spirit."[66]

Philosophical idealism remained quite popular among liberal Protestants well into the twentieth century. After about 1910, however, some mainline liberal Protestant thinkers—mostly academic theologians and philosophers of religion—began to join philosophers of a more secular orientation in expressing allegiance to a metaphysical realism predicated on the claim that many phenomena within the created order exist independently of consciousness. Those liberals commonly maintained that the way to establish theological certainty was to ground the claims of theology on a tough-minded critical analysis of the data of human experience. Shailer Mathews, the dean of the University of Chicago Divinity School, declared that the "substitution of scientific method for reliance upon authority is characteristic of our modern religious thought" and defined "modernism" itself as "the use of the methods of modern science to find, state and use the permanent and central values of inherited orthodoxy in meeting the needs of a modern world."[67] Employing that procedure, Mathews arrived at a conception of God as the source of the "personality evolving" and "personally responsive" constituents of the "cosmic environment with which we are organically related." He believed that it was appropriate to use the term "Christ-like" in describing that Deity.[68]

In contrast to Mathews and most other liberal mainline Protestants, who emphasized the continuity of their views with what they regarded as the essential elements of the Christian tradition, some metaphysical realists abandoned belief in the normative character of that tradition in favor of the idea that all theological propositions must be validated through the use of scientific modes of analysis.[69] The epicenter of the latter perspective was the University of Chicago, but proponents of a rigorously "scientific" approach to experience could be found elsewhere, as well. Douglas Clyde Macintosh, a professor of theology and the philosophy of religion at Yale, gave voice to the animus motivating those liberals when he declared that "if theology is to become really scientific it must be by becoming fundamentally empirical."[70] Exponents of what was often described as an "empirical theology" maintained that religious experience provided grounds for positing the existence of God; the task confronting theologians was to employ "scientific theological observation and experiment under the guidance of definite working hypotheses" in determining the nature of that God.[71] Protestants who embraced that position acknowledged that their approach to experience would not make religion itself a science, for religion involved not only a critical analysis of experience but also an effort "to find that adjustment to God which will yield most abundant life."[72] Nevertheless, they expressed confidence that use of the so-called scientific method in evaluating the data of experience would enable them to develop a theology consistent with the results of scientific inquiry.

Although many proponents of the idea that empirical analysis of experience rather than the Christian tradition should serve as the basis for religious affirmation held distinguished positions within American academic institutions, their position never commanded the allegiance of more than a small minority of liberal Protestant clergy and theologians. This is hardly surprising. In dissociating themselves from the Gospel tradition and in making their doctrinal claims utterly dependent on an understanding of empirical reality that was always evolving, they were implicitly endorsing an epistemological relativism that eschewed altogether any claims to religious certainty. In addition, though, many liberal Protestants resisted the notion that it was important or even useful to place theological discourse within the framework of a rigorous philosophical system. Some, in fact, were convinced that the commitment of theologians to determinate propositions concerning the nature of reality constituted the source of periodic tension between science and Christianity.

One of the most ardent proponents of that conviction was the historian and distinguished educator Andrew Dickson White. Since the 1860s, White had been lamenting the "interference with Science in the supposed interest of religion."[73] In 1896, he brought those concerns to a conclusion of sorts with the publication of his two-volume *A History of the Warfare of Science with Theology in Christendom*. Noting that he endorsed Matthew Arnold's belief in "'a Power in the universe, not ourselves, which makes for righteousness,'" White adamantly rejected the notion that science and religion were at odds. Rather, he asserted, the real enemy of scientific investigation was the "mass of unreason" associated with "Dogmatic Theology." It was White's hope that clearing away theological dogma would enable the "stream of 'religion pure and undefiled'" to "flow on broad and clear, a blessing to humanity."[74] He was notably vague, however, in specifying precisely what he actually meant by that.

Liberal Protestants responded somewhat more favorably to White's work than they had to the views of John William Draper. As Christians who regarded all knowledge as revelation from an immanent Deity, they were sympathetic to White's claims that "untrammeled scientific investigation . . . has invariably resulted in the highest good both of religion and of science."[75] Some liberal interlocutors also praised White's work for acknowledging both that theologians had often favored freedom of thought and that scientists had occasionally been on the "wrong" side in the hostilities.[76] However, in spite of the fact that commentators within the liberal tradition commonly invoked the trope that Christianity is a life and not simply a set of beliefs, they typically resisted White's notion that Christian theology itself could, or even should, be dissociated from religion. Theology, they emphasized, was simply thought concerning "God in His moral and spiritual relationship to man," and such thought was too important to abandon.[77] Convinced that theology and science "lie in two diverse planes of thought and fact," the Albany, New York, clergyman Walton Battershall insisted that it was not only possible but imperative that "every thinking man" bring to bear science and theology alike in grappling with the nature of human experience.[78]

If most mainstream liberals remained committed to the retention of theological discourse, by the turn of the century many of them had become just as convinced as White that it was eminently desirable to detach that discourse from concerns associated with science. However, in contrast to White, who made it clear that he wanted to free science from the trammels of dogmatic theology, most liberals were motivated to support the separation of religion from science because they were intent on exalting religion as a uniquely important constituent of human thought and behavior. One of the inferences that they drew from their belief in "the intrinsic and absolute superiority of man to nature," was that knowledge of the material world did little to minister to the human desire for meaning and that it was important to move beyond scientific understanding in attempting to discern that meaning.[79]

Many liberals who wished to separate science and religion were influenced by the work of the German theologian Albrecht Ritschl.[80] As a neo-Kantian, Ritschl maintained that theology could profit little from an alliance with metaphysics, for he was convinced that the nature of the God of Christianity lay beyond the "horizon" of human knowledge. The only knowledge of God available to human beings, he maintained, was that which God had chosen to reveal to them. The most important source of that divine revelation, he emphasized, was the teachings of Jesus as those teachings had been received and imparted by the "Christian community." Lying at the heart of the Christian message, he maintained, was "redemption through Jesus" and the concept of the "universal moral Kingdom of God." From Ritschl's vantage point, the essential nature of Christianity was to be found in history rather than metaphysics.

Ritschl's convictions concerning the limits of metaphysical reasoning conditioned his thinking about the relationship that existed between science and religion. He acknowledged the value of scientific investigation in disclosing truths about the natural world and insisted that Christian theology, rightly interpreted, was in harmony with its findings. He also maintained, however, that the human spirit, though "part of the world of nature," was also capable of transcending that world. One of the inferences that he drew from that conviction was that science and religion dealt with essentially separate realms of human experience. Whereas "theoretical science" was a cognitive enterprise that sought to obtain "impartial knowledge" of natural phenomena obtained through empirical analysis, religion was a perspective predicated on "value-judgments." Those value judgments, which differed "in kind" from knowledge embodied in the assessments made by science, enabled human beings to enjoy "dominion over the world vouchsafed him by God" and attain the "blessedness" resulting from "elevation above the world in the Kingdom of God."[81]

Ritschl's approach to the Christian Gospel, often conveyed through the work of Adolf von Harnack and other students, attracted widespread support among liberal Protestant thinkers in the United States after the turn of the century.[82] Although many of those thinkers did not share Ritschl's indifference to metaphysics and drew on an eclectic mixture of sources in articulating their theology, they typically shared his belief that science and religion should be seen as fundamentally different endeavors. William Adams Brown, the Roosevelt Professor of Systematic Theology at Union Theological Seminary and an enthusiastic proponent of Ritschl's views, gave voice to a position quite common among American liberal Protestants during the first three decades of the twentieth century when he asserted that science, "however loyally pursued," could not perform the task that was rightfully that of religion: it failed to address the central issue "of life itself—the issue of purpose and meaning."[83] Like Ritschl, Brown emphasized that "the world of values and of meanings is as real in its own way as the world of sense."[84] Harry Emerson Fosdick, the most eminent liberal clergyman in America during the first half of the twentieth century, similarly held that "science and religion move in different realms and should peacefully pursue each separate task in the interpretation of man's experience."[85] Fosdick, who credited his reading of Andrew Dickson White with showing him the absurdity of the doctrine of biblical inerrancy, regarded science as an important source of knowledge, and he stressed the importance of trying "not to be unscientific," but he insisted that "a scientific description never tells the whole truth about anything."[86] It was religion, he maintained, that provided humanity with "the appreciation of life's spiritual values and the interpretation of life, its origin, its purpose, and its destiny in terms of them."[87] It was also religion, Fosdick declared, that provided human beings with "power to release faith and courage for living, to produce spiritual vitality and fruitfulness."[88]

Mainline liberal Protestants who endorsed the separation of science and religion commonly maintained that it was not arguments derived from a consideration of the natural world but rather the fact of religious experience that constituted "the supreme evidence of the faith."[89] Nevertheless, they reasoned, theological understanding was no less grounded in experience than were the natural sciences. Both were essentially empirical approaches to understanding.

Not surprisingly, efforts to show that science and religion occupied "separate spheres" appealed to not only to many liberal Protestant clergy and theologians but also to practitioners of the sciences. While the terms of separation provided proponents of liberal theology with a realm of experience that lay beyond the realm of natural science, they accorded the task of describing natural phenomena to members of the scientific community, thereby giving scientists the autonomy that they desired. Given that fact, perhaps it is not altogether surprising that Robert A. Millikan, a Nobel Laureate in physics, sponsored one of the most influential expressions of the belief that science and religion were separate enterprises. In 1923, he succeeded in securing the signatures of more than thirty well-known scientists, theologians, and clergy to a "Joint Statement upon the Relations of Science and Religion" that affirmed that those enterprises "meet distinct needs" and thus "supplement rather than displace or oppose each other." Millikan's statement held that whereas scientific investigation sought "to develop, without prejudice or preconception of any kind, a knowledge of the facts, the laws, and the processes of nature," it was the role of religion "to develop the consciences, the ideals, and the aspirations" of humankind.[90]

During the late 1920s and early 1930s, a number of mainline liberal Protestants began to engage in a critique of their own tradition. Those critics resisted the idea that the redemptive elements of Christian theology could be established through "rationalistic intellectualism," and they placed greater emphasis than did most liberals on the centrality of the Word of God contained in the scriptures.[91] In keeping with their belief that the scriptures possessed singular importance in disclosing gospel truths, spokespersons for what often came to be described as "neo-orthodoxy" complained that proponents of mainstream liberalism had wrongly dismissed the importance of the supernatural elements of Christianity and were guilty of an "overemphasis on divine immanence."[92] Like the liberals whom they criticized, neo-orthodox commentators continued to accept the well-established results of scientific investigation. At the same time, they made the sharp distinction between science and religion an important element in their campaign to differentiate Christ from culture.

For their part, during the period between 1920 and 1960, many Protestants who remained committed to the more conventional elements of the liberal tradition made it clear that they would not accept ideas merely because they had been put forward in the name of science. Their belief in the privileged status of personality prompted virtually all of them to reject Freudianism as a distorted interpretation of human nature.[93] Similarly, they denounced behaviorism as an interpretation of the human condition that left "no place" for either a personal God or a meaningful concept of human personality.[94] In addition, liberals did not hesitate to criticize scientists who chose to "stand aloof" from religion while claiming expertise in evaluating the truth of its claims."[95]

For the most part, however, mainline liberal Protestant thinkers continued to look favorably on the work of practitioners of the natural sciences even while insisting that the concerns of those practitioners were essentially different from those of Christian theologians. Those thinkers sometimes differed in their descriptions of the precise nature of the difference between science and religion. Many endorsed the Ritschlian distinction between facts and values, while others of a more metaphysical inclination chose to emphasize that "the scientist seeks the HOW of things, while the theologian seeks the WHY of things."[96] Both of those strategies proved to be quite durable in the period prior to 1960.

Conclusion

Although mainline liberal Protestants were convinced that their openness to the established results of scientific investigation exonerated them from responsibility for conflict between science and religion, they recognized that not everyone was as committed as they to the importance of preserving harmony between the scientific and Christian worldviews. That conclusion seemed to be confirmed not only by assertions that appeared in the work of individuals antagonistic to Christian theology but also in the opposition that fundamentalists and other conservative Christians expressed to the theory of evolution. In view of those expressions of hostility, liberal mainline Protestants reluctantly concluded that notwithstanding their efforts to establish peace, acrimony between science and religion would continue to vex moderns for the foreseeable future.

ACKNOWLEDGMENTS

My thanks to Katie Moore and Alexis Buckley, who transcribed some of the primary source data that figured in this chapter. I am also indebted to Ron Numbers for providing his usual intelligent reading of two drafts and to members of the Boston Historians of Religion group for their perceptive comments and criticisms. Finally, thanks, as always, to Sharon (ILYS) and Jeff for their love and support.

NOTES

1. Unsigned, "Draper's Religion and Science," *Presbyterian Quarterly and Princeton Review*, n.s., 4 (1875): 165.

2. John William Draper, *History of the Conflict between Religion and Science* (1874; New York: D. Appleton, 1875), x; [Borden P. Bowne], "Draper's Religion and Science," *Independent* 27 (February 4, 1875): 10. The most comprehensive and perceptive treatment of the "military metaphor" in discussions of science and religion is still James R. Moore, *The Post-Darwinian Controversies: A Study of the Protestant Struggle to Come to Terms with Darwin in Great Britain and America, 1870–1900* (Cambridge: Cambridge University Press, 1979), 19–122.

3. Unsigned review of *History of the Conflict between Religion and Science* by John W. Draper, *Universalist Quarterly and General Review*, n.s., 12 (1875): 251–52 (quotation on 252).

4. Thomas H. Huxley, "The Origin of Species" [1860], *Darwiniana*, in *Collected Essays by T. H. Huxley* (1893; New York: Greenwood Press, 1968), 2:52.

5. E. L. Youmans, "Draper and His Critics," *Popular Science Monthly* 7 (1875): 232–33.

6. Charles Hodge, *What Is Darwinism?* (New York: Scribner, Armstrong, 1874), 126.

7. James T. Bixby, *Similarities of Physical and Religious Knowledge* (New York: D. Appleton, 1876), 7.

8. Hodge, *What Is Darwinism?*, 126–40.

9. Draper, *History of the Conflict*, vi.

10. Those denominations are conventionally regarded as "mainline Protestantism." For more extended discussions of the concept, see William R. Hutchison, "Protestantism as Establishment," in William R. Hutchison, ed., *Between the Times: The Travail of the Protestant Establishment in America, 1900–1960* (Cambridge: Cambridge University Press, 1989), 4; and Elesha A. Coffman, *The Christian Century and the Rise of the Protestant Mainline* (New York: Oxford University Press, 2013), 4.

11. Theodore T. Munger, *The Freedom of Faith* (Boston: Houghton, Mifflin, 1883) 8.

12. S. R. Calthrop, "Religion and Science," *Unitarian Review and Religious Magazine* 2 (1874): 313; J. H. Rylance, "Theological Re-adjustments," *North American Review* 138 (1884): 47.

13. Washington Gladden, *Burning Questions of the Life That Now Is, and of That Which Is to Come* (1890; New York: Century, 1891), 42.

14. William Newton Clarke, *An Outline of Christian Theology* (1898; New York: Charles Scribner's Sons, 1922), 51; Alexander Winchell, *The Doctrine of Evolution: Its Data, Its Principles, Its Speculations and Its Theistic Bearings* (New York: Harper & Brothers, 1874), 112.

15. George F. Moore, "The Modern Historical Movement and Christian Faith," *Andover Review* 10 (1888): 333.

16. Theodore T. Munger, *The Appeal to Life* (Boston: Houghton, Mifflin, 1887), v.

17. Bixby, *Similarities of Physical and Religious Knowledge*, 10–11.

18. Rylance, "Theological Re-adjustments," 49.

19. W. N. Rice, "The Darwinian Theory of the Origin of Species," *New Englander* 26 (1867): 608–9.

20. For a more extended treatment of this theme, see Jon H. Roberts and James Turner, *The Sacred and the Secular University* (Princeton, NJ: Princeton University Press, 2000), 27–31.

21. Rice, "Darwinian Theory," 608–9.

22. George Harris, *Moral Evolution* (Boston: Houghton, Mifflin, 1896), 426–27; Washington Gladden, *How Much Is Left of the Old Doctrines? A Book for the People* (Boston: Houghton, Mifflin, 1899), 174; Herbert Alden Youtz, *The Enlarging Conception of God* (New York: Macmillan, 1914), 23, 107–8.

23. Simon Newcomb, "Evolution and Theology. A Rejoinder," *North American Review* 128 (1879): 660.

24. Jon H. Roberts, "That Darwin Destroyed Natural Theology," in Ronald L. Numbers, ed., *Galileo Goes to Jail and Other Myths about Science and Religion* (Cambridge, MA: Harvard University Press, 2009), 163–69.

25. Harris, *Moral Evolution*, 185.

26. Washington Gladden, *Where Does the Sky Begin?* (Boston: Houghton, Mifflin, 1904), 21.

27. Lewis F. Stearns, "Reconstruction in Theology," *New Englander*, n.s., 5 (1882): 86–87; J. W. Chadwick, "The Basis of Religion," *Unitarian Review and Religious Magazine* 26 (1886): 256.

28. Newman Smyth, *The Religious Feeling: A Study for Faith* (New York: Scribner, Armstrong, 1877), 158; W. R. Benedict, "Theism and Evolution," *Andover Review* 6 (1886): 347.

29. Smyth, *Religious Feeling*, 33–35.

30. Smyth, *Religious Feeling*, 110–14, 124–26, 165.

31. Stearns, "Reconstruction in Theology," 86–87.

32. James C. Parsons, "Religious Experience," *Unitarian Review and Religious Magazine* 26 (1886): 461.

33. Smyth, *Religious Feeling*, 108.

34. Joseph Le Conte, "Man's Place in Nature," *Princeton Review*, n.s., 2 (1878): 794.

35. George T. Ladd, "History and the Concept of God," *Bibliotheca Sacra* 37 (1880): 597.

36. Joseph Le Conte, "Evolution in Relation to Materialism," *Princeton Review*, n.s., 7 (1881): 166.

37. Newman Smyth, *Old Faiths in New Light*, 2nd ed. (New York: Charles Scribner's Sons, 1879), 269.

38. James Douglas, "The Divine Immanency," *Bibliotheca Sacra* 45 (1888): 498.

39. Harris, *Moral Evolution*, 188–89.

40. George P. Fisher, "The Alleged Conflict of Natural Science and Religion," *Princeton Review*, n.s., 12 (1883): 38.

41. Fisher, "Alleged Conflict of Natural Science and Religion," 37; Smyth, *Old Faiths*, 76. See also R. Heber Newton, *The Right and Wrong Uses of the Bible* (New York: John W. Lovell, 1883), 109.

42. Washington Gladden, *Who Wrote the Bible? A Book for the People* (Boston: Houghton, Mifflin, 1891), 351.

43. Henry Ward Beecher, *Evolution and Religion* (New York: Fords, Howard, and Hulbert, 1885), 37.

44. Smyth, *Old Faiths*, 119.

45. John W. Chadwick, "The Revelation of God," *Unitarian Review* 27 (1887): 495.

46. Myron Adams, *The Continuous Creation: An Application of the Evolutionary Philosophy to the Christian Religion* (Boston: Houghton, Mifflin, 1889), 51.

47. Youtz, *Enlarging Conception*, 116–17.

48. William Newton Clarke, *Sixty Years with the Bible: A Record of Experience* (New York: Charles Scribner's Sons, 1912), 196.

49. Clarke, *Outline of Christian Theology*, 10–11, 50–51.

50. Gladden, *Where Does the Sky Begin?*, 21.

51. Frank Hugh Foster, "Evolution and the Evangelical System of Doctrine," *Bibliotheca Sacra* 50 (1893): 419.

52. David N. Beach, "The Reconstruction of Theology," *Bibliotheca Sacra* 54 (1897): 130–31.

53. George Harris, "The Function of the Christian Consciousness," *Andover Review* 2 (1884): 345; Francis A. Henry, "Reconstruction in Religious Thought," *Princeton Review*, n.s., 14 (1884): 24.

54. Foster, "Evolution and the Evangelical System," 418; Asher H. Wilcox, "The Ultimate Criteria of Christian Doctrine," *Andover Review* 8 (1887): 345; Harris, "Function of the Christian Consciousness," 339–40.

55. Lyman Abbott, *Reminiscences* (Boston: Houghton Mifflin, 1915), 452.

56. For a more extensive discussion of this idea, see Jon H. Roberts, *Darwinism and the Divine in America: Protestant Intellectuals and Organic Evolution, 1859–1900* (Madison: University of Wisconsin Press, 1988), 174–76.

57. See, for example, T. T. Munger, "Evolution and the Faith," *Century Magazine*, n.s., 10 (1886): 112–13; Clarke, *Outline of Christian Theology*, 223–25.

58. Henry Ward Beecher, "Progress of Thought in the Church," *North American Review* 135 (1882): 108; F. H. Johnson, "Coöperative Creation," *Andover Review* 3 (1885): 345; Le Conte, "Man's Place," 788–89, 777–78.

59. W. E. Parson, "The Materialist Heresy," *Lutheran Quarterly*, n.s., 18 (1888): 463. For a fuller discussion of this idea, see Roberts, *Darwinism and the Divine*, 176–79.

60. G. H. Howison, "The City of God, and the True God as Its Head," in Josiah Royce et al., *The Conception of God: A Philosophical Discussion concerning the Nature of the Divine Idea as a Demonstrable Reality* (1897; New York: Macmillan, 1902), 84.

61. Borden Parker Bowne, *Personalism* (Boston: Houghton Mifflin, 1908), 277–78.

62. George Trumbull Ladd, *The Philosophy of Religion: A Critical and Speculative Treatise of Man's Religious Experience and Development in the Light of Modern Science and Reflective Thinking* (New York: Charles Scribner's Sons, 1905), 2:59; G. H. Howison, "Is Modern Science Pantheistic?," *Journal of Speculative Philosophy* 19 (1885): 381.

63. Ladd, *Philosophy of Religion*, 2:79; Bowne, *Personalism*, vii.

64. Ladd, *Philosophy of Religion*, 2:59–60.

65. Borden P. Bowne, *Studies in Theism* (New York: Phillips and Hunt, 1879), 286.

66. Clarke, *Outline of Christian Theology*, 118.

67. Shailer Mathews, "Scientific Method and Religion," in Shailer Mathews, et al., eds., *Contributions of Science to Religion* (New York: D. Appleton, 1924), 381; Shailer Mathews, *The Faith of Modernism* (New York: Macmillan, 1924) 22–23 (quotation originally in italics).

68. Shailer Mathews, *The Growth of the Idea of God* (New York: Macmillan, 1931), 226 (quotation originally in italics).

69. Shailer Mathews, "A Positive Method for an Evangelical Theology," *American Journal of Theology* 12 (1908): 22; Kenneth Cauthen, *The Impact of American Religious Liberalism* (New York: Harper and Row, 1962), 27–30.

70. Douglas Clyde Macintosh, *Theology as an Empirical Science* (New York: Macmillan, 1919), 11.

71. Macintosh, *Theology as an Empirical Science*, 28–29 (quotation on 29). See also Henry Nelson Wieman, *Religious Experience and Scientific Method* (New York: Macmillan, 1926), 30–45.

72. Wieman, *Religious Experience*, 65, 381.

73. Unsigned, "First of the Course of Scientific Lectures—Prof. White on 'the Battle-Fields of Science,'" *New York Daily Tribune*, December 18, 1869, 4.

74. Andrew Dickson White, *A History of the Warfare of Science with Theology in Christendom*, 2 vols. (New York: D. Appleton, 1896), 1:xii, vi, ix, vi.

75. White, *History of the Warfare of Science with Theology*, 1: viii (quotation originally in italics).

76. Unsigned, "'The Book of the Wars of the Lord,'" *Outlook* 53 (June 20, 1896): 1153; John Merle Coulter, review of *A History of the Warfare of Science with Theology in Christendom* by Andrew Dickson White, *American Journal of Theology* 1 (1897): 239.

77. Walton Battershall, "The Warfare of Science with Theology," *North American Review* 165 (1897): 90.

78. Battershall, "Warfare of Science with Theology," 89–90.

79. Lewis French Stearns, *The Evidence of Christian Experience* (New York: Charles Scribner's Sons, 1890), 71–72.

80. Useful secondary accounts of the thought of Ritschl and his German followers include Claude Welch, *Protestant Thought in the Nineteenth Century: Volume 2, 1870–1914* (New Haven, CT: Yale University Press, 1985), 1–30, 146–82; and Gary Dorrien, *Kantian Reason and Hegelian Spirit: The Idealistic Logic of Modern Theology* (Malden, MA: Wiley-Blackwell, 2012), 315–77.

81. Albrecht Ritschl, *The Christian Doctrine of Justification and Reconciliation: The Positive Development of the Doctrine*, ed. H. R. Mackintosh and A. B. Macaulay, 3rd ed. (1888; New York: Charles Scribner's Sons, 1900), 16–17, 212, 2–7 (quotation on 6), 202. 237, 10–13, 224–25, 204–9, 211.

82. Gary Dorrien, *The Making of American Liberal Theology: Idealism, Realism, and Modernity, 1900–1950* (Louisville, KY: Westminster John Knox Press, 2003), 21–72; William Hutchison, *The Modernist Impulse in American Protestantism* (Cambridge, MA: Harvard University Press, 1976), 122–32.

83. William Adams Brown, *A Teacher and His Times* (New York: Charles Scribner's Sons, 1940), 299.

84. William Adams Brown, *Pathways to Certainty* (New York: Charles Scribner's Sons, 1930), 124–25.

85. Harry Emerson Fosdick, "Science and Religion," *Harper's Magazine* 152 (1926): 296.

86. Harry Emerson Fosdick, *The Living of These Days: An Autobiography* (New York: Harper and Brothers, 1956), 52; Harry Emerson Fosdick, "Yes, but Religion Is an Art!," *Harper's Magazine* 162 (1931): 132–33.

87. Fosdick, "Science and Religion," 296.

88. Harry Emerson Fosdick, "The Dangers of Modernism," *Harper's Magazine* 152 (1926): 408.

89. W. P. King, "Christian Faith and the New Apologetics," *Methodist Review Quarterly* 62 (1913): 338.

90. The Joint Statement is found in [Robert A. Millikan], "Science and Religion," *Science*, n.s., 57 (June 1, 1923): 630–31 (part of the quotations originally appeared in italics).

91. E. G. Homrighausen, *Christianity in America: A Crisis* (New York: Abingdon Press, 1936), 58; George W. Richards, *Beyond Fundamentalism and Modernism: The Gospel of God* (New York: Charles Scribner's Sons, 1934), 22–24 (quotation on 24).

92. Reinhold Niebuhr, *Does Civilization Need Religion? A Study in the Social Resources and Limitations of Religion in Modern Life* (New York: Macmillan, 1927), 204–5.

93. For a sustained discussion of this topic, see Jon H. Roberts, "Psychoanalysis and American Religion, 1900–45," in David C. Lindberg and Ronald L. Numbers, eds., *When*

Science and Christianity Meet: From Augustine to Intelligent Design (Chicago: University of Chicago Press, 2003), 225–44.

94. Georgia Harkness, *Conflicts in Religious Thought* (New York, Henry Holt, 1929), 98–99. See also Harry Emerson Fosdick, *Adventurous Religion and Other Essays* (New York: Cornwall Press, 1926), 104–5.

95. Gerald B. Smith, “Some Conditions to Be Observed in the Attempt to Correlate Science and Religion,” *Religious Education* 23 (1928): 306.

96. George H. Richardson, “Scientist and Theologian,” *American Church Monthly* 24 (1928): 336.

CHAPTER NINE

Protestant Evangelicals

BRADLEY J. GUNDLACH

Evangelicals have been among the most vocal and colorful exponents of a conflict between certain forms of science (especially naturalistic evolutionism) and gospel Christianity. On the surface, their use of what James Moore calls "the military metaphor" links them with outspoken secularists like Richard Dawkins.[1] Both parties recognize the other as an enemy and enjoin readers to choose between fundamentally opposed worldviews: one that admits the supernatural and another that does not. Both see them as engaged in an epic struggle between light and darkness—though each side sees itself as light and the other as darkness.

But this similarity masks significant difference underneath, especially historically. Secularist exponents of the conflict thesis envisioned religion and science as inherent enemies, differing fundamentally on the appropriate method of knowing and the appropriate reach of science. On this view, religion (or rather authoritarian religious leaders and institutions) had for centuries erected barriers against free inquiry and scientific study, but the victory of scientific knowledge over superstition and obscurantism was sure to come. By contrast, evangelicals did not posit an ultimate and inherent conflict between "science" and "religion" at all. In their view, the problem was temporary and avoidable, a matter of scientists erring or trespassing into religious matters.

To add to the complication, Andrew Dickson White, president of Cornell University and the greatest popularizer of the "warfare between science and theology," was neither an evangelical nor a secularist. A liberal Protestant, he saw science as a purifier of religion and wrote his history of religious persecution of scientists only to urge Christians to give science free rein. They could keep their religion if they modified it sensibly.

The point here is that there is no unitary conflict thesis—I am proposing, if you will, a complexity thesis about the conflict thesis, with a focus on the differences between the conflict theses of evangelicals, Protestant liberals, and out-and-out secularists. I propose, further, to distinguish a *conflict thesis* from the *military metaphor* (the use of martial language to describe a disagreement and to prescribe a strategy for victory) in order to understand the mentality and purposes of participants in the contest. Finally, I will argue that within the Protestant evangelical camp over time there were, again, several versions of a conflict thesis. The story of their development is fascinating and illuminating.

The term "evangelical" is notoriously difficult to define, and evangelicals themselves have been a notoriously fractious bunch, so any attempt here to survey the place of the conflict thesis in the panoply of evangelical groups would prove impossible.[2] Nineteenth-century evangelicals included Baptists, Methodists, holiness groups, Presbyterians, Congregationalists, Lutherans, the Dutch and German Reformed, Episcopalians, and others, and they varied in their sense of identification with the label. As the century wore on, some moved toward liberalism, while others in the same denominations steadfastly resisted such moves. In the twentieth century, evangelicals diversified even more, forsaking denominations, starting others, going independent, joining cooperative fundamentalist movements, joining charismatic renewal movements and Pentecostal churches, persevering in liberal-dominated churches, and vastly complicating the ecclesiastical terrain in America.

From across the ever-broadening spectrum of evangelical Protestants, many looked to Princeton for help in the intellectual defense of the faith against scientific abuse.[3] For over a hundred years, from 1825 to 1929, Princeton Seminary and College published weighty theological quarterlies that helped evangelicals navigate the many intellectual challenges of the day. Editors Charles Hodge, A. A. Hodge, and B. B. Warfield, head theologians at the seminary, marshaled their faculty and a host of like-minded contributors to offer an erudite conservative assessment of developments in not only theology, but science, philosophy, literature, education, politics (early on), and the churches.[4] The *Princeton Review* and its successor journals were widely read on both sides of the Atlantic, as were important monographs by college and seminary professors and alumni. At the college, President James McCosh championed the harmony of religion with modern science in a long list of influential works. Princeton stood for Calvinist orthodoxy, evangelical piety, the truthfulness of the Bible, and the intellectual defensibility of Christian faith.

Historians have long recognized Princeton's importance here. Herbert Hovenkamp's study of science and religion in antebellum America opens and closes with vignettes of Princeton. Glenn T. Miller identifies Princeton as the model of theological scholarship after the Civil War: it "set the standard for much of contemporary American theological education." And as the twentieth century dawned, Mark Noll observes, "the Presbyterians at Princeton flew the banner of conservative scholarship in the battlefields of the day pretty much by themselves"—so that while their influence in the academy at large was waning, their importance as heroic warriors for evangelicals intensified. Ernest Sandeen identified the Princeton doctrine of biblical inerrancy as one of two main "roots of fundamentalism" (the other being dispensational premillennialism). Molly Worthen finds that the spirit of Princeton theologian J. Gresham Machen "hovered over the neo-evangelicals" of the mid-twentieth century. Carl F. H. Henry, Bernard Ramm, Fuller Seminary, and the American Scientific Affiliation all looked to what Ramm called "the noble tradition" exemplified by the Princetonian approach to faith and learning, science and religion.[5]

Accordingly, I will treat the Princetonians as exemplary voices of nineteenth-century evangelical engagement with science, well aware that many others, to varying degrees, shared their approach. I will then move to consider fundamentalists and neo-evangelicals into the 1950s. There were of course other evangelical voices, but these have been the most influential and the most widely studied.

The Military Metaphor at Mid-Nineteenth-Century Princeton

Geoffrey Cantor has traced the roots of the warfare model to John William Draper's history (1868–70) of the American Civil War. Draper, a pathbreaking chemist and outspoken anti-Catholic, adapted William H. Seward's language of "irrepressible conflict" to the history of science and religion. Thereafter, in the 1870s, "the idea of a preordained and necessary conflict between two opposing worldviews" took hold. With this "strong version" of the conflict thesis, "science became a weapon to be wielded in public attacks on religion."[6]

A Civil War connection can be found a bit earlier than that in the pages of the *Biblical Repertory and Princeton Review*. In 1863, ten months after the famous sea battle between the *Monitor* and the *Merrimac*, in an article exploring "The Scepticism of Science" and suggesting a Christian response, Joseph Clark wrote, "An enemy who brings against us new and formidable weapons, must be met by weapons equal or superior. A contest against iron-clad ships can be sustained successfully only by iron-clad ships, or something better."[7] Indeed, imagery of war—sovereignties, territories, invasions, defense plans, weaponry, citadels, charges, and retreats—appeared repeatedly in the *Princeton Review* of the 1860s in connection with the relations of science and religion. It began shortly before the real War between the States, about 1860, but clearly the national calamity brought martial imagery to mind, and the Princetonians, with many others, found the military metaphor apt. It provided not only a dramatic rallying cry to the faithful but also a language with which to consider both the logical relations of science and religion and the approach religious thinkers should take to the challenges posed by science. I have argued elsewhere that historians do well not to shun but to embrace the military metaphor—not as a historical metanarrative for ourselves, but as the expression of the mindset of historical agents at the time, surely a key subject of our inquiries. In the Princetonians' case, an exploration of their use of the military metaphor reveals a surprisingly careful and consistent effort at *reconnaissance* and *strategic planning*, which they then implemented over the next several decades. Their famous semi-acceptance of evolution traces back to principles and plans for battle articulated in the 1860s. I have found the military metaphor to be an important key to unlocking not only their objections to evolutionary theory, but their qualified acceptance of it as well.[8]

Just here it is important to distinguish evangelical use of martial imagery from adherence to "the conflict thesis" usually associated with Draper and White. Draper borrowed from the Civil War the idea of "irrepressible conflict," an inevitable clash between two systems resulting from fundamentally different approaches to truth. The Princetonians and many fellow Protestants applied the Civil War analogy in a different way: they saw the science-religion conflict as a war *to save a union*—or even, to take an analogy from the Civil War South, a war against an *aggressive invader*.[9]

As is often the case with the use of martial language, these evangelicals perceived their situation as a matter of defense: Christian truth-claims and the religious experience grounded in those truth-claims were coming under attack. This was, to them, nothing new; the church militant had been defending the faith against enemy arguments since the days of the church fathers. One of the three founding purposes of Princeton Seminary was the defense of the faith against the cavils of rationalism, meeting the arguments of deists with reasoned arguments for gospel belief. But in the 1860s, the attack now came from the ranks of science, which until recently had been a great supporter of theism and Christian notions of human nature.[10] Lyman Atwater warned that "a positive and semi-positive school, with their allies, under the lead of such men as Huxley, Darwin, Spencer, and Mill," were "assailing the fundamental, moral, and religious convictions of men from the scientific side, with weapons claimed to be forged in the laboratories of physical science." Together with Romanism and German idealism, he said, they formed "a combined and fearful host arrayed against the faith once delivered to the saints." He urged, "Those set for the defence of the gospel must therefore gird on their armor. They must watch, detect, expose, confront, and overpower their foe."[11]

Science and religion were "twin daughters of heaven," each occupying a proper sphere and owing the other recognition as legitimate sovereign within her respective territory. The field of knowledge involved an ultimately harmonious union of these sovereign states. Religious leaders should not presume to dictate scientific theory, nor should scientists invade religious territory, particularly in matters concerning humankind: the unity of the human race, the reality of the soul, the place of humanity in the scheme of nature. Gospel Christianity involved doctrines that shared ground with science; in these areas Christians must be vigilant. These notions appeared again and again in the *Princeton Review* of the 1860s.

The Princetonians and their allies pressed the military metaphor further to include a strategy for not only defense but offense. Clark wrote, "In every attack which has been made upon Christianity by hostile human learning, from the days of the apostles to the days of Dr. Strauss, the assailing party have been thwarted and vanquished by the church seizing and mastering the weapons of attack. The sons of the church have become learned in the learning of their adversaries, and have not only sustained the attack, but have succeeded in bringing from every newly opened field of inquiry something to strengthen the citadel of their faith."[12]

To a remarkable degree Princeton put this strategy for the offense into practice. James McCosh, president of Princeton College, sent his most promising students to Europe to pursue graduate study under leaders in the very fields that threatened Christian understandings, bringing cutting-edge science back to Princeton to strengthen the defenses of Christianity. The most striking case was that of Charles Hodge's own grandson, William Berryman

Scott, who studied under none other than Darwin's bulldog himself, Thomas Henry Huxley—named by Atwater as leading a "fearful host arrayed against the faith." So real was Princeton's stance of fearless trust to the "book of God's works" (science fairly pursued) that these "bright young men" were thus commissioned to fulfill Clark's purpose to "seize and master the weapons of attack." Besides Scott, their ranks included the future director of the American Museum of Natural History, Henry Fairfield Osborn (another Huxley student), and James Mark Baldwin, a pioneer in evolutionary psychology who studied under Wilhelm Wundt.[13]

In this way, a conscious battle plan, articulated through the military metaphor, propelled Princeton to its thorough evidentialist engagement with "enemy" ideas. It became a hallmark of the Princeton quarterlies to digest controversial new books, exposing areas of fundamental disagreement, but often taking an apparently problematic fact or claim and "turning the guns" (Huxley's phrase, by the way) to show its actual support for orthodoxy. For example, Geerhardus Vos used Albert Schweitzer's *Quest of the Historical Jesus* to show that Jesus was a radical apocalypticist and supernaturalist—not at all in keeping with the historical Jesus imagined by the liberals.[14]

Princeton celebrated its famous leader on the evolution question, James McCosh, with imagery of war and peace, underscoring its alternative version of the conflict thesis. McCosh "averted a disastrous war between science and faith," wrote natural history professor George Macloskie (in the same year White published his *History of the Warfare of Science with Theology in Christendom*), "and at 'his' college, men have studied Biology without discarding their religion." The union was saved; science and religion could both pursue their own callings without antagonizing the other. Macloskie even ventured to announce in 1903 that the time for tentative harmonization, an opening of the borders between science and theology, had arrived.[15]

Princeton and White

It is clear that while this evangelical conflict thesis exemplified at Princeton shared some things with the Draper/White version, it also differed considerably. Like Draper and White, the Princetonians believed that science and reason, fairly pursued, would yield the truth. But while Draper and White pictured the church and religious leaders as obstructing the free pursuit of science, evangelicals pictured scientists and rationalists as attacking cherished and well-established spiritual truths that lay outside the reach of scientific expertise. Again, Draper and White pictured the relations of science and religion as a matter of sure progress; they struck a pose of concern over theology's threat to science but were dead sure that science would win, and soon. The evangelical conflict thesis displayed considerably more angst. Despite their postmillennialism, the Princetonians and many fellow evangelicals feared that science wrongly pursued might ungod the mental universe of many people.

Ironically, Andrew Dickson White tried to use the Princetonians as examples in his monumental *History of the Warfare*, setting up a contrast between college president McCosh as the liberal-minded, forward-thinking religious leader who embraced evolution with the conviction that it still left "room for religion," and seminary professor Charles Hodge, who flatly rejected evolution on religious grounds: "What is Darwinism? It is atheism." White was no Huxley; by his telling, science, rather than shattering Christianity, was in the process of purifying it. For him, McCosh was a veritable "*deus ex machina*" who "at once took his stand against teachings so dangerous to Christianity" as those of Hodge. White's conflict thesis was a tool for advancing liberal Protestantism. For all its talk of science versus theology, it advocated a particular theology in the name of science: one based on the experience of dependence on God rather than on orthodox doctrinal points like the deity of Christ and the substitutionary atonement, a theology immune to factual assault because it did not lay claim to facts. It was also a theology that scorned talk of hell. Observing that Princeton mathematics professor John T. Duffield had found the biblical Adam and Eve inextricably woven into the New Testament account of salvation through Christ, so that human evolution logically entailed a denial of the gospel and a future without God for those who followed that logic, White fumed. Duffield's statement, he said, showed "that a Presbyterian minister can 'deal damnation round the land' *ex cathedra* in a fashion quite equal to that of popes and bishops." Duffield clinched White's argument here: doctrine, White said, was serving as a bludgeon to beat back scientific advance.[16]

McCosh had criticized an earlier version of White's book twenty years before. Princeton's unsigned review of the final version in 1896 (almost certainly by B. B. Warfield) discreetly ignored White's representations of Princeton men and instead offered a pointed critique of his whole conflict thesis. The greatly enlarged work was "still an 'indiscriminate and uncritical agglomeration of facts' [McCosh's words], brought together for the support of a thoroughly one-sided and fatally misleading proposition"—namely, that the Bible's teachings, "so far as they concern matters with which science has to do, are simply a mass of mythological conceptions" blocking the advance of knowledge. In effect, "the sphere of religion is to be sharply limited and the whole sphere of science left undisturbed by its influence." This would never do. Anticipating today's historical critics of the conflict thesis, the reviewer noted, "[A]ll that [the book] proves is that some religious men have been slow to follow the advance of science; and the same could easily be shown of some scientific men: nay, it was largely not because they were men of religion, but because they held too tenaciously to the current ideas of men of science, that these men lagged behind the vanguard of advance."[17]

The real problem with the book lay not in its many errors of detail, but in the "fundamental conception" conveyed in its title. Nettled, the reviewer asked, "What, in all strictness of thought, can be meant by affirming a warfare between 'Science' and 'Theology'? There is no such thing as 'Science,' with a big S: sciences there are, but the *scientia scientarum*, which alone can be meant by 'Science,' is a philosophy, and can only be interpreted as a general world-view supposed to be involved in the results of the various sciences. But how can Science, with a big S, in this sense come into conflict with Theology? Theology is itself a science, one of the *scientiae*, of the results of which Science is the abstract expression."[18]

That White deployed his conflict thesis to further a liberal theological viewpoint is noteworthy. In another place, Warfield argued that liberalism, "the 'newer religious thinking,'" was "simply the old eighteenth-century problem in fresh form"—namely, whether Christianity was a natural religion or a supernatural one.[19] Deists thought of their rational religion as a purified Christianity; White did essentially the same thing. And his warfare model largely matched theirs: the advance of enlightenment (reason and science) would dispel traditional myths; science would conquer not religion but dogma.

Princeton contra Modernism

The Princetonians flatly rejected White's thesis of conflict between science and traditional dogma, as they had rejected Huxley's thesis of conflict between science and religion altogether. In both cases they blamed an underlying philosophical stance for the apparent conflict. By the 1890s, the language of battle was far less prominent in their encounters with evolutionary science than back in the 1860s. McCosh's irenic stance seems to have framed Princeton's interactions with evolutionary science in the 1880s and 1890s, when the scientists at Princeton College enjoyed what Macloskie called "a large and friendly toleration," a wide latitude in their explorations of evolution's mechanisms and effects under the general concept of theism.[20]

At the same time, however, serious conflict occurred *within* the Presbyterian Church—battles over the higher-critical views of Charles Briggs, confessional revision, and the inerrancy of scripture. The conflict involved not just persuasion but coercion, as the conservatives wrote into church law the Portland Deliverance on inerrancy and employed the church courts in three highly publicized heresy cases (Briggs, Henry Preserved Smith, and A. C. McGiffert).[21] It is interesting to note that the battle over inerrancy came only now, in the 1890s, in the wake of having averted a war between science and religion outside the church—as if to say that having secured the union of sovereign states in the public realm of ideas, it was not to be lost in the church itself by any capitulation to the idea that scientific investigation invalidated the trustworthiness of the Bible or rendered the old creed obsolete.

Thus while the military metaphor largely vanished from Princetonian discourse about organic evolution, it persisted, or perhaps resurfaced, in Princetonian discourse about evolutionary philosophy, the evolution of religion and the Bible, and the emerging liberal theological views that took such evolutionary notions as a guide or an excuse for making alterations to the faith. This use of martial language certainly did not entail the conflict thesis of Huxley or that of Draper and White. Rather it harked back to the old conflict between Christianity and unbelief. Henry Collin Minton, a former student of Warfield's and later a Stone Lecturer at Princeton Seminary, discerned "the cloven hoof of evolutionary philosophy" when the world process was equated with God. His language made the contest one between good and evil, God and Satan. William Brenton Greene, Princeton Seminary's professor of apologetics, dusted off the martial terminology of the 1860s for renewed battle with this new foe within the church, denouncing an apologetical method consisting in "concession rather than refutation; in crying 'Peace! peace!' rather than in vigorous warfare; in trying to show that there is little over which to fight, rather than in 'contending earnestly for the faith once delivered to the saints.'" New Testament professor George T. Purves urged against William Sanday, ". . . his irenicon will not serve to stay the battle which is waging, and must be waged to the finish, between belief in an infallible Scripture and dependence in religion as elsewhere upon the partially enlightened reason of man."[22]

George S. Patton captured the heart of the matter: "One cannot consistently empty Christianity of all content, and then claim to speak in its defense."[23] Here was a battle over the battle—over how to defend the faith. The party emerging as Protestant liberals were embracing evolutionary ideas about religion and the Bible, and doing so in an attempt to rescue Christianity from scientific refutation. The Princetonians were coming to see this liberal salvage effort as not the defender of Christian faith, but an enemy within.

Toward Fundamentalist Militancy

The Princetonians were of course not alone in their suspicions of theological liberalism as an enemy within. By the 1910s, the fundamentalist movement was getting under way both within and across evangelical denominations, impelled by just that suspicion of internal betrayal. As tensions mounted in the Presbyterian Church, U.S.A., Princeton New Testament professor J. Gresham Machen penned a book that many readers within and outside the church admired for its penetrating analysis: *Christianity and Liberalism* (1923).[24]

Already in the second sentence of the book, Machen invoked martial imagery, noting that many "prefer to fight their intellectual battles in what Dr. Francis L. Patton has aptly called a 'condition of low visibility.'" Machen called for a fair fight in an open, well-lit field. He proposed to bring up the lights on the actual terrain of battle, making the points of contention perfectly plain, calling all sides to face squarely the fact of a basic and ineluctable enmity. "In the sphere of religion, in particular, the present time is a time of conflict; the great redemptive religion which has always been known as Christianity is battling against a totally diverse type of religious belief, which is only the more destructive of the Christian faith because it makes use of traditional Christian terminology."[25]

The battle in question was not between Christianity and science, but Christianity and liberalism, as the liberal scheme for adjusting Christianity to modern science entailed the discarding of gospel doctrine. Machen was sure that science and history finally *would not* falsify Christianity's fundamental claims. The liberals, by contrast, argued that science and history *could not* falsify Christianity, because they did not intersect with it at all. "Admitting that scientific objections may arise against the particularities of the Christian religion . . . the liberal theologian seeks to rescue certain of the general principles of religion, of which these particularities are thought to be mere temporary symbols, and these general principles he regards as constituting 'the essence of Christianity.'" Here the military metaphor came back into play: "[A]fter the apologist has abandoned his outer defences to the enemy and withdrawn into some inner citadel, he will probably discover that the enemy pursues him even there. Modern materialism . . . is just as much opposed to the philosophical idealism of the liberal preacher as to the Biblical doctrines that the liberal preacher has abandoned in the interests of peace. Mere concessiveness, therefore, will never succeed in avoiding the intellectual conflict. *In the intellectual battle of the present day there can be no 'peace without victory'; one side or the other must win.*"[26]

Just a few years after the Treaty of Versailles, Machen invoked language from World War I—indeed, he invoked (negatively) a slogan of a fellow Princetonian, Woodrow Wilson. Just as "peace without victory" had proved impossible in 1919, so "the liberal attempt at rescuing Christianity" would fail. "[I]n abandoning the embattled walls of the city of God he has fled in needless panic into the open plains of a vague natural religion only to fall an easy victim to the enemy who ever lies in ambush there." Yielding up the core of Christian belief, the modern liberal had taken refuge in a religion bereft of the doctrines of grace, and would, Machen warned, sooner or later fall into outright unbelief.[27]

This tragedy, Machen argued, was unnecessary, based as it was on a faulty picture of the conflict. "But in showing that the liberal attempt at rescuing Christianity is false," he wrote, "we are not showing that there is no way of rescuing Christianity at all; on the contrary, it may appear incidentally, even in the present little book, that it is not the Christianity of the New Testament which is in conflict with science, but the supposed Christianity of the modern liberal Church, and that the real city of God, and that city alone, has defences which are capable of warding off the assaults of modern unbelief."[28]

Machen's version of the conflict thesis was a slight variation on the earlier Princetonian version, differing mainly in its emphatic affirmation of irreconcilable conflict between, not science and religion, but liberalism and Christianity—liberalism being a peculiarly modern form of the age-old foe, rationalistic unbelief. The battle with liberalism was a new engagement in the old war to save the union of science and religion. "Modern liberalism may be criticized (1) on the grounds that it is un-Christian and (2) on the grounds that it is unscientific."[29]

It is not hard to see how fundamentalists then and later, lacking Machen's nuance and erudition, could conflate the battle against liberalism with their battle against evolution. Insofar as liberalism embraced evolutionary views of biology, human society, and religion, the substitution of liberalism for science as the enemy made sense. The Princetonians and their many allies had always refused to peg science as a real enemy—the actual foe was "science, falsely so-called," and even there they chided those voices that were too quick to condemn particular scientific theories.[30] When William Jennings Bryan invited Machen to come to Dayton, Tennessee, to serve on a planned expert panel of theologians in the Scopes trial, Machen politely demurred. In the tradition of McCosh and Warfield, Machen had no problem with evolution "properly limited and explained" (that's McCosh's phrase), and with them and the other Princetonians he embraced what I call providential developmentalism in a surprising array of applications, including the historical progress of revelation and of Christian doctrine.[31] But the fundamentalists, identifying evolutionary science with unbelief, perceived "science falsely so-called" to be a main prop of Christianity falsely so-called, effectively compounding the enemies pictured in two versions of the conflict thesis—and the battle raged with double strength.

Fundamentalists and the Conflict Thesis

George Marsden famously defined fundamentalism as "militantly antimodernist Protestant evangelicalism," embedding a martial stance and active warfare into the very essence of the movement.[32] A complex phenomenon with fuzzy borders, fundamentalism's most prominent moment was the antievolution campaign of the 1920s, an effort that has come, somewhat misleadingly, to represent the whole. Machen, whom H. L. Mencken dubbed "Dr. Fundamentalis," opted out of the antievolution business, as we have already seen.[33] Still, the bulk of the movement threw its weight into a militancy that went beyond metaphor to active campaigning (in the courts and churches, it must be noted, and not in actual shooting).[34] Northern Baptists and northern Presbyterians endured internal battles between those who wanted to tolerate liberalism and those (the fundamentalist factions) who saw it, as Machen did, as a traitorous enemy within. Fundamentalists of many denominational stripes joined the World's Christian Fundamentals Association, headed by Baptist pastor William Bell Riley of Minneapolis, leaguing together in opposition to liberal theology, modern culture, and especially evolutionary science. The horrors of World War I, they believed, owed to Germany's putting into practice the ethics of applied Darwinism: "survival of the fittest" became "might makes right." They feared lest Christian civilization in America should fall under the influence of evolutionary ideas, now rife within the education system. These folks were quite willing to turn to power, the coercive power of the law or the denomination, to, in Marsden's words, "bind together once again the many frayed strands of evangelical America." He describes fundamentalism in the early 1920s as waging an "offensive on two fronts": denominational battles for traditional gospel doctrine, and a larger cultural battle to halt the teaching of evolution in the public schools.[35]

Mid-nineteenth-century evangelicals had frequently painted a horrible vision of a mental universe stripped of spiritual and moral truths. Now, after World War I, that vision was no longer a distant possibility foreseen by logical entailments; it seemed a very present threat in church, school, and the affairs of nations. Fundamentalists returned with gusto to the military metaphor, finding it splendidly useful for rallying the faithful. Its inherent drama stirred hearts to action. Its binary of friend or foe encouraged the simplification of issues and the posing of a stark choice, necessary to successful political mobilization. It fit well, too, with the debate format that came to characterize the antievolutionist circuit. And very importantly, the aggressive foe without was now accompanied by the insidious traitor within—the compromising liberal or modernist who recast Christianity to suit modern evolutionary science.

Fighting in such a circumstance became a virtue, peacemaking a vice. The issue became one of moral fiber. The gendered imagery of "muscular Christianity," dating from the 1890s, dovetailed nicely with the military metaphor. The hour called for courageous, resolute, and manly soldiers of the faith. It is no coincidence that Harry Rimmer, premier antievolutionary debater, was a former boxer. Billy Sunday issued the classic fundamentalist call to ready fighting when he said that Jesus himself "was no dough-faced, lick-spittle proposition. Jesus was the greatest scrapper that ever lived."[36]

But the fundamentalists' embrace of the military metaphor, like that of nineteenth-century evangelicals before them, expressly rejected the notion of a basic and inevitable conflict between science and religion. Fundamentalists' militancy arose always in the name of a peace threatened; it was aggressive, yes, but defensive, seeing this conflict as wrongfully forced upon them. Riley insisted, "Scripture and Science are harmonious and . . . any imaginary conflict between them is only the nightmare of uninformed minds." Rimmer wrote, "Science and the Bible are not in conflict—they are each supplemental to the other." God "is the author of both creation and record, therefore the two must agree; and they do agree in spite of what the ignorant may say to the contrary." Note here a difference from the earlier evangelicals, though: Riley and Rimmer struck a pose of superior knowledge, calling the proponents of science-religion conflict "uninformed" and "ignorant." Riley went so far as to insult Andrew Dickson White outright. Noting that "the path by which Science has traveled is strewn with the decaying structures of discarded theories," he asked, "why should not Andrew White have withheld his endeavor until specialists in biology, geology and paleontology are themselves convinced that evolution is something more than a theory?" The reason: "It is not unusual for smaller followers of great minds to far exceed their masters."[37]

Despite their free and at times extravagant use of the military metaphor, leading fundamentalist antievolutionists did not glorify actual armed conflict. Rather, they identified evolution with struggle and war, and Christianity (absent an aggressive foe) with peaceableness. William Jennings Bryan resigned his post as secretary of state when Woodrow Wilson led the nation to join the European conflict. John W. Porter, president of the Anti-Evolution League of America, stated flatly, "Evolution logically and inevitably leads to war." Darwin led to Nietzsche, who "crowned the Superman, glorified war, expressed contempt for Christ, and decried all rule of right and right living." While rallying believers to defend Christian civilization, these fundamentalists saw the Great War as a catastrophe and blamed German aggression on Darwinian ideals.[38]

We see, then, in the fundamentalist antievolution crusade a variation on the older evangelical conflict thesis. Like nineteenth-century evangelicals, fundamentalists perceived a conflict between not science and religion, but naturalism and supernaturalism. To that threat of unbelieving, materialistic science they now added theological modernism, a betrayal of the old-time gospel, compounding the sense of enmity. In the wake of World War I, they focused their concern on evolution as the wellspring of naturalistic science, doctrine-denying theology, and nationalistic militarism. They turned to the coercive power of the law and the courts—power over persuasion—even as they multiplied pamphlets arguing that evolution was unscientific. Claims of "no real conflict" became considerably more shrill and extreme than a generation before. Harry Rimmer used underlining and capital letters to shout his statement to his readers: "Wherever a <u>proved</u> fact of science is touched by any of the Scripture writers the two are in ABSOLUTE AGREEMENT."[39] Despite many continuities, the fundamentalists' ardent antievolutionism and strident tone was a far cry from the old evangelical defense of the faith exemplified by Princeton.[40]

The Conflict Thesis in a Time of Consensus

If the Civil War suggested to evangelicals the use of martial imagery in a call to save the union of science and faith, and if World War I prompted a vision of courageous conflict to rescue Christian civilization, can we find a similar parallel in World War II? After all, it was during and just after that war that a neo-evangelical movement, whose hallmark was recovery of the noble, pre-fundamentalist Protestant tradition of faith, learning, and social engagement, was born. Leading "new evangelical" organizations included the National Association of Evangelicals (founded in 1942), the American Scientific Affiliation (1941), and Fuller Theological Seminary (1947).[41] Other organizations would follow in the 1950s, most notably the establishment of a flagship magazine, *Christianity Today*, the neo-evangelical answer to the liberal *Christian Century*.

Perhaps an analogy to the war can be found in the fight against fascist demagoguery and extreme populistic appeals. In that case, the culture of the 1950s, with its emphasis on refinement and consensus, would have an explanation beyond the oft-remarked needs of the Cold War contest against communism in an atomic age. It does seem that mid-twentieth-century evangelical Protestants made a deliberate decision to renounce not only the military metaphor but rabble-rousing tactics. On both sides of the Atlantic, "new evangelicals" struck an irenic pose, even as they retained a firm belief in basic conflict between Christian and non-Christian worldviews. They sought to persuade thoughtful nonbelievers through reasoned argument; to shape the culture at large in a time when the Christian view of a sinful humanity seemed proven beyond a doubt; to bring about religious revival; and, significantly, to raise the intellectual tenor of fundamentalism. The neo-evangelical turn away from the military metaphor seems to have had to do not only with the Cold War horror of atomic warfare, but also with distaste for the extremism and emotionalism of America's enemies in World War II.[42]

The best example is Bernard Ramm, Baptist professor of philosophy and theology, whose *Christian View of Science and Scripture* (1954) earned commendation from Billy Graham and aroused spirited opposition from fundamentalists. Graham recommended Ramm's approach as a model for *Christianity Today*; the book also deeply influenced the evangelical American Scientific Affiliation toward moderate views on progressive creationism and theistic evolutionism. Opponents to that development flocked after

1961 to the reformulated flood-geology creationism of Whitcomb and Morris's classic, *The Genesis Flood*.[43]

Ramm called evangelicals to turn from what he termed the "ignoble tradition" prevalent among fundamentalists in recent decades but also running far back in time. "Sad has been the history of the evil that good Christian men have done in regard to science," he wrote, adding a quotation of Joseph Pye Smith from 1840: "'[Evangelical castigators of science] are unwittingly serving the designs of [Christianity's] enemies [and are] secret traitors to the cause of Christianity'" (brackets are all Ramm's). "The judgment of White is proved a thousand times that the cheap weapons of religious opposition to science are like 'Chinese gongs and dragon lanterns against rifled cannon.'"[44]

Surely no rebuke could sting worse than this—an evangelical quoting Andrew Dickson White to charge fellow believers as *traitors to Christianity*, the very charge they made against the liberals. Their "strong, outspoken" and "negative approach to science" was hurting the scientific respectability of scripture. Their "pedantic hyperorthodoxy" was not only continuing but widening "the great cleavage between science and evangelicalism which occurred in the nineteenth century." And to drive the point home, Ramm cited Maynard Shipley's anti-antievolutionist exposé, *The War on Modern Science: A Short History of the Fundamentalist Attacks on Evolution and Modernism* (1927). Evangelicals "cannot deny the very words of the hyperorthodox which Shipley quotes. There they are in all their extremisms, and even fanaticisms."[45]

Ramm squarely faced White's story of a historic battle between science and religion, recasting it as a "battle to keep the Bible as a respected book among the learned scholars and the academic world." He admitted with White that it "was fought and lost in the nineteenth century"—not that the Bible was disproven, but the larger intellectual world discounted its truth value. Instead of mounting a new attack, Ramm chose to ask, "*Why did the battle go as it did?* Why did the populace, the universities, and even much of the clergy yield to the critical and scientific attacks on the Bible?" His long list of reasons began with one of White's main themes: "the continuing revolt of man from the religion and authoritarianism of the Roman Catholic Church in its medieval expression." Echoing Machen, he cited a "deep-moving secularism—life without God, philosophy without the Bible, community without the Church"—which "was all in favor of the radical and the critic" and put "a premium on criticism and skepticism." This, he said, gave unbelief a huge psychological and social advantage. "It was far easier for the radical to draw blood than for the Christian to do so." But beside these issues of an uneven playing field were certain undeniable failures on the part of believers: their divisions over interpretation, their use of "sarcasm or vilification or denunciation" regarding the facts of science, and especially their "poorly developed philosophy of science" and their presenting "the problems of modern science as resolving down to (i) fiat, instantaneous creationism; or (ii) atheistic developmentalism"—a "gross oversimplification." Ramm called on evangelicals to abandon the tactics of extremism and to re-engage science and philosophy in a serious effort toward mutual understanding. There was indeed a conflict, not with science, but with scient*ism*. Echoing old Princeton (without naming any Princetonians), Ramm argued that if the theologian and the scientist had stuck to their respective duties, each learning the other side's positions carefully before pronouncing upon them, "there would have been no disharmony between them save that of the non-Christian heart in rebellion against God."[46]

And so Ramm made clear his aim "to call evangelicalism back to the noble tradition of the closing years of the nineteenth century," back before fundamentalism.[47] But there was a slight difference: Ramm factored in the soul's disposition toward or away from God. This was something the Princetonians and their ilk rarely did in their discussions of science and religion (though it did appear in other venues).[48] Their midcentury followers at Westminster Theological Seminary, though, devoted considerable attention to it, as did the Calvinists in the Dutch-American evangelical tradition, lionizing the work of Abraham Kuyper and Herman Bavinck.[49]

By midcentury, the already diverse ranks of America's evangelical Protestants were approaching a reconfiguration along lines dividing fundamentalists from neo-evangelicals on the matter of science and religion, but with plenty of other fault lines as well. Their attitudes toward and use of a conflict thesis varied widely. Large swaths of the Bible-believing Protestant population still sided with the fundamentalists in fierce battle against the menace of evolution, a story richly explored in Ronald Numbers's *The Creationists*. Members of the American Scientific Affiliation (practicing scientists and teachers of science) held the same doctrinal basics as the fundamentalists, but carefully avoided the conflict pose. They were "eager to advance scientific knowledge within the churches by dispelling unnecessary objections while affirming the harmony of science and Scripture," combining biblical fidelity with scientific credibility. With this limited purview, the ASA did not aim to wage war at all, leaving the theological and philosophical questions to others and concentrating on scientific particulars.[50] Meanwhile, the premier neo-evangelical educational institution, Fuller Seminary, was launched as a latter-day Princeton to wield current scholarship in the service of faith. Wilbur Smith boasted in 1950 that with a few new additions Fuller "will probably then have the greatest evangelical faculty that has been assembled on the North American continent since the days of Princeton's glory."[51]

Conclusion

Though they all stoutly refused the conflict thesis of Draper and White, variations on the conflict thesis appeared, faded, and reappeared among Protestant evangelicals in the century after Darwin's *On the Origin of Species*. Martial imagery had been native to the Christian tradition since biblical times, so it is no wonder that evangelicals found it natural to speak in the accents of the military metaphor. The particular nature of the conflicts envisioned—enemies, territories, key fields of battle, issues at stake, and tactics for victory—varied over time and often reflected national concerns during actual wars. But through all the variations, evangelicals agreed that the conflicts between science and religion were ultimately misconceived, for true science and true religion could never finally conflict.

NOTES

1. James R. Moore, *The Post-Darwinian Controversies: A Study of the Protestant Struggle to Come to Terms with Darwin in Great Britain and America, 1870–1900* (New York: Cambridge University Press, 1979), chaps. 1–4.

2. On defining evangelicalism, see Douglas A. Sweeney, *The American Evangelical Story: A History of the Movement* (Grand Rapids, MI: Baker Academic, 2005), chap. 1.

3. The literature on the Princeton theology and on Princetonian views of science and religion is considerable. Important works include Mark A. Noll, ed., *The Princeton Theology: Scripture, Science, and Theological Method from Archibald Alexander to Benjamin Warfield* (Grand Rapids, MI: Baker, 1983); D. G. Hart, *Defending the Faith: J. Gresham Machen and the Crisis of Conservative Protestantism in Modern America* (Baltimore, MD: Johns Hopkins University Press, 1994); David B. Calhoun, *Princeton Seminary: Faith and Learning, 1812–1868* and *Princeton Seminary: The Majestic Testimony, 1869–1929* (Edinburgh: Banner of Truth, 1994 and 1996); Mark A. Noll and David N. Livingstone, eds., *B. B. Warfield: Evolution, Scripture, and Science—Selected Writings* (Grand Rapids, MI: Baker, 2000); James H. Moorhead, *Princeton Seminary in American Religion and Culture* (Grand Rapids, MI: Eerdmans, 2012); Bradley J. Gundlach, *Process and Providence: The Evolution Question at Princeton, 1845–1929* (Grand Rapids, MI: Eerdmans, 2013).

4. Noll gives a handy overview of the Princeton journals in *Princeton Theology*, 22–24.

5. Herbert Hovenkamp, *Science and Religion in America, 1800–1860* (Philadelphia: University of Pennsylvania Press, 1978); Glenn T. Miller, *Piety and Profession: American*

Protestant Theological Education, 1870–1970 (Grand Rapids, MI: Eerdmans, 2007), 6; Mark A. Noll, *Between Faith and Criticism: Evangelicals, Scholarship, and the Bible in America* (San Francisco: Harper and Row, 1986), 51; Jon R. Roberts, *Darwinism and the Divine: Protestant Intellectuals and Organic Evolution, 1859–1900* (Madison: University of Wisconsin Press, 1988), 222; Ernest R. Sandeen, *The Roots of Fundamentalism: British and American Millenarianism, 1800–1930* (Chicago: University of Chicago Press, 1970); Molly Worthen, *Apostles of Reason: The Crisis of Authority in American Evangelicalism* (New York: Oxford University Press, 2014), 31; George M. Marsden, *Reforming Fundamentalism: Fuller Seminary and the New Evangelicalism* (Grand Rapids, MI: Eerdmans, 1987), 98; Christopher M. Rios, *After the Monkey Trial: Evangelical Scientists and a New Creationism* (New York: Fordham University Press, 2014), 54, 82; Bernard Ramm, *The Christian View of Science and Scripture* (Grand Rapids, MI: Eerdmans, 1954), 9.

6. Geoffrey Cantor, "What Shall We Do with the 'Conflict Thesis'?," in Thomas Dixon, Geoffrey Cantor, and Stephen Pumfrey, eds., *Science and Religion: New Historical Perspectives* (Cambridge: Cambridge University Press, 2010), 293–94.

7. Joseph Clark, "The Scepticism of Science," *Biblical Repertory and Princeton Review* 35 (1863): 65.

8. Gundlach, *Process and Providence*, esp. chaps. 2–3.

9. I am indebted to Chris Rios for this southern viewpoint.

10. See Charles Coulston Gillispie, *Genesis and Geology: A Study in the Relations of Scientific Thought, Natural Theology, and Social Opinions in Great Britain, 1790–1850* (1951; New York: Harper Torchbooks, 1959).

11. Lyman H. Atwater, "Herbert Spencer's Philosophy: Atheism, Pantheism, and Materialism," *Biblical Repertory and Princeton Review* 37 (1865): 569–70.

12. Clark, "Scepticism of Science," 67–68.

13. Gundlach, *Process and Providence*, chap. 5.

14. Geerhardus Vos, review of *The Quest of the Historical Jesus* by Albert Schweitzer, *Princeton Theological Review* 9 (1911): 132–41.

15. George Macloskie, quoted in William Milligan Sloane, *The Life of James McCosh* (New York: Charles Scribner's Sons, 1896), 124; Macloskie, "The Outlook of Science and Faith," *Princeton Theological Review* 1 (1903): 597–615. He delivered on that promise of harmonization the following year, with his article "Mosaism and Darwinism," *Princeton Theological Review* 2 (1904): 425–51.

16. Andrew Dickson White, *A History of the Warfare of Science with Theology in Christendom* (New York: Appleton, 1896), 1:79.

17. [B. B. Warfield (?)], review of *A History of the Warfare of Science with Theology in Christendom* by Andrew Dickson White, *Presbyterian and Reformed Review* 9 (1898): 510–12.

18. [Warfield (?)], review of *A History of the Warfare of Science with Theology*. This objection smacks of course of Charles Hodge's famous contention that theology is a science by virtue of its using the same inductive method as in the other sciences: the facts of scripture are to the theologian what the facts of rocks are to the geologist. See Charles Hodge, *Systematic Theology* (New York: Scribners, 1872–73; reprint, Grand Rapids, MI: Eerdmans, 1993), 1:10; cf. Theodore Dwight Bozeman, *Protestants in an Age of Science: The Baconian Ideal and Antebellum American Religious Thought* (Chapel Hill: University of North Carolina Press, 1977). Warfield here (if it is Warfield), instead of appealing to Baconian induction, defines Science with a big S as a "world-view"—but not the Kuyperian notion of worldview as a set of presuppositions. Rather, this Science or "world-view" is "the whole we have erected in our minds" out of the "results" of the various sciences, including theology.

19. Warfield, review of *"The Ascent of Man": Its Note of Theology* by Principal Hutton, *Presbyterian and Reformed Review* 6 (1895): 367.

20. George Macloskie, "Theistic Evolution," *Presbyterian and Reformed Review* 9 (1898): 3.

21. See Lefferts A. Loetscher, *The Broadening Church: A Study of Theological Issues in the Presbyterian Church since 1869* (Philadelphia: University of Pennsylvania Press, 1954), chaps. 5–8.

22. Henry Collin Minton, review of *Ideas from Nature* by William Elder, *Presbyterian and Reformed Review* 10 (1899): 546–47; George Macloskie, review of *Genetic Philosophy* by David Jayne Hill, *Presbyterian and Reformed Review* 6 (1895): 574; William Brenton Greene Jr., review of *The New Apologetic* by Milton S. Terry, *Presbyterian and Reformed Review* 8 (1897): 547; George T. Purves, review of *Bampton Lectures on Inspiration* by William Sanday, *Presbyterian and Reformed Review* 6 (1895): 181.

23. George S. Patton, review of *Moral Evolution* by George Harris, *Presbyterian and Reformed Review* 8 (1897): 539.

24. On the book and its reception, see D. G. Hart, *Defending the Faith: J. Gresham Machen and the Crisis of Conservative Protestantism in Modern America* (Baltimore, MD: Johns Hopkins University Press, 1994), chap. 3.

25. Machen, *Christianity and Liberalism* (New York: Macmillan, 1923), 1.

26. Machen, *Christianity and Liberalism*, 6 (emphasis added).

27. Machen, *Christianity and Liberalism*, 6–7.

28. Machen, *Christianity and Liberalism*, 7.

29. Machen, *Christianity and Liberalism*, 7.

30. Roberts, *Darwinism and the Divine*, chap. 2; cf. Gundlach, *Process and Providence*, 65.

31. Gundlach, *Process and Providence*, 297–98.

32. Marsden, *Fundamentalism and American Culture: The Shaping of Twentieth-Century Evangelicalism, 1870–1925* (New York: Oxford University Press, 1980), 4. For subsequent discussion of Marsden's definition, see William L. Svelmoe, "*Fundamentalism and American Culture*," in Darren Dochuk et al., eds., *American Evangelicalism: George Marsden and the State of American Religious History* (Notre Dame, IN: University of Notre Dame Press, 2014), 171.

33. H. L. Mencken, "Dr. Fundamentalis" (obituary for J. Gresham Machen), *Baltimore Evening Sun*, January 18, 1937, sec. 2, p. 15.

34. This is an important point in view of the move by historians and journalists since the early 1990s to extend the term "fundamentalism" to include militantly traditionalist Muslims and Hindus, some of whom do resort to physical violence as a matter of course. Representative of that move is Martin E. Marty and R. Scott Appleby, eds., *Fundamentalisms Observed* (Chicago: University of Chicago Press, 1991).

35. Marsden, *Fundamentalism and American Culture*, 161, 164. Christopher Rios explores the rise of antievolutionism in Britain in the years between the World Wars, comparing it to the American case. Rios, *After the Monkey Trial*, 29–34.

36. On Rimmer, see the introduction to *The Antievolution Pamphlets of Harry Rimmer*, ed. Edward B. Davis (New York: Garland, 1995; vol. 6 of *Creationism in Twentieth-Century America*, ed. Ronald N. Numbers), ix–xxviii; and Numbers, *The Creationists: The Evolution of Scientific Creationism* (Berkeley: University of California Press, 1992), 60–71. Sunday is quoted in William G. McLaughlin Jr., *Billy Sunday Was His Real Name* (Chicago: University of Chicago Press, 1955), 179; cf. David S. Gutterman, *Prophetic Politics: Christian Social Movements and American Democracy* (Ithaca, NY: Cornell University Press, 2006), 63; Josh McMullen, *Under the Big Top: Big Tent Revivalism and American Culture, 1885–1925* (New York: Oxford University Press, 2015), 66–68.

37. William Bell Riley, *Are the Scriptures Scientific*? (n.d.), in William Vance Trollinger Jr., *The Antievolution Pamphlets of William Bell Riley* (New York: Garland, 1995; vol. 4 of *Creationism in Twentieth-Century America*, ed. Ronald N. Numbers), 4; Harry

Rimmer, *The Harmony of Science and the Scriptures* (1927), in Trollinger, *Antievolution Pamphlets of Harry Rimmer*, 3; Riley, "The Theory of Evolution and False Theology," *Bible Student and Teacher* 9 (July 1908): 37, repeated in *Is Man a Developed Monkey?* (n.d.), in Trollinger, *Antievolution Pamphlets of William Bell Riley*, 7.

38. John W. Porter, *Evolution—A Menace* (1922), in Willard B. Gatewood Jr., ed., *Controversy in the Twenties: Fundamentalism, Modernism, and Evolution* (Nashville, TN: Vanderbilt University Press, 1969), 129.

39. Rimmer, *Harmony of Science and the Scriptures*, 3.

40. It would, however, be a mistake to suppose that all fundamentalist leaders were flashy showmen making populistic appeals. Most of the top leaders of the fundamentalist movement retained an air of respectability and restraint, emphasizing the philosophical and scientific issues, even if they treated those issues somewhat naively.

41. See Joel A. Carpenter, *Revive Us Again: The Reawakening of American Fundamentalism* (New York: Oxford University Press, 1999), chap. 8; Rios, *After the Monkey Trial*, chap. 2; Marsden, *Reforming Fundamentalism*, chap. 3.

42. On the neo-evangelical movement, see David F. Wells and John D. Woodbridge, eds., *The Evangelicals: What They Believe, Who They Are, Where They Are Changing* (Nashville: Abingdon Press, 1975); Carpenter, *Revive Us Again*; Marsden, *Reforming Fundamentalism*; and Worthen, *Apostles of Reason*. Carpenter treats the Cold War connection beginning on 188; Worthen treats it in many places.

43. Numbers, *Creationists*, 184–85; Marsden, *Reforming Fundamentalism*, 158–59.

44. Ramm, *Christian View of Science and Scripture*, 9, 27–28.

45. Ramm, *Christian View of Science and Scripture*, 28. Shipley was founder of the Science League of America.

46. Ramm, *Christian View of Science and Scripture*, 17–23 (italics in the original). On the conflict with scientism, clearly retaining White's vocabulary if not White's identification of the warring parties, see chap. 2, "An Analysis of the Conflict between Theology and Science."

47. Ramm, *Christian View of Science and Scripture*, 9.

48. See Paul K. Helseth, *"Right Reason" and the Princeton Mind: An Unorthodox Proposal* (Phillipsburg, NJ: Presbyterian and Reformed, 2010).

49. For starters, see Hendrik Hart et al., eds., *Rationality in the Calvinian Tradition* (Lanham, MD: University Press of America, 1983).

50. Rios, *After the Monkey Trial*, 46.

51. Wilbur Smith, quoted in Marsden, *Reforming Fundamentalism*, 119.

CHAPTER TEN

Jews

NOAH EFRON

Behind seventeen-year-old Maximilian Theodore Rosenberg sat thirty-odd grandees on the stage of the yawning New York Academy of Music, as a renowned band played comic opera from the pit. Rosenberg stood at the dais to deliver a "philosophical oration," his contribution to the commencement ceremony of the 1878 class of the New York City University. Behind him sat the university chancellor, the newly appointed secretary of the Smithsonian Institution, the president of the New York Medical Society, the city park commissioner, the newly installed editor of the *Jewish Messenger*, a retinue of alumni professors, and a raft of alumni clergymen.[1] Of the personages on stage, there was one in particular whom Rosenberg addressed, although he did not mention him by name: Professor John William Draper.

Four years earlier, Draper had published his famous *History of the Conflict between Religion and Science*, and Rosenberg spoke on "The Influence of Science on Religion." "True religion and science are never and can never be in conflict," Rosenberg told the crowd. "They are the two mighty forces that control civilization. True religion has nothing to dread from science. Nay, it even finds its strongest confirmation in scientific discovery." But what holds for "true religion," Rosenberg continued, cannot be said of the poor substitute often mistaken for religion: "Religion, depending on unauthorized tradition for its popular support, upon unfounded speculation and upon degrading ideas of the Almighty, alone dreads science. . . . For a thousand years the Church placed a ban upon science; yet science throve in secret. In vain did the Inquisition threaten the daring investigator; in vain did the fate of Giordano Bruno impend over his head. . . . Science did not assail the Church. She assailed merely the patristicism that it upheld. . . . [T]hese two mightiest of forces, science and revelation, unite all mankind in the grand chorus of "Glory to God in the highest and on earth, peace, good-will toward men."[2]

It was a good day for Rosenberg. He received a university fellowship and the prestigious Butler Prize, and was chosen to represent the university in an intercollegiate Latin contest. And Rabbi Abram S. Isaacs, who received an honorary doctorate on that same day, would publish Rosenberg's philosophical oration in the *Jewish Messenger.*[3]

That the *Jewish Messenger*, a traditionalist New York weekly, printed Rosenberg's excoriation of "patristic religion" (on page one) is less surprising than it may seem.[4] In Rosenberg's oration, one finds themes that were popular among American Jews in his day and for the next three-quarters of a century. Rosenberg asserted that in America a war raged between science and religion; however, it was a pointless war between science misapprehended and religion mispracticed. Beating drums for this war was the "Church," not in its generic sense, meaning all religion, but in a narrower sense of what Andrew Dickson White, the president of Cornell University and an impassioned advocate of secular education, was already then describing as repressive "Christendom."[5] Further, Rosenberg argued, if science and "patristicism" were at battle, science must win. These assertions were commonplace among Jews in Rosenberg's day and remained so for generations. They were held by "traditionalist" Jews (whom we would today call "Orthodox"), and they were held by "reformers."

Max Rosenberg's future was one of high respectability. He graduated from Columbia University Law School, became lead partner at a posh Jersey City law firm, and held a variety of honorific positions in society and the legal profession. Eventually, Rosenberg moved to an exclusive Central Park West address in Manhattan, where his neighbors included owners of department stores and factories. By the time he died in 1936, after being struck by a truck as he exited Hudson Terminal, Rosenberg had achieved influence, wealth, and respectability.[6] Rosenberg made his way in the world during years of enormous change for American Jewry, and for America itself. Over these years, many Jews registered, as Rosenberg had, that a boisterous and fearsome sort of Christians were at war with science. It was a fight that they followed warily but closely, because while they saw Jews as bystanders, the struggle between Christianity and science was a proxy war, standing in for a battle over whether America would become a place in which Jews could thrive. At stake in the "conflict between religion and science" that Draper had described was, in the eyes of Rosenberg and many other American

Jews of his day, neither religion nor science, but rather, as Rosenberg said, whether America could "unite all mankind" with "peace [and] good-will toward men."[7]

Witnesses to War

Fifteen years before Max Rosenberg's oration, and more than a decade before John Draper published his *History of the Conflict between Religion and Science*, the *Jewish Messenger* had already approvingly quoted Draper at length from a lecture he delivered before the New York Academy of Medicine in December 1863. In that lecture, Draper described the superstition that reigned in Christian Europe: "There were fairies in the moonshine, ghosts in the darkness, apparitions in the twilight, goblins in the kitchen, spectres in the garret." He also noted the horrid price that "cosmopolitan" Jewish physicians paid for the benightedness of their Christian patients, who saw in doctors' efficacy a "sinister" magic.[8]

By the time the *Messenger* quoted Draper, seeing conflict between religion and science as *Christian* conflict, misappropriated at times by misinformed Jews to tragic effect, was already a mainstay of traditionalist Jewish thought. Writing in 1867 in the traditionalist, Philadelphia-based Jewish monthly he founded and edited, the *Occident*, Isaac Leeser decried the circumstances of his day, in which some believed that science cast doubt on Judaism and others believed that Judaism cast doubt on science:

> Neither of the parties had, indeed, a correct appreciation of the value of science. The new men thought . . . that natural sciences and philosophy would themselves prove that the Bible taught erroneous things, and was therefore of no value and binding force. At the same time, those who wished to preserve Judaism dreaded to be brought in contact with an education which necessarily had such pernicious effects. . . . The evident errors of Christian dogma, which we have always rejected, these pretenders affected to think were inherent in Judaism also, and they thus made themselves merry over the simplicity of their fathers and less enlightened companions who still adhered to the requirements of Scriptural Judaism.[9]

Leeser had long argued that Judaism, unlike Christianity, was utterly at ease with science.[10] In an 1841 sermon titled "Religious Education," he insisted that "sciences can well be blended with the study of religion."[11] It was only when their faith was tinctured with Christianity, Leeser asserted, that Jews became convinced that their religion and rational science were at odds.[12]

This assertion fit well with Leeser's long campaign against Jewish reform. By arguing that what enmity existed between religion and science was, in fact, a product of false Christian dogma, and that Judaism and science had always coexisted peaceably, Leeser was able to bolster his contention that reform of Judaism was unneeded and ignorant. Unneeded, because traditional Judaism was by its nature well-suited to absorb all that was good about modernity, from belle lettres to "the progress of science."[13] Ignorant, because those who saw conflict between Judaism and science confused Judaism with Christianity. For Leeser, the reform project, then, was based on the same error that allowed Jews to think that their religion (like Christianity) was at war with science. In refashioning Judaism in the mold of American Protestantism (forsaking Hebrew prayer for English, celebrating the Sabbath on Sunday instead of Saturday, building organs in sanctuaries, and more), reformers would draw Jews *away* from rationality and toward dogmatism, *away* from peaceful embrace of science toward the belligerent obscurantism of dogmatic Christianity. The lesson Leeser learned from the warfare between Christianity and science was that Jews must stick steadfast with the Judaism of their fathers and grandfathers, which was at ease with science.

Reformers saw matters differently, of course, although some elements of Leeser's views recurred in their own. Isaac Mayer Wise—who, as a founder of the Reform rabbinical school, Hebrew Union College, and founding editor of the *Israelite*, was arguably the most influential Reform Jew in nineteenth-century America[14]—wrote in his newspaper in 1873 that Christianity is belligerent toward "philosophy, science, and criticism," while Judaism is "in profound peace" with them. Indeed, the "war of the Church against philosophy and science, as old as the Church herself," amounts to a battle "against the onward march of progressive and irresistible liberty in all parts of the civilized world."[15] Still, in its doomed war on science and philosophy, Wise argued, Christianity did not represent *religion*. Religion without dogma and without "priestly arrogance" would survive and thrive. "This is Judaism," Wise wrote. "The religion of future generations will be Judaism."[16]

In all this, Wise and Leeser agreed. What they disagreed about was whether Judaism required reform to keep from succumbing to the dogmatism that characterized Christianity at its worst. Leeser believed that reform would spur such dogmatism, not diminish it. Wise felt otherwise, believing that there was a rigidity to traditionalist Judaism that was ill-suited to America and ill-suited to modernity. The men agreed that there should be no antagonism between their religion and science; they disagreed about what sort of Judaism was needed to prevent such antagonism from taking root among Jews. Historian Marc Swetlitz smartly observed that disagreements about matters of science "among American Jews were often debates about the nature of Judaism"; in the case at hand, *agreement* about the easy compatibility of science and Judaism offered opportunity for debate about the nature of Judaism.[17]

At the end of the nineteenth century and start the twentieth, the differences between reformers and traditionalists grew greater and more varied. The great migrations of eastern European Jews brought to America, among others, observant Jews whose impact was polarizing. Influenced by the new arrivals, traditionalists became more orthodox, while reformers grew more adamant about the need for change. As historian Hasia Diner has explained, the immigrants represented "the demon feared by the reformers: an old-style, decidedly unmodern eastern European Judaism transplanted to America."[18]

It was in recoil from such old Jews that "radical reformers" like Rabbi Kaufman Kohler dominated the 1885 Pittsburgh Conference of Reform rabbis, determined to define, once and for all, the nature of their creed, and preaching, with rising urgency, the progressivism of Judaism and its compatibility with reason. Sharing Kohler's outlook, the *American Israelite* (the name was changed in 1874) decried the orthodox immigrant synagogues proliferating in Philadelphia for "fostering only superstitions and absurd notions and customs, and holding services which are far removed from dignity and propriety."[19] It was in hope of extinguishing superstitions and absurd customs that the "Pittsburgh Platform" declared that "all such Mosaic and rabbinical laws [that] regulate diet, priestly purity, and dress originated in ages and under the influence of ideas entirely foreign to our present mental and spiritual state." In the same spirit, the platform declared: "We hold that the modern discoveries of scientific researches in the domain of nature and history are not antagonistic to the doctrines of Judaism . . . [which is] a progressive religion, ever striving to be in accord with the postulates of reason."[20]

For reformers, retiring such superstitious practice as dietary laws, Hebrew prayer, and traditional Jewish dress was of a piece with acknowledging the compatibility of science and Judaism. For traditionalists, retiring age-old practices was an unnecessary apostasy because Judaism had always been compatible with science, and required no modernization. Kohler told his congregation that Jews of their day must choose one of two directions: "Which are we to espouse? The one that turns the dials of the time backward, or the one that proudly points to the forward move of history?"[21] Traditionalists insisted that they too pointed forward, without sacrificing Jewish practice. So it was that nearly all American Jews of the age concluded that their religion had always been, and remained, companionably disposed toward science. Like "the Crusades and the Thirty Years War, the Auto-da-fees [*sic*] and the Inquisition," the battle between religion and science was a Christian affair.[22] "Cultivated Europe should blush to her very finger-nails," wrote Emma Lazarus in 1883, "at her ignorance of [the] pacific and philosophical maxim hidden in the neglected lore of the Jewish sages . . . the harmony existing between sublimated Judaism and scientific philosophy."[23]

This was the context in which Jews praised the works of John Draper and, in time, Andrew Dickson White. Draper's *History of the Conflict between Religion and Science* was praised by the traditionalist *Messenger* for its "accurate scholarship," and for delivering a "crushing blow at the Papacy": "The errors and the crimes committed by Church and State in the interest of religion are pointed out so forcibly and so impartially, that the reader is irresistibly impelled to admit the full extent of the wrong done to civilization in impeding the pursuit of knowledge."[24]

These church wrongs, the reviewer continued, contrast with "the services rendered to science by the Jews at Alexandria, at Rome, in Spain, and in Arabia" that Draper recounts.[25] The paper soon reported that readers could find in *Popular Science Monthly* "President White's admirable address on 'The Warfare of Science,' whose thesis does not differ widely from Dr. Draper's in his recent book."[26] When White's own *History of the Warfare of Science with Theology in Christendom* was published, the *American Hebrew* praised it as an "interesting and most instructive book" and reviewed its two "principal theses"—that schools need to be free from church control and that science is not opposed to "religion," but rather "dogmatic theology": "As to the second thesis, however, we should suggest a slight modification. Neither religion nor theology in the true sense is opposed to science; and the adjective 'Dogmatic' used in the preface, and the qualifying words 'in Christendom' of the title show that the author himself recognizes that too sweeping an assertion has been made. . . . Jewish theologians have not been altogether innocent of opposition to science, but on the whole, as Dr. White's book makes clear, such opposition has been insignificant, and the influence of Jewish thought has been largely with the party of progress."[27]

Such was the dominant view among both traditionalists and reformers as the nineteenth century drew to a close: that the warfare between science and religion was waged by Christians on a Christian battlefield. While, on rare occasions, deluded and misguided Jews might stumble upon this battlefield, for the most part Jews peaceably, and without sacrifice of their pieties and principles, enlisted in the party of progress, the party of science.

The nineteenth-century responses to the putative conflict between religion and science soon served as patterns for similar responses in the first decades of the twentieth century. Pittsburgh's Rabbi Rudolph Coffee wrote in 1900 that during the Middle Ages, "all ideas and thoughts were sifted through the channels of the Established Church. . . . Science was utterly crushed, and the man who attempted to prove truths contrary to accepted teaching was either compelled to cease, as Roger Bacon, or put to death, as Bruno." The position of Jews "in this vexing question of religion and science" is clear: "They are not forces in direct antithesis to one another, . . . nor is it at all necessary to make a sharp line of demarcation between the two."[28] Two years later, Charleston Rabbi Barnett Elzas wrote, "Draper has written a work on 'The Conflict Between Science and Religion,' but his book discloses no such conflict. It merely describes the conflict between science and the Roman Catholic theologians—which is a very different matter. *Religion and science have never been in conflict* [emphasis in the original]."[29]

In 1908, the *American Hebrew* could declare flatly that "the 'Conflict between Religion and Science' has no terrors for us."[30] In 1912, Far Rockaway's Rabbi Ephraim Frisch wrote that "the warfare between religion and science, once so bitter, is over."[31] In 1915, Rabbi Nathan Kras of New York's Central Synagogue wrote that the "whole quarrel between religion and science is outside the walls of Judaism; for us there can be no conflict."[32] Yet despite the similarity between these statements, attitudes among American Jews toward science were changing. Like new wine in old vessels, the rejection of the notion of a war between religion and science came to take on new meaning and new urgency as the nineteenth century gave way to the twentieth.

Gilded-Age Greenhorns, Progressive Proletarians, and the Politics of Science and Religion

The last decades of the nineteenth century, and the first decades of the twentieth, were years of rapid and profound change for Jews in America. The immigration of millions of Jews was the greatest transformation, and an engine of further changes. To the approximately 230,000 Jews who lived in the

United States in 1880, almost two million immigrants had been added by 1914.[33] Before 1880, Jews did not aim to exert parochial influence on American politics and society.[34] With the rapid appearance of so many Jews, mostly from Russia and eastern Europe, whether they wished to be or not, Jews were increasingly seen and heard.

When writer Hutchins Hapgood began frequenting New York's down-and-out Lower East Side for his 1902 book about immigrant Jews, he found three types: old religious Jews seeking to recreate in New York what they had left behind in the Pale of Settlement, hustlers driven to make a good buck, and boisterous intellectuals, "the anarchists, the socialists, the editors, the writers; some of the scholars, poets, playwrights and actors, . . . the many men of 'ideas' who bring about in certain circles a veritable intellectual fermentation."[35] Hapgood's tripartite schema oversimplifies matters but captures something.[36] For most Jewish immigrants, finding economic footing was of first concern. But alongside economic drive, a great reformatory urge swelled within Jewish communities, finding expression in print, in the formation of Jewish social organizations, in unions, and in other efforts to change the status quo. As historian Hasia Diner explains, "Many American Jews—and not just the socialists and labor activists among them—advocated that the state enter into the economic life of the nation on behalf of the workers and the poor. . . . They lambasted the United States for the pervasiveness of racism and implored the country to live up to its creed."[37]

It is unsurprising that reform captured the imaginations and the evenings of poor and ambitious immigrants who still bore the imprint of the eastern Europe from which they had come, itself a dynamo tirelessly producing schemes for reform and revolution. It is in that context that one must understand the commitment of Jews of the day to the project of the "De-Christianization of American public culture," as historian David A. Hollinger has so memorably put it.[38] This project took a variety of forms. It was at this time, as Diner observes, that Jews "began to insist to the overwhelmingly Christian majority that it did not have the right to claim America as a Christian nation."[39] More practically, it expressed itself in efforts (as the 1913 charter of the then-new Anti-Defamation League put it) to "secure justice and fair treatment to all citizens alike and to put an end forever to unjust and unfair discrimination against and ridicule of any sect or body of citizens."[40] The American Jewish Committee lobbied the New York legislature to ban advertisement of hotels and other establishments that discriminated on the basis of race, color, or creed, and in 1913 a ban was made law.

Ad hoc organizations like Brooklyn's Jewish Protective Association arose to see that neighborhoods, like hotels, ceased to discriminate based on race, background, or belief.[41]

Soon, schools too became a tool for challenging Protestant hegemony and a perpetual site of conflict. Public schools were important because they were seen to offer a way out of poverty and a way into American boardrooms and clubrooms, serving as what historian Arthur Goren called "the bridge to the new society and the key to self-improvement."[42] For schools to do this, they had to be of a certain character. They had to be hospitable to Jews, and they had to train these Jews in the ways of America. Further, they had to teach an ideal, not just to Jews but Christian children to, of indifference to creed. Jews sought from public schools not just skills needed to get their own children good middle-class jobs, but also the ideological indoctrination of the children of their Christian neighbors, so that one day they would be willing to hire Jews. They looked to public schools not simply to remain mum about all matters of religion, but to advance the ideal that religion ought not have a place in public life. And it was for this reason that schools became and remained sites of continuing conflict. When Fred F. Harding, the principal of P.S. 144 in Brownsville, Brooklyn, told a 1905 Christmas assembly of his mostly Jewish charges that "you all have the feeling of Christ in you," he spurred Jews of all persuasions to protest.[43] In the following year, protestors returned to the Board of Education to challenge the celebration of Christmas in public school assemblies; when the board failed to act, twenty thousand Jewish children boycotted, staying home the day before Christmas, 1906.[44] Dozens of similar actions were launched with the grim certainty that Jewish children would find places as equals in the public schools only to the degree that expressions of Christianity—prayer, Bible study, religious homilies, celebration of Christmas and Easter, crucifixes and crèches—were kept out of the schools.

Disparate trends among the rapidly growing Jewish community—an urge for reform at the workplace and polling place, a vision of a public square free from religion, a wish to end discrimination and defamation, a desire for schools accessible to all—cohered into a complex of progressivist commitments. These commitments were shared (to varying degrees) by immigrant and native-born Jews, secular and religious,[45] Reform, Conservative, Traditionalist and Orthodox, Germans, Russians, Poles, and all the rest, by political radicals, liberals, and conservatives. Beneath the differences in background, position, language, ideology, belief, wealth, and prospects was a shared wish to establish for Jews an enduring place in America.

Science was part of the complex of progressivist commitments many American Jews shared. Mary Antin, who gained fame with her 1912 immigrant's autobiography, *The Promised Land*, wrote that "the Progressive movement is an attempt to record . . . the progress of humanity since the founding of the American Republic. Now the greatest achievements of this interval have been in the accumulation of scientific knowledge. But all this knowledge is useless until it is applied to daily life, both public and private. . . . The inspiration of the legislatures shall be the laboratory, not the counting-room. . . . The law of the land, in every instance, shall be the law of nature."[46] "The law of nature," Antin believed, would form a society more fair and decent than tradition-bound, discriminatory laws of man. There were reasons for this belief. Scientific studies produce a new sort of scholar and citizen, argued Yale physicist Henry Bumstead in 1910, speaking as if for science itself: "they think more of facts and less of words; . . . they are more careful to guard against being prejudiced by external circumstances and implications."[47] The new sorts of institutions, and new sorts of intellectuals, associated with science fit well with the progressivist views embraced by many Jews. Science came to be seen by many American Jews as advancing values they cherished. Science promised to refashion America, rattle the complacency of exclusionary elites, dissolve sectarianism, and expand universalist education. It promised to make fact and data the basis of social policy, rather than tradition and prejudice.

And it was in this context, that resistance to science—and surely any putative *warfare* waged by church zealots against laboratory savants—was taken as resistance to the changes that many Jews believed they needed to thrive in America. Warfare was still seen as a Christian affair, but now it was understood less as a matter of comparative theology than of the politics of the public square. Christian warfare against science was seen as a proxy war against the entry of Jews into American public life. Its soldiers counted nativists seeking to staunch the immigration of Jews from Europe, fundamentalists seeking to bring Christ into classrooms, segregationists seeking to maintain a caste system, and anti-Semites seeking to keep Jews at the margins of America: a sectarian coalition threatening to prevent Jews from thriving on American soil.

This was the perspective with which many Jews reacted to the Scopes trial. The *American Israelite*'s first report of the trial made no pretense of objectivity: "The extent to which ignorance, intolerance and sectional bigotry may flourish in this country is almost beyond belief."[48] On Rosh

Hashanah, 1925, rabbis in New York of all denominations attacked the trial in sermons.[49] Rabbi I. Mortimer Bloom best captured the prevailing mood:

> America, once a land of light and liberty, bids fair soon to be shrouded in a pall of ignorance and illiberalism that will extinguish the torch of learning and bring back the darkness of medieval night. Contemplate the multiplying efforts and proposals to compel Bible reading in the public schools, to outlaw the teaching of organic evolution in schools and colleges, to elaborate and make the Blue Laws more rigorous, to introduce censorship of the stage, the cinema, the novel. What are all these but opening guns in a cunningly contrived campaign to control education, to establish statutory morality, to convert government into the secular arm of the Church and institute a state religion with religious tests for office, for suffrage, for citizenship and, ultimately, for property-owning and for residence.[50]

Bloom saw in Scopes a war launched by church partisans against science and saw in this war on science the opening salvos of a wider war against all that is liberal and enlightened, high and holy in American life. Bloom feared that the Christian attack on science, if successful, could refashion America as a place in which Jews were denied a vote, a job, and a home.

This fear was not born in Dayton, Tennessee. In the years before Scopes, the position of Jews in the United States had grown stronger: their numbers increased, their wealth grew as many rose from hardscrabble immigrant privation to middle-class security, a complicated network of communal organizations took form, a cadre of extravagantly rich Jewish entrepreneurs devoted themselves to philanthropy and community service, and a small number of Jewish luminaries—men like Supreme Court Justice Louis Dembitz Brandeis, American Federation of Labor founder Samuel Gompers, and Pulitzer Prize–winning novelist Edna Ferber—signaled increased integration of Jews into American politics and society. During these same years, though, the position of Jews in the United States had also grown weaker. Six years before Scopes, Henry Ford bought a provincial weekly, the *Dearborn Independent*, and repurposed it as a populist voice of disgruntled nativism. Beginning in May 1920, the paper ran article after article about the woeful influence of Jews on America. It procured a copy of the notorious Russian anti-Semitic hoax, *The Protocols of the Elders of Zion*, and set out to Americanize it.[51] The *Independent*'s circulation rose to almost 700,000, an exorbitant reach in an American population that numbered 120 million,[52] and its essays about Jews were also collected in a popular book.

Contemporary Jewish leaders tried at first to ignore the paper, but found that they could not. Louis Marshall, the president of the American Jewish Committee, telegraphed Ford to protest "the insult, the humiliation, and the obloquy" inflicted by his paper.[53] Princeton historian William Starr Myers wrote in 1924 that Ford spurred the Ku Klux Klan: "[T]here is no doubt that the anti-Jewish campaign of Henry Ford and his *Dearborn Independent* is a direct cause of much of this growth of mental aberration."[54] Indeed, the Klan was suddenly everywhere. Two and a half weeks after the Scopes trial ended, thirty thousand Klansmen braved driving rain to march from the Capitol to the Lincoln Memorial, in Washington DC.[55]

The crass anti-Jewish agitation of the KKK was all the more fearful because it joined a more gentile anti-Semitism, of a sort Jews discerned in the growingly severe restrictions Congress placed on immigration. In 1917, it set in place literacy requirements. In 1920, the House of Representatives passed a bill (which never became law) banning all immigration for fourteen months. The following year, a bill was signed into law reducing immigration, and another, far stricter law (denounced by Louis Marshall as an embodiment of "the ideology of the Ku Klux Klan") was passed in 1924.[56] Jewish immigration from Europe dwindled to a fraction of what it had been only a few years before.

In the same years, many American Jews came to fear that public schools, too, were becoming less hospitable. In the year of the Scopes trial alone, legislatures in New York, New Jersey, California, Colorado, Ohio, Wisconsin, and Tennessee all acted to ensure and expand Christian religion in local schools. Various means were pressed into service: mandating or sanctioning religious study in the schools, banning evolution, introducing daily chapel, mandating the "Lord's Prayer," introducing formal Bible study, and more. As contemporary Jews saw it, all these measures stripped public schools of the nonsectarian public culture they avidly sought. Reform, Conservative, Orthodox, and secular Jewish leaders united to fight these religious initiatives. Louis Newman, who edited a 1925 book called *The Sectarian Invasion of Our Public Schools*, wrote: "The shrine of our common Americanism is the public school. Our children enter it not as Catholic, Protestant, Jew or unbeliever, but as Americans all. Nothing should ever be done, as the Bible amendment seeks to do, to split the student community into warring sectarian camps."[57] The united Orthodox Jewish Congregations of Cleveland issued a statement protesting "solemnly against the . . . attempts to divide the children of the public schools into separate racial, national or religious groups in which they belong."[58] Rabbi Stephen Wise, the founder of the "Free Synagogue" movement and a cofounder of the NAACP, dryly remarked that "when the Church infringed on the State the Jews would be the first to suffer."[59]

Concern that schools were becoming less welcoming to Jews was stoked by the constriction of Jewish entry into universities and professional schools, a reversal of two generations of rapid increase in the number of Jews in American higher education. When the twentieth century began, Jews accounted for half the student body of City College and Hunter College in New York. By the end of the Great War, almost nine in ten of the students in these schools were Jewish. Almost half the students at New York University were Jewish, a quarter at Columbia, a fifth at Harvard, and an eighth at Yale. As their numbers grew, complaints multiplied against the gruff incivilities of Jewish students, particularly the children of eastern European immigrants. A 1922 article in *Outlook* demanded that quotas be set to limit the damage:

> [I]t is one of the severest and most distressing tasks of college authorities today to exercise that discrimination which will keep college ideals and atmosphere pure and sound and yet not quench this eager spirit. In particular, among these alien youths—alien in spirit but not in body—are many who have their origin in eastern Europe, a majority of whom are Jews. The fact that in their endeavor to maintain their standards the wholesome discrimination exercised by college authorities may exclude a very large proportion of these particular aliens . . . ought to be understood as a natural and inevitable consequence of the immigrant tide. . . . [T]he effort to maintain standards against untrained minds and spirits is not oppression or prejudice.[60]

This was the attitude of most university administrators who adopted quotas limiting Jews. With the exceptions of City College and the University of Chicago, America's elite universities all established some sort of quota system. The percentage of Jews allowed admission differed greatly from university to university, with Columbia College allowing Jews to constitute 21 percent of its student body in 1918; Harvard, 12 percent; Yale, 7 percent; and Princeton, less than 4 percent.[61]

When the rabbis of New York arose on Rosh Hashanah in 1925 to defend John Scopes, they did so with a sick sense that a broad field of forces was arrayed against American Jews: Ford and his *Dearborn Independent*, Klansmen on Capitol lawn, sentry gates at Ellis Island, Bible-thumpers on school boards, tweedy university admissions officers, and more.[62] Common to all these was a shared view of American Jews as unpatriotic, unassimilable, uncivil, uncreative, noncontributing, and parochial. This was the view that the rabbis—and Jewish political leaders—were at pains to disprove. It is against the backdrop of their enduring efforts to refute this view that an emerging Jewish enthusiasm for science flourished, and it is against this background that it must be understood.[63] And it was against this backdrop that the fundamentalist attack on science they witnessed in Dayton was experienced by many Jews as an attack on their own, and on the conditions needed if they were to thrive, or even merely survive, in the United States.

War to End all Warfare: A Summary Reflection

In the years that separated young Max Rosenberg's chiding of John Draper from the stage of the New York Academy of Music from his violent end outside Hudson Terminal, Jewish perceptions of the ostensible "warfare" between science and religion had changed. Still, certain themes remained constant. The greatest of these themes was that such combat as there was, was someone else's affair. Most Jews who considered the question concluded that enmity between what is revealed in the laboratory and what is revealed in scripture might develop among Christians, but not among Jews. In the last decades of the nineteenth century, the relationship between science and religion was viewed, for the most part, as a theological matter. Jewish writers invoked the issue most often to differentiate their beliefs from those of the Christians among whom they lived. Occasionally, they did so to argue that liberal Christians, those who accepted science, were reframing their own religion in a more Jewish way (some even suggesting that in the epoch of science, Christians would return to the Judaism their forebears too hastily abandoned). As the nineteenth century moved toward its end, and the rift between Reform and traditionalist Jews grew greater, some Jews suggested that their own openness to the scientific theories of the day was a sign of the superiority of their own worldview over that of Jews who saw things differently. This argument was asserted, but only very rarely. Rarer still were outright rejections of science: one may scour the archives in search of a contemporary Jew who described his or her religion as being at odds with science, *tout court* and return empty-handed.[64]

With the great changes that American Jewry underwent at the end of the nineteenth century and the start of the twentieth—the mass immigrations from eastern Europe and all that came in their wake—the ostensible warfare between Christianity came to be viewed in a more political light. It was seen as part of a complex of fundamentalist urges—an expression of a roiling

Christian Id—to re-Christianize (or desecularize) America that expressed itself in rare, sporadic outbursts against science (of which the Scopes trial was a bracing example); in unending fights over prayer and Bible study in public schools; and in enduring efforts to tamp down immigration, enforce quotas, and otherwise preserve the Protestant atmosphere of colleges and universities and the workplaces of their graduates, courthouses, newspapers, radio, and almost every other aspect of civic life.

By this time, science had become associated in the minds of many American Jews with the America they longed for. This was an America of blind meritocracy, with a secular public square and secular public schools, and without discrimination based on background or belief. Many Jews saw any challenge to science as a challenge to the modernist, liberal, and democratic values that characterized the America they believed in, if not always the American they lived in.

Conclusion

It may seem odd that generation after generation of Jewish clergymen, journalists, and scholars wrote earnest essays insisting that, for Jews, war was unthinkable between their companionable religion and the science their people had admired for millennia, each time as though their insight was fresh. They did so, however, because the notion that Christians were engaged in a Verdun pitting their religion against their science, was *useful*. It was a metric, one of many, by which a Jewish minority distinguished itself from the Protestant majority within which they lived, and one by which they distinguished the Protestants with whom they could make common cause from those with whom they could not.

"Taste classifies," wrote sociologist Pierre Bourdieu, "and it classifies the classifier."[65] The penchant of American Jews for science, and their rehearsed recoil from those (real and imagined) who, in the name of religion, attacked science, were (among other things) acts of classification. They served to identify Jews with a complex of ideas that they and their contemporaries sometimes called "modernist" or "progressivist." They served to identify Jews with a broad cultural and political agenda that included secularizing American civic life; dismantling the sometimes-formal, sometimes-informal mechanisms of discrimination on the basis of race, ethnicity, and creed; opening America's borders to immigrants; establishing institutions (like public libraries) to help those of meager means but ample talents to succeed on their merits; loosing fetters on free expression; and more. Any effort to limit science in the name of religion was seen as an affront to this entire modernist complex of commitments.

For American Jews, then, the notion that war rages between religion and science was an idea that would not die, because it had become a symbol and signifier of cultural and political battles that remained very much alive, at least until after World War II, and retained urgency for Jews aching to be equal citizens on American soil.

NOTES

1. "Closing the College Year," *New York Times*, June 21, 1878.

2. Maximilian Theodore Rosenberg, "The Influence of Science on Religion," *Jewish Messenger*, July 12, 1878, 1.

3. "Closing the College Year."

4. The term "traditionalist" refers to what we today might call "Orthodox" or "Modern Orthodox." The denominations into which American Jews divided themselves have long been in flux and ill-defined at the edges. In nineteenth-century America, most Jews called themselves either Reform or Traditional. As the nineteenth century gave way to the twentieth, a variety of descriptors like "Orthodox," "Conservative," "Reconstructionist," "Secular," "Liberal," "Ethical Culture," and more, were pressed into service. This essay does not discuss Jewish immigrants to America who might be called "ultra-Orthodox" or *Haredi* (although neither of those appellations were in use during the period considered herein). This community was small in the first decades of the twentieth century, and its influence on organized Jewish life in America was limited. We still await a good scholarly account of their attitudes toward science and technology in the years before the Word War II.

5. Andrew Dickson White, *The Warfare of Science* (New York: Appleton, 1876), 52.

6. "M. T. Rosenberg Dies," *New York Herald Tribune*, April 9, 1936. "M. T. Rosenberg, 75, Is Dead," *New York Times*, April 9, 1936.

7. This chapter, like the book, focuses on the history of Jews in the United States. Although a great deal of what follows reflects circumstances that were unique to America, Jews in other places embraced and integrated science into their worldviews and theologies in ways that were similar to what is described here. For a discussion of some of these other places, especially Russia (and the Soviet Union) and Mandatory Palestine (and Israel), see Noah J. Efron, *A Chosen Calling: Jews in Science in the Twentieth Century* (Baltimore, MD: Johns Hopkins University Press, 2014).

8. "Literature and Art," *Jewish Messenger*, February 26, 1864.

9. Isaac Leeser, "Progress," *Occident*, 5, 24, no. 11 (February 1, 1867).

10. Lance J. Sussman, *Isaac Leeser and the Making of American Judaism* (Detroit, MI: Wayne State University Press, 1996).

11. Isaac Leeser, *Discourses, Argumentative and Devotional: The Subject of the Jewish Religion*, 2nd ser. (Philadelphia: C. Sherman, 1841), 3:230–31.

12. See, for instance, Isaac Leeser, *Discourses, Argumentative and Devotional*, 2 vols. (Philadelphia: Haswell and Fleu, 1836), 2:53; Isaac Leeser, *Discourses on the Jewish Religion*, 3rd ed. (Philadelphia: C. Sherman & Co., 1867), 6:282–83; Isaac Leeser, "Our Motto," *Occident and American Jewish Advocate* 1, no. 3 (June 1843): 110.

13. Isaac Leeser, "Scriptural Predictions," in *Discourses on the Jewish Religion*, 6:264.

14. Wise's influence is difficult to overestimate. As historian Hasia Diner put it, the period between 1855 and the 1880s "can be considered the 'age of Wise.'"; Hasia R. Diner, *The Jews of the United States, 1654 to 2000* (Berkeley: University of California Press, 2004), 122.

15. Isaac Mayer Wise, "The Religion of Future Generations," *Israelite*, November 28, 1873, 4.

16. Wise, "Religion of Future Generations," 4.

17. Marc Swetlitz, "American Jewish Responses to Darwin and Evolutionary Theory, 1860–1890," in Ronald L. Numbers and John Stenhouse, eds., *Disseminating Darwinism: The Role of Place, Race, Religion, and Gender* (Cambridge: Cambridge University Press, 1999), 234. Swetlitz shows that attitudes toward Darwin differed a great deal among Jews, and they changed with the passing of the years. But even among Darwin's most extreme Jewish critics, rejection of evolution rarely led to rejection of science.

18. Diner, *Jews of the United States, 1654 to 2000*, 123.

19. Maggid, "Correspondence: Philadelphia," *American Israelite*, February 26, 1886, 1.

20. Quoted in Michael A. Meyer, *Response to Modernity: A History of the Reform Movement in Judaism* (Detroit, MI: Wayne State University Press, 1995), 270.

21. Kaufmann Kohler, *Studies, Addresses and Personal Papers* (New York: Bloch Publishing, 1931), 203.

22. The events within the quotation marks are part of a litany of "horrors which fanaticism has inflicted upon man" cited by Isaac Mayer Wise. Isaac M. Wise, "The Eternal Progression of Humanity," *American Israelite*, August 21, 1874.

23. Emma Lazarus, "An Epistle to the Hebrews," *American Hebrew*, January 5, 1883, 88.

24. "Church and Science," *Jewish Messenger*, December 18, 1874, 4.

25. "Church and Science," 4.

26. "Literature of the Day: The February Magazines," *Jewish Messenger*, January 28, 1876, 5.

27. "Science and Theology," *American Hebrew*, December 11, 1896, 164.

28. Rudolph I. Coffee, "The Conflict between Religion and Science," *American Hebrew*, August 3, 1900, 318.

29. Barnett A. Elzas, "The Conflict between Science and Religion," *American Israelite*, January 30, 1902, 4.

30. "The Charity Jew," *American Hebrew & Jewish Messenger*, July 3, 1908, 200.

31. Ephraim Frisch, "The Position of Judaism in Present Day Religion," *American Hebrew & Jewish Messenger*, August 23, 1912.

32. "The Golden Gem of Judaism," *American Hebrew & Jewish Messenger*, August 13, 1915, 374.

33. Simon Kuznets, "Immigration of Russian Jews to the United States," *Perspectives in American History* 9 (1975): 35–124.

34. See, for instance, Mark Twain, "Concerning the Jews," *Harper's Magazine*, September 1899, 532–33. See also Hasia Diner, "The Encounter between Jews and America in the Gilded Age and Progressive Era," *Journal of the Gilded Age and Progressive Era* 11, no. 1 (January 1, 2012): 7.

35. Hutchins Hapgood and Jacob Epstein, *The Spirit of the Ghetto: Studies of the Jewish Quarter in New York* (New York: Funk and Wagnall's, 1902), 40–42.

36. Lucy S. Dawidowicz writes colorfully of "three types of greenies": the observant Jews, the *shvitsers* (who, with success, become complacent *alrightniks*), and the radical freethinkers. Dawidowicz, *On Equal Terms: Jews in America, 1881–1981* (New York: Holt, Rinehart and Winston, 1982), 47–54. Jonathan D. Sarna recently used the same categories in his magisterial *American Judaism: A History* (New Haven, CT: Yale University Press, 2004), 158–59.

37. Diner, "Encounter between Jews and America in the Gilded Age," 7–8.

38. In this, Hollinger rightly points out, Jews were hardly alone, finding common cause with Catholics (for "De-Christianization," in early-twentieth-century America, amounted principally to "De-Protestantization"), secular Protestants, many mainline Protestants, and "freethinkers" of a great variety of backgrounds. David A. Hollinger, "Jewish Intellectuals and the De-Christianization of American Public Culture in the Twentieth Century," in *Science, Jews, and Secular Culture: Studies in Mid-Twentieth-Century American Intellectual History* (Princeton, NJ: Princeton University Press, 1996), 17–41.

39. Diner, "Encounter between Jews and America in the Gilded Age," 8.

40. Leon L. Lewis and Sigmund Livingston, "Anti-Defamation League," *American Israelite*, November 20, 1913, 4.

41. John Higham, "Social Discrimination against Jews in America, 1830–1930," *Publications of the American Jewish Historical Society* 47, no. 1 (September 1, 1957): 16–18.

42. Arthur A. Goren, *New York Jews and the Quest for Community: The Kehillah Experiment, 1908–1922* (New York: Columbia University Press, 1979), 98.

43. "The Harding Case Decided: Board of Education Modifies Report of Local School Board—Mr. Harding Censured," *American Hebrew & Jewish Messenger*, June 22, 1906, 74.

44. Rafael Medoff, *Jewish Americans and Political Participation: A Reference Handbook* (Santa Barbara, CA: ABC-CLIO, 2002), 86.

45. It is worth noting that, in the first decades of the twentieth century, the great majority of Jewish *scientists* were themselves secular, and very few were Orthodox (a state of affairs that would change only after World War II). Many secular Jewish scientists, and many secular Jews more generally, shared with religious Jews of various sorts the impulse to distinguish a more "rational" and open-minded Jewish tradition, ostensibly at ease with science, from the more aggressively antiscientific views they observed among fundamentalist Christians.

46. Mary Antin, "A Woman to Her Fellow-Citizens," *Outlook*, November 2, 1912, 287.

47. H. A. Bumstead, "The Functions of a University Laboratory," *Science*, n.s., 31, no. 793 (March 11, 1910): 365.

48. "The Scopes Trial: Intolerance and Fanaticism Have Their Day in Court," *American Israelite*, July 16, 1925, 7.

49. For further detail, see Efron, *Chosen Calling*, 12–14.

50. Quoted in "Rabbis Assail Law against Evolution," *New York Times*, September 20, 1925.

51. Historian Norman Cohn wrote that Ford's publications "probably did more than any other work to make the *Protocols* world-famous." Cohn, *Warrant for Genocide: The Myth of the Jewish World Conspiracy and the "Protocols of the Elders of Zion,"* 2nd ed. (London: Serif Publishing, 2006), 159.

52. Victoria Woeste, *Henry Ford's War on Jews and the Legal Battle against Hate Speech* (Stanford, CA: Stanford University Press, 2012), p950.

53. Quoted in Robert S. Rifkind, "Confronting Antisemitism in America: Louis Marshall and Henry Ford," *American Jewish History* 94, nos. 1/2 (March 1, 2008): 72.

54. William Starr Myers, "Know Nothing and Ku Klux Klan," *North American Review* 219, no. 818 (January 1, 1924): 7. The link between antievolutionism and the Klan was emphasized in a 1926 article in the *American Israelite* called "Still Fighting Science." *American Israelite*, June 17, 1926, 4.

55. A connection between the Scopes trial and Klan power was made by many Jews of the day. See, for example, Harold Berman, "The Comedy Has Become a Tragedy," *American Israelite*, August 20, 1925, 1.

56. "Leading Jews Protest Immigration Restriction," *American Israelite*, January 10, 1924.

57. Louis I. Newman, "Vote 'No' on Amendment No. 17, " General Committee, District Lodge No. 4, Independent Order of B'nai B'rith, 1926, photocopy in author's possession.

58. Louis I. Newman, *The Sectarian Invasion of Our Public Schools, and Resolutions and Opinions Opposing the Miller Bill, No. 128 in the California Assembly and Similar Legislation throughout the United States* (San Francisco: N.p., 1925), 47.

59. In Newman, *Sectarian Invasion of Our Public Schools*, 46.

60. "Exclusion from College," *Outlook*, July 5, 1922, 407.

61. Dan A. Oren, *Joining the Club: A History of Jews and Yale*, 2nd rev. ed. (New Haven, CT: Yale University Press, 2001), 230; Robert A. McCaughey, *Stand, Columbia: A History of Columbia University in the City of New York, 1754–2004* (New York: Columbia University Press, 2003), 268; Heywood Broun and George Britt, *Christians Only, a Study in Prejudice* (New York: Vanguard Press, 1931).

62. All of these should be seen in the broader context of the rise of American Nativism, about which no better introduction exists than the justly canonical John Higham, *Strangers in the Land: Patterns of American Nativism, 1860–1925* (New York: Atheneum, 1963).

63. This enthusiasm for science also led to ever-growing numbers of Jews seeking to devote their careers and their lives to sciences. For a description of the Jewish entry into science, see "Holding High the Torch of Civilization: American Jews and Twentieth Century Science," in Efron, *Chosen Calling*, 12–38.

64. The archives produce a bounty of rejections by American Jews of Darwinian evolution, but none of these texts questioned the legitimacy of science.

65. Pierre Bourdieu, *Distinction: A Social Critique of the Judgement of Taste*, trans. Richard Nice (Cambridge, MA: Harvard University Press, 1984), 6.

CHAPTER ELEVEN

Muslims

M. ALPER YALÇINKAYA

Recent scholarship has provided students of science and religion with many new perspectives and case studies that highlight the shortcomings of the conflict narrative and the warfare metaphor. Among the key contributions to this scholarship have been works that contextualize dialogues about science and religion in terms of specific histories and geographies in a way that opens up new pathways for scholars by raising new questions: Is there one "conflict narrative" that simply travels from one location to another? Did the metaphors introduced by the polemical supporters of science John William Draper and Andrew Dickson White in nineteenth-century America inspire disputes in different parts of the world? Did different cultural traditions and social structures give rise to other ways of representing science and religion and evaluating the relations between them? Can there be a variety of conflict narratives rather than one? This chapter asks similar questions about the debates regarding the relationship of science and religion in Muslim societies and argues that while Muslim debates were influenced by European and American discourses, local conditions and concerns powerfully shaped them. The Muslim narratives about religion and science—even if they ultimately resembled those taking place elsewhere—were products of the interactions between international and local political conditions and intellectual climates.

Muslim societies are certainly familiar with narratives of conflict between religion and science. Indeed, these narratives have a peculiar and quite colorful history in the Muslim world, and the positions different parties have taken in the "science and religion" debates in Muslim societies are by no means mere mirror images of those taken by their counterparts in European or American cases. Ideas from Europe and the United States *were* introduced into the debates, but Muslims read those ideas from the point of view of the debates they themselves were having.[1]

Probably the fundamental difference between European or North American arguments on science and religion and those arguments that emerged in Muslim societies is that in the former there was little concern regarding authenticity. That is, even if a nineteenth-century European "man of science" claimed that religion and science were in conflict, he rarely, if ever, represented the conflict as between something native to European culture and something foreign to it. For Muslim authors of the same period, however, the question of native versus foreign influence was not only a very significant issue, it was *the* issue. If science had progressed so much in the Christian world, did that progress have to do with Christianity? Was Islam a religion that impeded progress? Would adopting the new sciences of the Europeans mean adopting a new identity and creed? Owing to the magnitude of these questions, the themes of authenticity and imitation were (and arguably still are) central to the debate in Muslim societies. Muslim authors rarely treated the "science and religion" question as primarily an epistemological one, and represented it, instead, as an overtly social and political issue.[2]

In this chapter, I focus on some of the positions adopted by participants in the Muslim debates on "conflict" in the late nineteenth and early twentieth centuries. As noted above, understanding the local context is crucial in the analysis of any "science and religion" debate, and it would be rather too simplistic to comment on the entire "Muslim world" in a brief chapter such as this. Therefore, I consider primarily the Muslims in the Ottoman Empire, but also refer to several examples from other predominantly Muslim societies. I discuss the stakes involved for those who made a case about the harmony or conflict between science and religion and the concerns that motivated them.

Contextualizing the Debates

In discussions about science and religion in the Muslim world, a popular approach is to list examples demonstrating that no conflict between Islam and science actually exists. An entire apologetic literature attests to this tendency, and we see such approaches in contemporary erudite academic studies as well. In the current global political setting where the "true nature" of Islam is a matter of much controversy, such responses are not particularly surprising. Studies written in this vein rightly caution against assuming that the conflict narrative that became so prominent in Europe should also sooner or later emerge in identical fashion in other parts of the world, no matter how different institutional and cultural contexts may be. But while scholars should not assume that the conflict narrative was an inevitable development in Muslim societies, neither should they assume that it was irrelevant to those societies. For many Muslims, the conflict narrative resonated precisely because it seemed to speak to important aspects of their cultural and institutional realities.

The "science and religion" debate evolved within Muslim societies in an era defined by the social and cultural changes brought about by European colonialism as well as the articulation of a global capitalist order. For many nineteenth-century Muslims, "the wonders of science" was but one of the many new themes of political and intellectual discourse, and the institutional changes associated with the new sciences were but part of much broader programs of structural transformation that colonialism and capitalism helped to effect. As a result, the "science and religion" debate emerged as a series of responses to these significant social and cultural changes in Muslim societies, changes that included new ideas on progress and the place of science within it, the emergence of new elites with a new type of cultural capital, and changes in class and status configurations, which in some cases were also related to ethno-religious identity.

The Intellectual Context: Progress, Islam and Science

Many Muslim travelers, bureaucrats, diplomats, military officers, students, and intellectuals visited and wrote about Europe in the nineteenth century. These Muslim visitors commonly maintained in their private or public writings that Europe owed its progress to scientific inventions and that Muslims had neglected science at their peril. Indeed, this discourse that posits humanity on a path of progress and defines each nation's (or civilization's) part in progress as its scientific contributions probably influenced the Muslim debate on science and religion more than any other. Muslim intellectuals were familiar with European works on the history of science (or civilization) written in the eighteenth and nineteenth centuries that highlighted these notions. Encountering such works precisely at a time when they themselves were interrogating the causes of the growth of European economic and military might at the expense of the Muslim world, Muslim readers developed ways of thinking about what their own ancestors had contributed to humanity.

It is important to note that a theme Muslim readers commonly encountered in European texts was that Islam was an obstacle against progress. Representations of Islam as superstition or a faith based on a lethargic acceptance of simple doctrines are not hard to come by in the literature of this period. Hence, a peculiar, but particularly strong version of the "conflict narrative" in European texts—indeed, one of the two "conflict narratives" that can be observed in the case of Muslim societies—primarily asserted that the cause of the decline of the so-called Orient was Islam's conflict with progress. As is well known, arguments about the supposed backwardness of Muslim societies, and in particular, the Ottoman Empire, were not only becoming popular in history books but also as a part of a defense of colonialism, often made in connection with disputes about the future of Ottoman territories in Europe and of European colonies with Muslim populations, such as India.[3]

But this was not the only strand in this discourse. As is also well known, the European historiography of civilization in this period frequently inquired about the missing link between the achievements of ancient Greece and Rome and those of the European Renaissance. Many authors argued that Muslims, by adopting and building on the legacy of the ancients, had made the awakening of Europe possible. Most authors favoring this view highlighted the importance of the Moors of Spain, who were, so to speak, the representatives of Islam in Europe.

In his *History of the Decline and Fall of the Roman Empire* (1776–89), Edward Gibbon notes the contributions of Muslims to astronomy, chemistry and medicine.[4] Similarly, Victor Cousin's *Cours de l'histoire de philosophie* (1829) devotes several pages to exalting the contributions of scholars like Avicenna and Averroës.[5] Alexander von Humboldt's *Cosmos* (1845–62) is filled with remarks about Arabs' contributions to science.[6] "The Arabians are, we repeat, to be regarded as the proper founders of the physical sciences," he asserts.[7]

A perhaps most instructive combination of these two strands of the progress narrative—Islam as obstacle and Islam as link—could be found in William Whewell's *History of the Inductive Sciences*: "The Arabs bring to the cultivation of the science of the Greeks their own oriental habit of submission, their oriental love of wonder; and thus, while they swell the herd of commentators and mystics, they produce no philosopher. Yet the Arabs discharged an important function in the history of human knowledge, by preserving and transmitting to more enlightened times, the intellectual treasures of antiquity."[8] Although several Ottoman texts refer to Gibbon and Cousin, it is undoubtedly difficult to know how familiar Muslim readers were with each of these texts. But we can safely assume that Muslim intellectuals who traveled to Europe gained acquaintance with such ideas, particularly if we take into account how commonly they interacted with Orientalists both in Europe and in the Muslim world. Indeed, many nineteenth-century texts by Muslim travelers allude to the idea that the contributions of early Muslim scholars identified by Europeans themselves proved that Islam was not, and could not be, in conflict with science.

One of the earliest Muslim writers to emphasize Muslims' contributions to science was Rifa'a al-Tahtawi (1801–73), an Egyptian intellectual and statesman who discussed his views on science in the travelogue he published after returning from Paris in 1831. In this work, which was also translated into Ottoman Turkish soon afterward, al-Tahtawi argued that Europeans were simply the current bearers of the torch of learning that had traveled from one civilization to another throughout history. Europeans had received it from Muslims, and Muslims were now to repossess it.[9] The Ottoman bureaucrat Mustafa Sami made similar arguments in his *Treatise on Europe*, which he published in 1840 after returning from diplomatic posts in Vienna and Paris. Sami stated that while it was true that science had progressed most rapidly in Europe in recent periods, the origins of that progress lay in the works Muslims had produced during Europe's Middle Ages. These sciences are our "true heritage," Sami wrote.[10]

A particular outcome of Muslims' encounters with the new European historiography of science was a rediscovery of early Islamic scholars. In a sense, nineteenth-century Muslim intellectuals reimagined the history of their societies with a new emphasis on the so-called contributions to civilization or, more specifically, to science. Names such as Ibn Sina (Avicenna), al-Farabi, and al-Haytham started to be mentioned commonly in texts in this new genre, and the Moorish kingdom in Spain was raised to almost mythic status, as it was a Muslim state within Europe and signified the zenith of Islamic contributions to humanity via science.

While the history of the Islamic state in Spain was rarely discussed in Ottoman texts until this period, it became a source of inspiration for Ottoman authors after the 1850s. Ziya Pasha (1825–80), an influential man of letters and a leading member of the Young Ottoman movement that pushed for a constitutional government in the Ottoman Empire, published sections from Louis Viardot's *Essai sur l'histoire des Arabes et des Mores d'Espagne* (1833), with additional material from other sources, under the title *The History of Andalusia* in 1859—a work containing a long section devoted to the Moors' contributions to science. Following the publication of this work, it became a commonplace rhetorical move to state that whoever mistook Islam for a religion that is in conflict with science should study the history of Islamic Spain.

The proclamations of the litterateur Namık Kemal (1840–88), another prominent member of the Young Ottomans, addressed to European audiences, is a case in point: "[Y]ou still declare our religion an obstacle to progress. . . . Wasn't it Islam that preserved the glories of civilization after the decline of the Romans? Wasn't it Islam that advanced and revived rational knowledge? Some wise men among you state 'The Arabs of Andalusia were the teachers of knowledge to Europe.' Weren't they Muslim?"[11] Kemal made similar points in many of his essays, especially in those he wrote for *Hürriyet*, the newspaper Young Ottomans published during their self-imposed exile in Paris. For instance, in one piece, after complaining that Ottoman bureaucrats hoping to gain the favor of Europeans had also started to believe that their own religion was the reason behind the Ottoman Empire's weakness, he stated, "[I]f the glory, prosperity and erudition of the Muslim world in Damascus, Baghdad, Egypt and Andalusia [were] taken into account," it would be impossible to argue that Islam was in conflict with progress.[12]

Hence, by the 1870s, the narrative about Islam's harmony with science and progress had stabilized. Muslim authors had access to a relatively well-defined set of ideas, names, dates, and arguments in order to "defend" their religion against European critics. Needless to say, this new apologetics also entailed constructing a new understanding of Islam itself—an understanding that portrayed Islam as a distinct civilization based on specific tenets, and with unique contributions to humanity.

This representation of Islam as a religion in harmony with science probably found its best articulation in the public dispute set off by Ernest Renan's (1823–92) lecture titled "Islam and Science." Indeed, the war of words that followed the lecture was perhaps the most widely popularized (and definitely the most commonly referenced) debate over Islam's relationship to science. A renowned French scholar and author of a study on the influence of Averroës on European thought, Renan gave his lecture at the Sorbonne, and his arguments reached a wide audience when the *Journal des débats* published the lecture on March 29, 1883. Renan started this short lecture by making it clear that his purpose was to demonstrate that there was no such thing as

"Arabic" or "Islamic" science, thus directly and deliberately challenging the harmony narrative that had become so dominant in Muslim societies. In fact, according to Renan, it did not require much work to demonstrate the conflict between Islam and science:

> Every person . . . sees clearly the actual inferiority of Mohammedan countries, the decadence of states governed by Islam, and the intellectual nullity of the races that hold, from that religion alone, their culture and their education. All those who have been in the East, or in Africa, are struck by the way in which the mind of a true believer is fatally limited, by the species of iron circle that surrounds his head, rendering it absolutely closed to knowledge. . . . This bent of mind inculcated by the Mohammedan faith is so strong that all differences of race and nationality disappear by the fact of conversion to Islam.[13]

In his writings, Renan commonly argued that Semitic races as a whole did not possess the qualities needed for intellectual creativity, and among Muslims he considered only Persians to have had the ability to contribute to science. Besides the racial reasons, this Persian science had been possible according to Renan because Persians were Shi'ites and therefore in some sense "immune" to Islamic orthodoxy. Thus, for as long as Arabs were the dominant group in the intellectual world of Islam, no contribution had taken place. So-called Islamic science had arisen only after the conquests, thanks to which Persians, Christians, Jews and members of the more skeptical groups within Islam had come to play a part in intellectual production, culminating in the translation of Greek works into Arabic. Yet this scientific accomplishment was a temporary aberration that could not be sustained in a society dominated by Islam, especially after "the birth of the European genius in the twelfth and thirteenth centuries." Renan continued, "The liberals who defend Islam do not know its real nature. Islam is the close union of the spiritual and the temporal; it is the reign of a dogma; it is the heaviest chain that humanity has ever borne. . . . Islam has been liberal in its day of weakness, and violent in its day of strength. Do not let us honor it then for what it has been unable to suppress."[14]

As France and French intellectuals constituted the chief source of inspiration for the new Muslim elites, and as many Muslim students were attending French schools and academies in this period, Renan's remarks garnered unprecedented attention. The best-known response it provoked is that of the activist intellectual Jamaluddin al-Afghani (1838–97). In his reply, which was published in the same journal as Renan's lecture and soon afterward, al-Afghani conceded that religions could indeed hinder scientific thought, and Islam was no exception. Yet he also argued that history demonstrated that after adopting Islam, Arabs had "developed, extended, clarified, perfected, completed and coordinated [the Greek legacy] with a perfect taste and a rare precision and exactitude."[15] What Muslims had to do, al-Afghani asserted, stating the fundamental principle of what later came to be known as Islamic modernism, was to revive this spirit and radically reform their understanding of Islam.

As the literary theorist Edward Said has noted, in al-Afghani's text we observe that "the native, using terms defined in advance by Renan, tries to 'disprove' the European's racist and culturally arrogant assumptions about his inferiority."[16] A desire to speak on the same level, within the same discourse as the European scholar, shaped al-Afghani's response. It is also in this sense that the debate over Islam's relationship to science cannot be properly understood without taking into account the context shaped by colonialism and orientalism.[17] The consequence of al-Afghani's admittance of the terms imposed by his adversary becomes all the more apparent when one examines Renan's reply to him in which he states, "Sheikh Jemmal-Eddin [*sic*] is an Afghan, entirely emancipated from the prejudices of Islam; he belongs to those energetic races of the Upper Iran, near India, where the Aryan spirit still flourishes so strongly, under the superficial garb of official Islamism. . . . [He] is the finest case of racial protest against religious conquest that could be cited."[18] Thanks to his arguments and style, al-Afghani proved to be a Muslim who had transcended Islam.

Undoubtedly, at stake in these disputes over Renan's lecture was not only the interpretation of Muslim history, but Muslims' very status in the present era. After all, Renan's contention that the Islamic religion precluded contemporary Muslim societies from progress was a legitimization of European domination of them. In his response to Renan, the Ottoman intellectual Namık Kemal emphasized that the dispute was at least as much about the future as about the past: "When the fact is that Europeans reached their current state by resisting the fiery persecution exerted by the clergy, and thanks to the Arab learning that they adopted, why does M. Renan deny us the chance to [make progress] by availing ourselves of the sciences of Europe? He should know very well that Muslims are obligated by their religion to educate themselves, and that among them there exist neither clerics who are enemies of learning nor institutions of Inquisition that would have scientific and philosophical works burnt."[19]

The effort to demonstrate to Renan and other critics the "virtues" of Islam can also be seen clearly in the work of the Indian intellectual Sayyid Ahmad Khan (1817–98). A well-educated man from a noble family, Khan sought for ways in which to promote cooperation between the colonial regime and the Muslims of India. Noting that Muslims were indeed in a wretched intellectual state, he argued that what needed to be done was to purify Islam of the superstitions by which it had been invaded and to restore "true Islam," which was entirely in agreement with reason and science. Khan's project was clearly one of objectifying Islam—transforming it into a well-defined, bounded entity abstracted from practice and history. Khan argued, "The person that states Islam to be true must also state how he can prove the truth of Islam. . . . [T]he only criterion for the truth of religions . . . is whether the religion [in question] is in correspondence with the natural disposition of humankind, or with nature." Thus according to him, the best way to understand religion was to look at it as an outsider and to see it as a complex of tenets the truth of which could be proven or disproven.[20] Islam was not "life itself"; it could be seen neither as a set of practices nor as "mere belief." Islam had to be plausible and proven to the "third person," that is, to the colonizer or the native students who were trained in Western schools opened after colonization.

Thus, for Muslim intellectuals, challenging the Eurocentric argument about the conflict between Islam and science was a matter of finding room for their ancestors in the so-called history of humanity and, perhaps more important, demonstrating their "right" to exist in the contemporary world. Just as urgent, however, was the task of preventing new generations from feeling alienated from their own community, a subject to which this chapter now turns.

The Social Context: Conflict between "Men of Science" and Society?

These intellectual encounters with orientalist perspectives, the Eurocentric history of human progress, and arguments about the conflict between Islam and science helped produce popular representations of Islam as a pro-science religion (which, perhaps, needed only to be reinterpreted). But equally important, local political and social changes of significance also led to the emergence of a different set of ideas about potential conflicts having to do with science and religion—namely, the narrative that presents Muslim men of science as people potentially in conflict with their own society. Although less commonly studied, this discourse constitutes a distinctive and very prominent conflict narrative in Muslim societies. The context that produced this narrative is of course multilayered, but three important transformations stand out.[21]

Among the most consequential changes that Muslim societies experienced in the nineteenth century was the emergence of new types of educational institution—schools where the new sciences were taught to Muslim youth. In addition to learning science, the young men attending these schools typically also learned French, and many of them also received some training in Europe and/or worked as diplomats and bureaucrats in European capitals. Because of their knowledge of the French language and their exposure to European society, a number of these students developed tastes, interests, and attitudes that distinguished them from more traditionally educated elites. Using a sociological term, it can be argued that these men had a new type of *habitus*—embodied dispositions that generate practices and understandings. This differentiation gave rise to a split in the bureaucratic middle class, as the case of the Ottoman Empire illustrates.[22] The new *habitus* in question involved not only some familiarity with the new sciences and, generally speaking, "European ways," but also tastes and lifestyles that resembled those of the non-Muslim members of the Ottoman middle class and of the Europeans living in Istanbul.

This was a significant development at a time when status concerns were heightened among Muslim Ottomans as a result of several factors. For one, some members of non-Muslim communities were able to reap benefits from the articulation of the empire's economy into the global capitalist system, while Muslim Ottomans' participation in the new economic order remained disproportionately low. Furthermore, the strategic support that the Great Powers offered to non-Muslim Ottomans further created a wedge between Muslims and non-Muslims, and led to perceptions within the Muslim community that the Ottoman state was but a puppet of European powers, and oblivious to the concerns of the Muslim community. This reaction was directed primarily at the new bureaucrats who came to be commonly represented as Muslim Ottomans who had abandoned their ties to the Muslim community and become imitators of Europeans.

An additional transformation that contributed to the development of this narrative about "alienated elites" was the growing influence of missionary schools across the empire over which the Ottoman state had little control. These schools tended to be better organized and equipped than Ottoman schools and gradually acquired significant popularity, attracting Muslim youths and leading to concerns within the Ottoman government. Particularly in the 1880s and 1890s, Ottoman bureaucrats produced numerous reports expressing their anxiety about the potential "corruption" of Muslim youth in these schools.

The heightened tension within the Muslim middle class together with the Ottoman state's concerns about the loyalty of the educated youth gradually gave rise to a discourse that cast young men trained in the new sciences as alienated from their own society, hostile to its beliefs, unlikely to defend the interests of Muslim Ottomans, and, in sum, as a potential liability to the Ottoman throne and its security. *This* is the conflict that Ottoman texts, especially those written after the 1860s, alluded to.

A case in point is an article published in an issue of the journal *Kasa* that was devoted entirely to mocking the popular scientific journals published in late-nineteenth-century Istanbul. Criticizing the "arrogance" of the editors of such journals, the writer proclaims:

> While our youths try to learn clownishness at balls, dancing in theaters, the latest fashions of Paris, *alla franca* haircuts, . . . the qualities that adorn the moral virtues of the Ottomans are lost entirely. It is true that learning is the water of life for any nation. . . . [But] we want to progress in the knowledge of civilization while maintaining the morals of our community; we do not need the vileness and degeneracy of Europe in the guise of civilization. We sincerely hope that we will see in our country a lot of progress with respect to the philosophy and experiments of the natural sciences, . . . roads, railways, new weapons; but if our fops will excuse us, we do not want to and will never see the . . . balls, dances, hubbubs and fashions of Europe. . . . We will progress within Islam, we will not be Franks in fezzes.[23]

In his well-known poem, Ziya Pasha makes a similar point:

> The zealous man is now accused of fanaticism;
> Finding wisdom in the irreligious, this is now the fashion
> Islam, they say, is what prevents the progress of the state;
> This story was not known before, now it is the fashion.
> Forgetting our religious loyalty in all our affairs,
> Allegiance to Frankish ideas is now the fashion.[24]

An example that is more direct in its references to science and religion concerns a series of articles written by the prolific author Ahmed Midhat (1844–1912) in which a broadly materialistic theme can be detected. In these articles, which he published in his journal *Dağarcık* (The knapsack) in 1872 and 1873, Midhat discusses how matter cycles continuously in nature, blurring the distinction between the organic and the inorganic. Additionally, he refers to Lamarckian ideas on evolution, and while he does insist that these ideas could not explain how humans had acquired their current form, his discussion on the topic does not include references to a Creator. The response to these essays came from a religious cleric—as may have been expected—but not in a form that can be reduced to a simple "science versus religion" framework. Harputlu Ishak (1803–92), the cleric, wrote in his response that authors like Midhat are reminiscent of those who had participated in the Paris Commune—the enemies of order and tradition that even European states themselves had been trying to expel. Discussing these issues is thus to Ishak ultimately a matter of defining good citizenship and patriotism: "This country's thirty-five million People of the Book, Muslim and non-Muslim, . . . the Protector of the state and religion and the possessor of the throne of the Caliphate, and the great men of the state share the same view and faith with us, whereas [Midhat] has nothing other than his journal and his walking stick."[25]

Fifteen years after this debate between Midhat and Ishak, a much more heated and famous controversy took place. This time the protagonist was a young military officer named Beşir Fuad (1852–87), who published numerous pieces arguing that science was incomparably superior to any other activity and that Ottoman history contained very few names of which Ottomans could be proud in this respect. In the ensuing debate, which escalated rapidly, Fuad's opponents accused him of being a naive materialist and a simple worshipper of anything European. Especially after he committed suicide in the midst of the debate, Fuad was transformed into the symbol of all that was wrong with the new science enthusiasts. An article that framed Fuad's suicide as an act caused by weak morals stated, "If a man of science is regarded by the public as immoral, it does not matter in the least whether that person knows the sciences or not."[26] Ahmed Midhat, who had, in a sense, learned his lesson from his own earlier controversies and become an unofficial representative of the throne in the Ottoman press, contributed to the villainization of Fuad. In a series of essays, he argued that although Fuad was a very intelligent and promising youth, he read only European works and was ignorant of the Islamic legacy and the Qur'an. Thus, Midhat claimed, Fuad had failed to "write about things that would be in harmony with the philosophy as well as interests of Muslims and the Ottoman Empire."[27]

The "dangers" that materialist science aficionados posed to social order was not a topic discussed only in Istanbul. Jamaluddin al-Afghani condemned "materialist" authors influenced by Darwin's ideas in his highly popular writings that circulated throughout the Muslim world. While his own attitude toward Darwinism would get increasingly more sympathetic near the end of his life, al-Afghani accused authors such as Sayyid Ahmad Khan and Shibli Shumayyil (1850–1917), a physician and popular author who had translated Ludwig Büchner's lectures on Darwin into Arabic, of blindly imitating the West and potentially weakening the bonds among Muslims. For Shumayyil, who was a Christian Arab, Darwin's and Büchner's views were important precisely because they indicated the need to limit the role of religion in public life and for creating a new civic identity that would unite all Ottoman citizens regardless of faith. In other words, while al-Afghani and Shumayyil were on different sides of the debate over materialism, both men's concerns about science and religion were closely linked with political projects and social issues.[28] "Science and religion" was not an issue about an unavoidable existential clash between alternative views on nature; it was about broader social and political concerns.

Harmony: Universal Science, Muslim Scientists

The culmination of these debates over the relationship between Islam and science was the formation of a discourse that accentuated harmony to unprecedented degrees. European claims that there was no room for Muslims in the civilized world, growing missionary influence, the colonial threat, and finally, the anxiety regarding the affiliations and loyalties of new generations transformed the idea of a conflict between science and Islam itself into a menace to be combated.

The Ottoman sultan Abdulhamid II (reigned 1876–1909) was particularly concerned with the loyalty of the educated youth and encouraged the publication of works that demonstrated that Islam and science were in harmony. The Lebanese theologian Hussein al-Jisr (1845–1909) was rewarded by the sultan in 1888 for his *Hamidian Treatise on the Truthfulness of the Islamic Religion and the Praiseworthy Path of Muhammad*, which made the argument that Islam, unlike Christianity, was in agreement with scientific findings. Similarly, Ahmed Midhat, once again writing as the sultan's mouthpiece, produced a particularly significant work in this genre of harmony: the translation of Draper's *History of the Conflict between Religion and Science* into Ottoman Turkish (four volumes, 1895–1900). In this work, Midhat not only translated Draper's text, but inserted his own lengthy comments on Draper's arguments.[29] According to Midhat, it was possible to interpret Draper's work as a solution to the pressing problems of the empire. If read "correctly," Draper's work would show "confused men of science," such as Beşir Fuad, that Islam was the most scientific of religions and thereby prevent them from falling under the influence of missionaries or European thinkers. In Midhat's view, a proper understanding of Draper provided a path for creating a docile, patriotic youth. It was only such a youth that could be trusted at a time when the empire and the entire Muslim world were under immense European pressure.

Midhat was not necessarily misrepresenting Draper's arguments. Indeed, many of Draper's remarks about Islam and Christianity resembled beliefs commonly held by Muslims or that fit very well with the discourse about the harmony between Islam and science. For instance, Draper emphasized repeatedly that Christianity had embraced pagan ideas while Islam had remained intact—a belief central to Muslim faith itself. Draper also noted that Muslims' fatalism had enabled them to develop the idea of laws of nature much more easily than Christians, who trusted far more in miracles. According to Draper, Islamic philosophy had transcended, especially in the works of Averroës, anthropomorphic conceptions of God and even developed approaches resembling evolutionary thinking. As a result, the staunchly anti-Catholic Draper found it apt to refer to Islam as the "Southern Reformation." For Midhat, the conclusion to be drawn from Draper's remarks was simple: when they adopted a "clear and impartial" perspective, even European (or American) scholars could see that while science and Christianity were undoubtedly in conflict, Islam was a religion in harmony with science.

Reading the Qur'an with the purpose of demonstrating its harmony with science became increasingly popular in this period, giving rise to what is commonly referred to as scientific exegesis—an approach that still has many adherents in the Muslim world. Another modernist, the Egyptian author Muhammad Abduh (1849–1905), authored perhaps the best-known work of scientific exegesis, in which he analyzed Qur'anic verses as expressions of scientific facts. He argued that passages on the "jinn" (invisible beings created by God), for instance, could essentially be about microbes. Similarly, he interpreted the way the Qur'an describes the emergence of humans as in harmony with recent embryological findings.[30] This hermeneutical approach would also be taken up by the Kurdish theologian Said-i Nursi (1877–1960) and, later, by the French physician Maurice Bucaille (1920–98).

Yet another approach that emerged in the late nineteenth century and remains popular gave rise to the literature on the "scientifically proven benefits" of Islamic rituals. The microbiologist Huseyin Remzi (1839–96) and the physician Mehmed Fahri (1860–1932) were among the Ottoman authors who wrote on how fasting and praying as Islam instructed had health benefits that contemporary medicine had made manifest. In fact, even Harputlu Ishak—the staunch critic of Midhat's essays on evolution—used this approach in his book against missionaries and argued that fasting and ablution, contrary to the claims of Protestant missionaries, were beneficial for health as well.[31]

Conclusion

For the Muslims of the late nineteenth and early twentieth centuries, the political nature of the "science and religion" debate was simply too obvious to ignore. The idea that the new era, and the right to dominate the world, belonged to those who contributed to science (as defined by European discourses on progress) made the question of "Islam and science" vital for Muslim intellectuals. This not only meant the adoption of the terms of the European debate, but a far-reaching reimagination of what Islam itself was. Regardless of sectarian background—al-Jisr and, most likely, al-Afghani had Shi'i origins, while Abduh and Midhat were Sunni, for instance—these intellectuals imagined Islam as a discrete, unified entity that was already in existence in perfect form or that needed to be reinterpreted, but they nevertheless shared the assumption of its essential harmony with science. In the meantime, as a result of the anxieties highlighted above, an entire apologetic literature arose, composed of works written with not only European critics or missionaries but also young Muslim readers in mind. The task for many a Muslim author was to demonstrate to all of these groups that Islam was in harmony with science and that stating the opposite could not but lead to the annihilation of Muslim societies. The power of Islam as the cement holding together a people whose destiny seemed increasingly precarious remained unchallengeable, even in the eyes of authors who at times came close to arguing that Islam had, indeed, impeded scientific progress, like al-Afghani or the Ottoman intellectual and political activist Abdullah Cevdet (1869–1932).[32] As a result, conflict in the epistemological sense was rarely the dominant theme in the Muslim world. Instead, the debate on science and Islam turned inward and became one about the lifestyles, interests, and loyalties of young generations of educated Muslims. Emphasizing that Islam and science were in harmony was a way to discipline the young Muslim men of science. The conflict was social through and through.

NOTES

1. James Secord, "Knowledge in Transit," *Isis* 4 (2004): 654–72; David Livingstone, "Science, Text and Space: Thoughts on the Geography of Reading," *Transactions of the Institute of British Geographers* 4 (2005): 391–401.

2. John Evans and Michael Evans, "Religion and Science: Beyond the Epistemological Conflict Narrative," *Annual Review of Sociology* 34 (2008): 87–105.

3. Consider the arguments of the historian E. H. Nolan: "The limitation of Mohammedan power and influence must be one of the results of the reconstitution of British authority in India, and such a change must affect the whole social condition of that country. Mohammedanism and a high degree of civilisation cannot co-exist among the same people. The Koran is not only the Bible of the Mussulman, it is his book of science and of government. . . . On all scientific subjects its contents are absurd, puerile, and superstitious. . . . It is true that when the light of science does find entrance to the mind of the Mohammedan his religion is destroyed, for if the Koran be confuted in one point, it is confuted in its entirety. Infallible in its pretensions on all subjects, as soon as it is found to be in error, its authority perishes. The public schools in India, and the missionaries, have infused just philosophical notions among the better classes of Mohammedan youth, and where this has been the case they have invariably become sceptics to their creed. A perception of this fact has roused the fanaticism of all Mohammedan India against the English." Nolan, *History of the British Empire in India and the East* (London: James Virtue, 1858), 1:iv.

4. Edward Gibbon, *The History of the Decline and Fall of the Roman Empire* (London: Cadell and Davies, 1813), 10:47–50.

5. Victor Cousin, *Cours de l'histoire de philosophie* (Paris: Pichon et Didier, 1829), 1:352–54.

6. Alexander von Humboldt, *Cosmos: A Sketch of the Physical Description of the Universe* (London: Longman, 1848), 2:200–228

7. Humboldt, *Cosmos*, 2:212.

8. William Whewell, *History of the Inductive Sciences: From the Earliest to the Present Time* (Cambridge: John W. Parker, 1837), 1:364–65. See also Lucien Leclerc, *Histoire de la médecine arabe* (Paris: Ernest Leroux, 1876), 1:1–4; Gustave Le Bon, *La civilization des Arabes* (Paris: Firmin-Didot, 1884).

9. See John W. Livingston, "Western Science and Educational Reform in the Thought of Shaykh Rifaa al-Tahtawi," *International Journal of Middle East Studies* 4 (1996): 543–64. Note that, as Livingston also underlines, al-Tahtawi also emphasized that Muslims needed to be very cautious during this process in order to be able to preserve their identity.

10. Mustafa Sami, "Avrupa Risalesi," in Fatih Andı, ed., *Bir Osmanlı Bürokratının Avrupa İzlenimleri: Mustafa Sami Efendi ve Avrupa Risalesi* (Istanbul: Kitabevi, 1996), 80–81.

11. Namık Kemal, untitled article, *Hürriyet* 11 (20 Cemaziyelevvel 1285 / September 7, 1868): 8.

12. Namık Kemal, "Âzâr-ı mevhume," *Hürriyet* 35 (10 Zilkade 1285 / February 22, 1869): 8.

13. Ernest Renan, "Islamism and Science," in Bryan Turner, ed., *Readings in Orientalism* (London: Routledge, 2000), 1:200.

14. Renan, "Islamism and Science," 209. For a comprehensive study of the intellectual milieu that Renan was addressing, see B. Harun Küçük, "Islam, Christianity, and the Conflict Thesis," in Thomas Dixon et al., eds., *Science and Religion: New Historical Perspectives* (Cambridge: Cambridge University Press, 2010), 111–30. On Muslim responses to Renan, see also Cemil Aydın, *The Politics of Anti-Westernism in Asia: Visions of World Order in Pan-Islamic and Pan-Asian Thought* (New York: Columbia University Press, 2007), 47–52.

15. Sayyid Jamal al-Din al-Afghani, "Answer to Renan," in Charles Kurzman, ed., *Modernist Islam: A Sourcebook* (New York: Oxford University Press, 2002), 109.

16. Edward Said, *Culture and Imperialism* (New York: Alfred A. Knopf, 1993), 263.

17. In particular, orientalism as Said defines the term: ". . . the corporate institution for dealing with the Orient—dealing with it by making statements about it, authorizing views of it, describing it, by teaching it, settling it, ruling over it: in short, Orientalism as a Western style for dominating, restructuring, and having authority over the Orient." Edward Said, *Orientalism* (New York: Vintage, 1979), 3.

18. Ernest Renan, "Appendix to the Preceding Lecture," in Turner, ed., *Readings in Orientalism*, 213.

19. Namık Kemal, *Renan Müdafaanamesi* (1910; Ankara: Güven, 1962), 56.

20. Sayyid Ahmad Khan, "Lecture on Islam," in Kurzman, ed., *Modernist Islam*, 295–96.

21. I discuss these transformations and their consequences in Yalçınkaya, *Learned Patriots: Debating Science, State, and Society in the Nineteenth-Century Ottoman Empire* (Chicago: University of Chicago Press, 2015).

22. See Carter Findley, *Bureaucratic Reform in the Ottoman Empire: The Sublime Porte, 1789–1922* (Princeton, NJ: Princeton University Press, 1980); and Findley, *Ottoman Civil Officialdom: A Social History* (Princeton, NJ: Princeton University Press, 1989).

23. "Tiyatro ve Ahlak," *Kasa* 3 (1291/1874–75): 53–55. Unless otherwise noted, all translations of passages quoted in the text are mine.

24. Ziya Pasha, "Terkib-i Bend," in Kenan Akyüz, ed., *Batı Te'sirinde Türk Şiiri Antolojisi*, 3rd ed. (Ankara: Doğuş, 1970), 48.

25. "Varaka-i Cevabiyye," *Basiret* (9 Muharrem 1290 / March 9, 1873): 2. Note that walking sticks were associated with the "dandies" of Istanbul.

26. "Terbiye—Tahsil," *Tercüman-ı Hakikat* 2621 (11 Cemaziyelahir 1304 / March 7, 1887): 2.

27. Ahmed Midhat Efendi, *Beşir Fuad* ([1305/1888]; Istanbul: Oğlak, 1996), 14–15.

28. Marwa Elshakry, *Reading Darwin in Arabic* (Chicago: University of Chicago Press, 2014), esp. chap. 3.

29. M. Alper Yalçınkaya, "Science as an Ally of Religion: A Muslim Appropriation of the 'Conflict Thesis,'" *British Journal for the History of Science* 44 (2011): 161–81.

30. On Abduh and other authors who wrote this type of exegeses, see Elshakry, *Reading Darwin in Arabic*.

31. Harputlu Ishak, *Cevab Veremedi [Diyâ-ül Kulûb]*, 36th ed. (1876; Istanbul: Hakikat, 2009), passim. This annotated edition also contains the following sentence, possibly added later: "I wonder if [missionaries] really believe that dirty water used in baptisms doesn't transmit diseases, and that, in contrast, Muslims' fasting and ritual ablution are unhealthy. Or are they making such unscientific, irrational, ugly allegations because of the money they are paid by Protestant associations?" (236).

32. On Cevdet, see Şükrü Hanioğlu, "Blueprints for a Future Society: Late Ottoman Materialists on Science, Religion and Art," in Elisabeth Özdalga, ed., *Late Ottoman Society: The Intellectual Legacy* (London: Routledge, 2005), 28–116.

CHAPTER TWELVE

The New Atheists

RONALD L. NUMBERS AND JEFF HARDIN

In recent years, especially since the September 11, 2001, attacks on the United States by Islamic terrorists, no group has more loudly insisted on the warfare between science and religion than the so-called New Atheists.[1] In stark contrast to "Neo-Harmonists," who have argued for harmony between science and religion (see chapter 13 in this volume), the New Atheists have been uncompromising critics of religion, proclaiming that religion is the implacable foe of science. Led by the "big four"—Sam Harris, Richard Dawkins, Daniel C. Dennett, and the late Christopher Hitchens, each of whom wrote at least one best-selling book—from time to time the New Atheists have also asserted that religion has *always* impeded the progress of science. Significantly, they rarely explore those past relations. Given their vehement indictment of religion, their indifference to the history of the relationship between science and faith is a curious omission and one that has not gone unnoticed by their critics.

Intellectual Forebears of the New Atheism

The backdrop to the New Atheism is rooted in the nineteenth and early twentieth centuries and is certainly influenced by the spirit of, if only infrequently by explicit reference to, the chief architects of the "warfare thesis," John William Draper (1811–82) and Andrew Dickson White (1832–1918). Most leading atheists of the late nineteenth and early twentieth centuries, including such thinkers as Friedrich Nietzsche (1844–1900) and Bertrand Russell (1872–1970), rarely addressed the historical relations between science and religion. A notable exception among the atheists was the apostate Catholic priest Joseph McCabe (1867–1955), who went on to become a famous atheist. In his mass-produced booklet *The Conflict between Science and Religion* (1927) he optimistically reported that "religion is in decay all over the earth."[2]

In the late twentieth century, the foremost scientific popularizer—and promoter—of the conflict thesis was the American astronomer and secular humanist Carl Sagan (1934–96). Sagan opened his well-known Public Broadcasting System television show *Cosmos*, first aired in 1980, with the memorable quip, "The cosmos is all there is or ever was or ever will be." However, as Stephen P. Weldon has pointed out, despite Sagan's repeated portrayals of a war between science and religion, he did not think that science conflicted with all forms of religion, "only with a certain type of religion: fundamentalism and orthodoxy. Religious beliefs based on rationalism and humanism could be very good. They were the sources of our inspiration." Indeed, in *The Demon-Haunted World* (1996), published the year he died, Sagan presented a softening view of religion: "Of course many religions, devoted to reverence, awe, ethics, ritual, community, family, charity, and political and economic justice, are in no way challenged but rather uplifted, by the findings of science. There is no necessary conflict between science and religion. . . . The religious traditions are often so rich and multivariate that they offer ample opportunity for renewal and revision, again especially when their sacred books can be interpreted metaphorically and allegorically. Conflict was thus common but not necessary."[3]

Other famous atheists of Sagan's generation, such as the molecular biologist Francis Crick (1916–2004) and the cosmologist Stephen Hawking, likewise fostered acceptance of unbelief in the public sphere. The only historian of science to play a prominent role in the promotion of atheism was William B. Provine (1942–2015), a distinguished professor at Cornell University and an expert on evolution. "Let me summarize my views on what modern evolutionary biology tells us loud and clear," he once announced loudly and clearly. "There are no gods, no purposes, no goal-directed forces of any kind. There is no life after death. . . . There is no ultimate foundation for ethics, no ultimate meaning to life, and no free will for humans, either." Though evangelistic in his promotion of atheism, he never imposed his antireligious views on his history, and Provine himself maintained a cordial personal relationship with a well-known proponent of Intelligent Design, Phillip Johnson during frequent speaking engagements together.[4]

Implicit Rather than Explicit: The New Atheism and the Conflict Thesis

While these forerunners may be regarded as "spiritual advisers" to the New Atheists, most of the New Atheists themselves simply assume that science and religion, past and present, are at war. Significantly, most of them ignore the alleged historical conflict in any explicit sense. The most famous of the leading four, the British evolutionary biologist and longtime atheist Richard Dawkins, years ago diagnosed religious faith "as a kind of mental illness" and described religious faith as "one of the world's great evils, comparable to the smallpox virus but harder to eradicate." In *The Blind Watchmaker* (1986) he stressed the role of blind natural selection in creating organized complexity and denied any role to the Abrahamic God. His best-selling book, *The God Delusion* (2006), which sold over two million English-language copies in its first four years in print, depicted the God of the Old Testament as "arguably the most unpleasant character in all fiction: jealous and proud of it; a petty, unjust, unforgiving control-freak; a vindictive, bloodthirsty ethnic cleanser; a misogynistic, homophobic, racist, infanticidal, genocidal, filicidal, pestilential, megalomaniacal, sadomasochistic, capriciously malevolent bully." He dismissed those who claimed that science and religion were converging: "Convergence? Only when it suits. To an honest judge, the alleged convergence between religion and science is a shallow, empty, hollow, spin-doctored sham."[5] However, Dawkins avoided retelling the tales of past scientific martyrs.

Equally lacking a historical dimension are the writings of Daniel C. Dennett, an influential philosopher of science at Tufts University, and Sam Harris, a relatively young neuroscientist who has been an outspoken New Atheist since September 11. In *Darwin's Dangerous Idea* (1995)—blurbed on the cover by Sagan and Dawkins as well as by the historian of science and religion James Moore—Dennett mentions the "apparent clash between science and religion," but he does not describe it or its historical inputs. For his part, Harris has described the "conflict between religion and science" as "inherent and (very nearly) zero-sum. . . . [T]he maintenance of religious dogma always comes at the expense of science." But he too makes no historical case.[6]

The least scientific of the four major New Atheists, the late journalist and literary critic Christopher Hitchens (1949–2011), wrote the most about the historical relationship between science and religion. Infuriated by September 11 and other religious atrocities, he focused on what he saw as the evil—and stupidity—of religion. In *God Is Not Great: How Religion Poisons Everything* (2007) he referred (without documentation) to "the terror imposed by religion on science and scholarship throughout the early Christian centuries." Elsewhere he conceded that some scientists, such as the great natural philosopher Isaac Newton (1643–1727), had "been religious, or at any rate superstitious"—but nevertheless concluded, "All attempts to reconcile faith with science and reason are consigned to failure and ridicule." He anachronistically dated the end of scientific appeals to a Creator to the meeting of the French mathematician and astronomer Pierre-Simon Laplace (1749–1827) with Napoleon Bonaparte (1769–1821) in the early nineteenth century, when, in response to the latter's inquiry about how God fit into Laplace's hypothesis about the natural development of the solar system, Laplace allegedly quipped, "I have no need of that hypothesis." In most respects Hitchens agreed with his fellow New Atheists, but he took issue with Dawkins and Dennett "for their cringe-making proposal that atheists should conceitedly nominate themselves to be called 'brights.'"[7]

Some of the most ardent New Atheist critiques featuring the warfare thesis have appeared in the articles and blogs of Jerry A. Coyne and PZ Myers. Coyne, a respected evolutionary biologist at the University of Chicago, has severely criticized several historians of science for what he views as hopeless accommodation. Coyne has been a particularly harsh critic of Ronald L. Numbers, whom he has characterized (despite Numbers's self-proclaimed agnosticism), as "one of the biggest accommodationists among historians of science." One contributor to Coyne's blog went so far as to describe Numbers as "the Neville Chamberlain of accommodation," while another suggested that Numbers denies conflict between science and religion by ignoring "the actual, historical conflicts."[8] In a review in the *New Republic* titled "The Never-Ending Attempt to Reconcile Science and Religion, and Why It Is Doomed to Fail," Coyne himself repudiates efforts to find compatibility between science and religion, writing that "a true harmony between science and religion requires either doing away with most people's religion and replacing it with a watered-down deism, or polluting science with unnecessary, untestable, and unreasonable spiritual claims. . . . It would appear, then, that one cannot be coherently religious and scientific at the same time." As he explained in an essay for *USA Today*, "Science nibbles at religion . . . relentlessly consuming divine explanations and replacing them with material ones. Evolution took a huge bite a while back, and recent work on the brain has shown no evidence for souls. . . . Science is even studying the origin of morality. So religious claims retreat into the ever-shrinking gaps not yet filled by science." Coyne harbors no doubts that science and Christianity have been and continue to be at war.[9]

Myers at times has used even stronger language than Coyne. Although Coyne once collaborated with Myers, he later broke with his colleague, describing Myers's blog as "devoted to drama, rage, and recrimination." Myers has compared religion in general to "a tapeworm" and has described Western conceptions of God in yet more derogatory terms. Like Coyne, Myers has been a harsh critic of recent historians of science and religion. For him, science and religion are implacable foes: "One is a method of analysis and experiment; the other is pretense and lies." Like his New Atheist peers, however, Myers pays little attention to the historical relationship between science and religion, although he insists that they "are incompatible in all the ways that count: Science works. Religion doesn't."[10]

If most New Atheists, including the "big four," have tended to ignore history, a few have not.[11] The most historically engaged of the self-identified atheists is the late physicist and former Catholic Victor J. Stenger (1935–2014), whose writings Dawkins has "strongly" recommended, Hitchens has introduced, and Harris has blurbed.[12] In Stenger's *God: The Failed Hypothesis—How Science Shows That God Does Not Exist* (2007), which briefly appeared on the *New York Times* list of best sellers, he claims to have "demonstrated that the absence of evidence that should be there is now sufficient to conclude beyond a reasonable doubt that the God worshipped by Jews, Christians, and Muslims does not exist."[13]

Stenger's greatest use of history appears in *The New Atheism* (2009) and in *God and the Folly of Faith: The Incompatibility of Science and Religion* (2012). In the former, Stenger devotes a brief section to answering the question (printed in bold capital letters) "Do science and religion conflict?," in which he introduces the nineteenth-century histories of Draper, a professor of chemistry, polemical historian, and vocal advocate for the acceptance of scientific rationality over religious dogma, and White, the cofounder of Cornell University and critic of denominational educational institutions, before going on to explain "why the conflict remains." Dismissing a number of reputable historians of science and religion as harmonizers and compromisers, he points out that "[s]ome of the most prominent scientific figures of the last century have acknowledged their strong atheism: Steven Weinberg, Stephen Hawking, Steven Pinker, and the later Stephen Jay Gould, just to mention the Steves. Others . . . include James Watson, Francis Crick, Carl Sagan, Richard Feynman, Edward O. Wilson, and, despite tales to the contrary, Albert Einstein."[14]

In *God and the Folly of Faith*, which features the provocative epigraph "Science flies us to the moon. Religion flies us into buildings," Stenger offers the most sophisticated—though tendentious—New Atheist treatment of the history of science and religion. "The notion that science and religion have been long at war with each other is widespread," he writes, "but, as we will see, somewhat of an oversimplification." Tracing the warfare model back to Draper and White, he notes, "Modern historical scholars, some with ideological motives of their own, have severely criticized the accuracy of Draper's and White's accounts, saying they oversimplified what was a far more complex relationship." He describes the author of the popular "complexity thesis," the distinguished Oxford historian John Hedley Brooke as mistaken and the Johns Hopkins historian Lawrence M. Principe as having a "strong proreligion bias [that] comes out no matter how hard he tries to hide it and to appear even-handed." In the end, Stenger concludes that "some historians have not been particularly careful or accurate in their criticisms of Draper and White. . . . Nevertheless, while history cannot be neglected because of its effect on the present, the incompatibility between science and religion that we see today arises primarily from current conflicts, not from ancient history." Ignoring John Heilbron's statement that "[t]he Roman Catholic Church gave more financial and social support to the study of astronomy for over six centuries . . . than any other, and probably all, other institutions"—and Michael H. Shank's extension of the Heilbron thesis to all of natural philosophy—Stenger sums up the case thus: "The totality of evidence indicates that, on the whole, over the millennia the Christian religion was more of a hindrance than a help to the development of science."[15]

It might be asked, given the ahistorical accounts of science and religion found in most New Atheist books, whether any historians of science count themselves as members of this group. The answer seems to be no, with one exception: Richard Carrier, who earned a doctorate in ancient history at Columbia University for his dissertation, "Attitudes toward the Natural Philosopher in the Early Roman Empire (100 B.C. to 313 A.D.)." There is no evidence that Carrier has had significant interactions, written or otherwise, with other historians of science or natural philosophy. In his most significant publication to date on the history of science and religion, "Christianity Was Not Responsible for Modern Science," he refutes the assertions of such scholars as the Benedictine priest-physicist Stanley Jaki (1924–2009) and the sociologist-Christian apologist Rodney Stark, who have attributed the rise of modern science to Christianity. Describing Stark as "incompetent" and "delusional" and his argument as "rubbish," Carrier finds Stark's assertions as "so *embarrassingly false* it's a wonder he stays employed" at the Baptist Baylor University. Whereas other historians certainly have similar disagreements with Jaki and Stark on this point (for additional analysis of Stark in particular, see chapter 13 in this volume), Carrier expands his critique beyond these two to claim that "when Christianity came to power *it did not* restore those scientific values [associated with the ancient Greeks], but instead sealed the fate of science by putting an end to all significant scientific progress for almost a thousand years. . . . Had Christianity not interrupted the intellectual advance of mankind . . . the Scientific Revolution might have occurred a thousand years ago." To support his claims about early Christianity's "hostility to curiosity, dethronement of empiricism, and disinterest in scientific progress (and eventual warming to these ideas over a thousand years later)," Carrier cites, among other authorities, Peter Harrison, who has never expressed similar views.[16]

Responses from Non-Theists: Gould, Ruse, and Wilson

To no one's surprise, the New Atheists' writings have provoked a lively response, occasionally bearing on the conflict thesis. Critics span a wide ideological spectrum, from fellow skeptics to Christian fundamentalists. Among the earliest in the former category was Stephen Jay Gould (1941–2002), a Harvard-based paleontologist who has had a reader base as large as Dawkins's. Even before the post–September 11 outbursts, Gould had publicly taken issue with what he called the "Darwinian fundamentalism" of Dawkins and Dennett, dubbing the latter "Dawkins's lapdog." In contrast to "pluralists," such as himself, who embraced multiple mechanisms to explain evolution, the "fundamentalists" reveled "in the belief that one overarching law—Darwin's central principle of natural selection—can render the full complexity of outcomes (by working in conjunction with auxiliary principles, like sexual reproduction, that enhance its rate and power)."[17]

As early as 1999, in *Rocks of Ages: Science and Religion in the Fullness of Life*, Gould, a self-described agnostic, expressed concern that these Darwinian fundamentalists were damaging science by touting "their private atheism (their right, of course, and in many ways my own suspicion as well) as a panacea for human progress against an absurd caricature of 'religion,' erected as a straw man for rhetorical purposes." Dismissing the notion of inevitable conflict, Gould sided with the "people of goodwill" who "wish to see science and religion at peace, working together to enrich our practical and ethical lives." His solution was to think of science and religion as occupying two "non-overlapping magisteria" (NOMA), that is, cultural spaces, according to which science would try "to document the factual character of the natural world, and to develop theories that coordinate and explain these facts," while religion would restrict itself to the "realm of human purposes, meanings, and values." Reviewing the Galileo affair as a misleading "model of inherent warfare between the magisteria," he went on to devote a long section to "Historical Reasons for Conflict," in which he provided a historically informed survey from Draper and White to those who would soon be known as the New Atheists and from the myth of a flat earth to the misinterpretations of young-earth creationism. He concluded by pronouncing his "anathema upon those dogmatists and 'true believers' who, usurping the good name of religion for their partisan doctrines, try to suppress the uncomfortable truths of science." He pointedly criticized the John Templeton Foundation for its "syncretist program" to foster dialogue between science and religion.[18] For his part, Dawkins dismissed Gould's writing as an example "of bad poetry in evolutionary science." In *The God Delusion*, he devoted an entire section to a scathing critique of Gould's NOMA, accusing Gould of carrying "the art of bending over backwards to positively supine lengths."[19]

The philosopher and historian of science Michael Ruse, also expressed exasperation with the rhetoric of the New Atheists. The author of *Can a Darwinian Be a Christian?* (2000) and numerous other books on science and religion, and a self-proclaimed "agnostic or skeptic . . . who is atheistic about traditional religions," Ruse denounced the "knee-jerk atheism" of Dennett and Dawkins, scolding them for their unwillingness to grapple seriously with the issues: "neither of you are willing to study Christianity seriously and to engage with the ideas—it is just plain silly and grotesquely immoral to claim that Christianity is simply a force for evil." Because the United States Constitution requires the public schools to maintain religious neutrality, Ruse feared that Dawkins and Dennett's "really dangerous" argument that Darwinism was essentially atheistic would provide "a legal loophole" to ban the teaching of evolution.[20] For "telling Christians how they can reconcile their theology with the facts science," as Coyne described Ruse's message, the Chicago evolutionist denounced Ruse for his "faitheism." For trying to play the role of peacemaker, Dawkins consigned Ruse to the "Neville Chamberlain School of Evolutionists." Dennett went further, telling Ruse that he feared the Florida State philosopher was "being enlisted on the side of the forces of darkness."[21]

Another skeptical critic of the New Atheists has been David Sloan Wilson. "Richard Dawkins and I share much in common," he wrote. "We are both biologists by training who have written widely about evolutionary theory . . . We are both atheists in our personal convictions who have written books on religion." However, Wilson found that in *The God Delusion* Dawkins failed to "get the facts about religion right"; indeed, he regarded "Dawkins' diatribe against religion . . . [as] deeply misinformed." In Wilson's estimation, Dawkins was "just another angry atheist, trading on his reputation as an evolutionist and spokesperson for science to vent his personal opinions about religion." Needless to say, Dawkins did not appreciate the criticism.[22]

Responses from Theists: Historians and Theologians

If non-theists such as Gould, Ruse, and Wilson have found the New Atheists disappointing, the critique from theists has been much more forceful. Historians who are explicitly Christian, for example, have tended to find the New Atheist approach to be either a relic of Victorian ideological battles or the direct descendant in spirit of White and Draper.[23] The Notre Dame historian Mark Noll, himself an evangelical, finds striking aesthetic similarities between Andrew Dickson White and the New Atheists that have more to do with marketing than any obvious historical thread:

> The United States does not have a monarchy, but we do have talk radio. . . . We do not have a divine right of kings, but we do have expert marketers who know how to promote books with titles like *The God Delusion* and *Defeating Darwinism*. We have many religiously-inspired statements about science from people with little or no scientific expertise. We have many statements by people claiming to speak for science who have little or no philosophical expertise. Here is the critical point: if misinformed people making such statements are more persuasive for a particular audience [, then] . . . the uninformed exercise a coercive authority over the informed.[24]

No Roman Catholic has responded to the New Atheists more fully and thoughtfully than John F. Haught, a theologian at Georgetown University. "As a Roman Catholic," he once explained, "I learned from an early age that there can be no genuine conflict between scientific truth and religious faith." Thus he found the martial language—and religious examples—of Dawkins and Dennett jarring and unpersuasive. He attributed "most of the apparent skirmishes between science and religious belief" to the (Protestant) "penchant" for reading the Bible literally. The New Atheists, he complained, tended to read the scriptures "at the same information-hungry level" as fundamentalist creationists. The atheists shared "with their religious opponents an anachronistic literalism that demands from ancient texts nothing less than a body of reliable contemporary scientific information." Haught, who embraced evolution as "a fresh way to understand the Psalmist's sense of the divine present in the depths of nature," recommended "an ongoing dialogue of science with religion."[25]

One of the most historically informed critiques of the New Atheists' promotion of the conflict thesis has come from theologian Alister McGrath, an ordained Anglican priest trained as a molecular biologist, who has written extensively about science and religion. In *Dawkins' God: Genes, Memes, and the Meaning of Life* (2005) McGrath rebuts Dawkins's claim that "[t]o an honest judge, the alleged convergence between religion and science is a shallow, empty, hollow, spin-doctored sham," describing it as "an interesting point of view, but [one that] belongs to another century. In recent years, . . . [i]ntensive historical scholarly research has demonstrated that the popular notion of a protracted war between church and science which continues to this day is a piece of Victorian propaganda, completely at odds with facts." "Today," he observes, Dawkins's "stereotype of the 'warfare of science and religion' lingers on in the backwaters of Western culture. Yet the idea that the natural sciences and religion have been permanently at war with each other is now no longer taken seriously by any historian of science."[26]

Critics of the New Atheism are not restricted to Christian theologians. Jonathan Sacks, moderate philosopher and from 1991 to 2013, Chief Rabbi of the United Hebrew Congregations of the British Commonwealth, wrote *The Great Partnership: Science, Religion, and the Search for Meaning* (2011), in part in response to a 2009 advertising campaign in London funded by the British Humanist Association, in which buses were festooned with the slogan "There's probably no God. So stop worrying and enjoy your life." Sacks sees no inherent conflict between science and religion; this idea for him "draws heavily on Greek myth, specifically the myth of Prometheus." Evolution is compatible with Abrahamic faith for Sacks, because "God . . . is a gardener, not a mechanic, one who plants systems that grow." As part of a thorough analysis of the history of ideas from his Jewish perspective, Sacks proffers the distinctive thesis that the fusion of Greek and Jewish thought in the theology of the Apostle Paul "contained one assumption that would eventually be challenged from the seventeenth century to today, namely that science and philosophy on the one hand, and religion on the other, belong to the same universe of discourse." Sacks views science and religion as "left- and right-brain modes of thinking" that should be kept separate but enhance one another: "*Science takes things apart to see how they work. Religion puts things together to see what they mean*" (his italics). For Sacks, New Atheism "is fundamentalism . . . the attempt to impose a single truth on the plural world."[27] For his part, Dawkins has described Sacks as "nice," but that attacking him was "like attacking a wet sponge."[28]

Responses from Theists: Scientists

A final group of critics of the New Atheist assumption of conflict are theistic scientists, who also span a wide spectrum of theological views. Kenneth Miller, a Brown University evolutionary biologist and Roman Catholic, sees the idea of irreconcilable conflict as misguided. In *Finding Darwin's God* (1999), he criticizes his atheistic colleagues' "triumphant excess," which he sees as resulting in part from their inability to conceive of a religious system that would rival atheistic materialism. Miller dismisses any notion of essential conflict between science and religion. In one blog post he recounts numerous Christians (including Catholics) who made key contributions to modern science. For him, the culprit is not conflict but ideology.[29]

The molecular immunologist Denis Alexander, a prominent British evangelical and director emeritus of the Templeton-funded Faraday Institute at St. Edmund's College, Cambridge, has come in for heavy criticism by Jerry Coyne, who once described Alexander as "one of the prize thoroughbreds in their [the Templeton] stable, so willing is he to twist science into supporting God." Alexander has criticized the New Atheists, especially Richard Dawkins, for fueling what he calls a "clash of fundamentalisms."[30]

Francis Collins, director of the National Institutes of Health (NIH) in the United States, founded the BioLogos Foundation, a leading American evangelical organization devoted to demonstrating that science and Christian faith can exist in harmony. Collins has elicited intense criticism from many New Atheists for his book *The Language of God: A Scientist Presents Evidence for Belief* (2006). For them, he is a walking—and particularly perplexing—contradiction, given their assumption of an ineluctable conflict between science and religion. When Collins was nominated in 2009 to be NIH director, Jerry Coyne, while granting that Collins "isn't a straight-up wackaloon creationist," expressed concern that "the guy is deeply, deeply superstitious, to the point where, on his website BioLogos and his book *The Language of God*, he lets his faith contaminate his scientific views. So I can't help but be a bit worried." Sam Harris, in an op-ed piece for the *New York Times*, expressed himself even more straightforwardly: "There is an epidemic of scientific ignorance in the United States. This isn't surprising, as very few scientific truths are self-evident, and many are counterintuitive. . . . But few things make thinking like a scientist more difficult than religion. . . . Francis Collins is an accomplished scientist and a man who is sincere in his beliefs. And that is precisely what makes me so uncomfortable about his nomination."[31]

Although Collins in *The Language of God* does not draw any direct connection back to the nineteenth-century purveyors of the conflict thesis, he portrays Dawkins and his fellow New Atheists as the intellectual progeny of Enlightenment materialism and Freudian psychology. Much of Collins's book affirms evolutionary biology but seeks to develop a constructive rapprochement between science, particularly evolution, and Christian faith. He introduces a neologism, "BioLogos," from the Greek *bios* and *logos*, to capture his view—and also to provide a name for his foundation. In Collins's view, the picture of irreconcilable conflict offered by the New Atheists was in part spawned the Intelligent Design (ID) movement: "ID could be thought of ironically as the rebellious love child of Richard Dawkins and Daniel Dennett." In a follow-up book with the Christian physicist Karl Giberson, Collins briefly mentions the books of Draper and White and their contribution to the warfare thesis: "Prior to the appearance of these books, science and religion, except for the occasional skirmish like the Galileo affair, got along fine. . . . And even the Galileo affair was nothing like its urban legend."[32]

Intelligent Design and Young-Earth Creationism

While numerous theists operating from within the scientific consensus criticized the New Atheism for historical shortsightedness, other theists have shown significantly less interest in dispelling the historical shortcomings of the New Atheism. The so-called Intelligent Design movement emerged in the late 1980s, motivated by an antipathy to "methodological naturalism" generally and to evolution specifically. One of its earliest books, *Darwin on Trial* (1991), came from a Berkeley law professor, Phillip E. Johnson, who had picked up Dawkins's *The Blind Watchmaker* and discovered, as he put it, that the argument for evolution was more rhetorical than factual. In *Darwin on Trial* and a sequel, *Reason in the Balance: The Case against Naturalism in Science, Law and Education* (1995), he sought not only to expose the weaknesses of Dawkins's blind watchmaker thesis but to discredit the privileged status of naturalism in science, which ruled out any consideration of theistic factors.[33] Attacks by Dawkins and other New Atheists brought the ID camp much-appreciated publicity. As one evangelical observed, Dawkins probably "single-handedly makes more converts to intelligent design than any of the leading intelligent design theorists." William Dembski, a leading ID theorist, sent the following note of appreciation to Dawkins: "I know that you personally don't believe in God, but I want to thank you for being such a wonderful foil for theism and for intelligent design more generally. In fact, I regularly tell my colleagues that you and your work are one of God's greatest gifts to the intelligent-design movement. So please, keep at it!"[34]

Although ID advocates understandably saw themselves engaged in an ideological war—one design theorist termed the New Atheists' effort to destroy religion as a "call to jihad" against religion—they seldom addressed the alleged historical conflict between science and religion. A notable exception was Alvin Plantinga, a leading Protestant philosopher of religion and ID sympathizer who had praised Johnson's *Darwin on Trial*. His book *Where the Conflict Really Lies: Science, Religion, and Naturalism*, based on his series of Gifford Lectures in 2005, is devoted to this question. His "overall claim," he announces in the preface, is to show that "*there is superficial conflict but deep concord between science and theistic religion, but superficial concord and deep conflict between science and naturalism*." Tracing the "simplistic" notion of warfare back to White, Plantinga seeks to correct the record using the historical studies of John Hedley Brooke and others. He admits that there had been "conflicts between science and particular religious beliefs," but only ones "that are not part of Christian belief as such: belief in a universal flood, a very young earth, etc." He finds that "[t]he scientific theory of evolution as such is not incompatible with Christian belief; what is incompatible with it is the idea that evolution, natural selection, is *unguided*," a view widely held in the scientific community. "In addition to those dancing on the lunatic fringe such as Richard Dawkins and Peter Atkins [another New Atheist], there are perfectly reasonable scientists who reject the idea of special divine action in the world." But, in Plantinga's opinion, that is precisely where the conflict lay. "If my argument is cogent, it follows that there is deep and serious conflict between naturalism and evolution, and hence deep conflict between naturalism and science."[35]

Occupying the far right wing of the science-and-religion spectrum are the so-called young-earth creationists (YECs), shunned by atheists and moderate theists alike. Although the New Atheists paid them little attention—indeed the two parties scarcely disagreed about the conflict between science and religion—YECs found in atheism an attractive target. The venerable

Institute for Creation Research (ICR), established in 1972 by Henry M. Morris (1918–2006), repeatedly lashed out against the atheists, especially their "anti-God" ads that appeared in Dallas, home of the ICR.[36]

Even before the death of Morris, the ICR had ceded leadership of the YEC movement to Answers in Genesis (AiG), led by a dynamic Australian-born schoolteacher and Christian apologist, Ken Ham. In 2007, Ham opened a $27 million Creation Museum in northern Kentucky, and shortly thereafter laid plans for a $150 million Ark Encounter, featuring a "life-size" replica of Noah's ark, scheduled to open in July 2016. The year he opened his museum, he penned a screed again the New Atheists, asking his followers for their "prayers and support" to stop the progress of this powerful movement. Like Morris and the ICR, he said little about the atheists' promotion of a war between science and religion; instead, he railed against their anti-Christian propaganda. To counter atheists billboards, AiG launched its own campaign featuring billboards addressed "To All of Our Atheist Friends—Thank God You're Wrong."[37]

Conclusion

Although the New Atheists have spilled substantial ink in their writings arguing that organized religion is antithetical to science, the somewhat surprising finding of this brief survey is that the New Atheists themselves rarely see the need to access any sort of historical analysis to support their position. There are several possible reasons for this rather striking omission. One is rather facile: none of the major New Atheists is a historian. A charitable interpretation of this observation is that none of the New Atheists is comfortable offering such an analysis. However, given that Dawkins in particular feels no compunction about wading into theological or philosophical waters in his polemical writing, this explanation does not seem particularly likely.

A second possible reason for the New Atheists' neglect of historical analysis is that the past is largely irrelevant for them. In their minds the conflict is so obvious that any nuanced attempt at historical analysis is, as Jerry Coyne so frequently labels it, "accommodationist."[38]

A third possibility is that such omission is intentional, because the current picture of the historical situation in the nineteenth through the twentieth-first centuries developed by historians of science fails to support a simplistic and counterfactual story of conflict. If, as Colin A. Russell quipped, the conflict thesis is "excellent drama but impoverished history," then the historical narrative that has emerged is ill suited to such drama.[39] This third possibility then represents a strategic decision by New Atheists to protect the cleanliness of a story of conflict as a polemical stratagem. Many analyses of the New Atheism, from that of the early critic Mary Midgley to those already mentioned, see the propagation of a coherent mythology as crucial to New Atheist writing.[40] Whether omission of recent historical work is part of a deliberate "sanitation project" is unclear. In any case, it is notable that those with such a large stake in perpetuating the conflict thesis commonly fail to examine its origin and propagation.

ACKNOWLEDGMENTS

We would like to thank Stephen Weldon, Salman Hameed, and Michael Ruse for their helpful suggestions.

NOTES

1. On the impact of September 11, see, e.g., Richard Dawkins, quoted in "Has the World Changed?—Part Two," *Guardian*, October 11, 2001, http://www.theguardian.com/world/2001/oct/11/afghanistan.terrorism2; and Victor J. Stenger, *The New Atheism* (Amherst, NY: Prometheus Books, 2009), 11. For the New Atheists and the conflict thesis, see Stephen D. Snobelen, "Science, Religion, and the New Atheism," in Susan Harris, ed., *The New Atheism* (Charlottetown, PEI: St. Peter Publications, 2013), 109–44, an essay brought to our attention by Ted Davis.

2. Joseph McCabe, *The Conflict between Science and Religion*, Little Blue Book no. 1211 (Girard, KS: Haldeman-Julius Publications, 1927), 64. See also Peter H. Denton, *The ABC of Armageddon: Bertrand Russell on Science, Religion, and the Next War, 1919–1938* (Albany: State University of New York Press, 2001), chap. 5.

3. Carl Sagan, *Cosmos* (New York: Ballantine Books, 1985), 1, based on the TV television series, which first aired in 1980; Stephen P. Weldon, draft MS of "The Scientific Spirit of American Humanism," chap. 10, p. 29 (in press, used with permission); Carl Sagan, *The Demon-Haunted World: Science as a Candle in the Dark* (New York: Random House, 1996), 277. See also Keay Davidson, *Carl Sagan: A Life* (New York: John Wiley and Sons, 1999), 409–10.

4. Francis Crick, quoted in Margaret Wertheim, "After the Double Helix: Unraveling the Mysteries of the State of Being," *New York Times*, April 13, 2004, D3; Stephen Hawking and Leonard Mlodinow, *The Grand Design* (New York: Bantam Books, 2010), 180; William B. Provine, "Darwinism: Science or Naturalistic Philosophy? A Debate between William B. Provine and Phillip E. Johnson at Stanford University," *Origins Research* 16, no. 1 (1994): 9.

5. Richard Dawkins, *The Selfish Gene*, new ed. (New York: Oxford University Press, 1989), 198 (mental illness); Dawkins, "Is Science a Religion?" *Humanist* 57 (January–February 1997): 26 (smallpox); Dawkins, *The Blind Watchmaker* (New York: W. W. Norton, 1986); Dawkins, *The God Delusion* (Boston: Houghton Mifflin, 2006), 31 (unpleasant character); "How Many Copies of *The God Delusion* Have Been Sold?," Patheos.com, February 3, 2010, http://www.patheos.com/blogs/friendlyatheist/2010/02/03/how-many-copies-of-the-god-delusion-have-been-sold; Richard Dawkins, *A Devil's Chaplain:*

Reflections on Hope, Lies, Science, and Love (Boston: Houghton Mifflin, 2003), 150–51 (convergence).

6. Daniel C. Dennett, *Darwin's Dangerous Idea: Evolution and the Meanings of Life* (New York: Touchstone, 1995), 22; Dennett, *Breaking the Spell: Religion as a Natural Phenomenon* (New York: Penguin, 2006), 263–64; Sam Harris, "Science Must Destroy Religion," *Huffington Post*, June 2, 2006, http://www. huffingtonpost.com/sam-harris /science-must-destroy-religion, reprinted in John Brockman, ed., *What Is Your Dangerous Idea?* (New York: Harper Perennial, 2007), 148–51. See also Harris, *The End of Faith: Religion, Terror, and the Future of Reason* (New York: W. W. Norton, 2004), 15, 165; Harris, *Letter to a Christian Nation* (New York: Alfred A. Knopf, 2006), 62–68; and Harris, *The Moral Landscape: How Science Can Determine Human Values* (New York: Free Press, 2010), 6, 24, 160–74.

7. Christopher Hitchens, *God Is Not Great: How Religion Poisons Everything* (New York: Twelve, 2007), 5 ("brights"), 64–67 (Newton, attempts to reconcile, Laplace), 260 (terror).

8. Jerry Coyne, "Faitheist T.V.: Historian of Science Joins Young-Earth Creationist in an Old Fashioned Coyne-and-Dawkins Roast," *Why Evolution Is True* (blog), July 25, 2009, https://whyevolutionistrue.wordpress.com (where he describes Numbers as "pusillanimous"); Coyne, "More Denialism of a Science-Religion Conflict," *Why Evolution Is True* (blog), December 10, 2012, https://whyevolutionistrue.wordpress.com (accommodationist). For an example of Coyne's hypocrisy, see also Paul Nelson, "Jerry, PZ, Ron, Faitheism, Templeton, Bloggingheads, and All That—Some Follow-up Comments," Uncommon-Descent.com, July 26, 2009, http://www.uncommondescent.com/intelligent-design. For Numbers on the New Atheists, see his "Aggressors, Victims, and Peacemakers: Historical Actors in the Drama of Science and Religion," in Harold W. Attridge, ed., *The Science and Religion Debate: Why Does It Continue*? (New Haven, CT: Yale University Press, 2009), 15–53. Neville Chamberlain was the British prime minister who sought to appease Adolf Hitler in 1938 by letting him take parts of Czechoslovakia.

9. Jerry A. Coyne, "Seeing and Believing: The Never-Ending Attempt to Reconcile Science and Religion, and Why It Is Doomed to Fail," *New Republic*, February 4, 2009; Coyne, "Science and Religion Aren't Friends," *USA Today*, October 11, 2010; Coyne, "Michael Ruse Advises the Faithful on How to See the Evolution of Humans as Inevitable," *Why Evolution Is True* (blog), September 12, 2012, https://whyevolutionistrue.wordpress .com/2012/09/12 (war). See also Coyne, *Faith vs. Fact: Why Science and Religion Are Incompatible* (New York: Viking, 2015); and Coyne, "NY Times Editor Proclaims That Science and Religion Are Compatible," *Why Evolution Is True* (blog), February 16, 2016, https://whyevolutionistrue.wordpress.com/2016/02/16.

10. Leo Buzalsky, "PZ Myers—Religion as a Parasitic Worm Analogy," YouTube.com, November 7, 2011, http://www.youtube.com/watch?v=t1CPJOFAEfE. For Myers's severe critique of Ronald L. Numbers, see PZ Myers, "Ron Numbers, Another Tool of the Religious Establishment," *ScienceBlogs*, January 2, 2007, http://scienceblogs.com/pharyngula; Myers, *The Happy Atheist* (New York: Pantheon, 2013), 146 (pretense and lies), 160 (incompatible). On the Coyne-Myers spat, see "Jerry Coyne's Blog Attacks Myers," *Shadow to Light* (blog), December 2, 2014, http://shadowtolight.wordpress.com/2014/12/02.

11. For passing reference to a historical conflict, see, e.g., Pascal Boyer, *Religion Explained: The Evolutionary Origins of Religious Thought* (New York: Basic Books, 2001), 320–21. Most books by New Atheists say nothing about history. See, e.g., Scot Atran, *In Gods We Trust: The Evolutionary Landscape of Religion* (New York: Oxford University Press, 2002). Despite its title and occasional historical references, Peter Atkins, *Galileo's Finger: The Ten Great Ideas of Science* (Oxford: Oxford University Press, 2003), says next to

nothing about the alleged war between science and religion; his position, however, is clear (237).

12. Dawkins, *God Delusion*, 118; Christopher Hitchens, foreword to Victor J. Stenger, *God: The Failed Hypothesis—How Science Shows That God Does Not Exist* (Amherst, NY: Prometheus Books, 2007), 1–6, which contains dust jacket blurbs by both Dawkins and Harris. Harris also contributed a cover blurb to Stenger, *New Atheism*.

13. Stenger, *God: The Failed Hypothesis*, 12. Like Harris, he gives a pass to naturalistic Eastern spiritual practices.

14. Stenger, *New Atheism*, 73–75.

15. Stenger, *God and the Folly of Faith: The Incompatibility of Science and Religion* (Amherst, NY: Prometheus Books, 2012), 1–38; John Heilbron, *The Sun in the Church: Cathedrals as Solar Observatories* (Cambridge, MA: Harvard University Press, 1999), 3; Michael H. Shank, "Myth 2: That the Medieval Christian Church Suppressed the Growth of Science," in Ronald L. Numbers, ed., *Galileo Goes to Jail and Other Myths about Science and Religion* (Cambridge, MA: Harvard University Press, 2009), 21. Historical sources for this introductory chapter include works by David C. Lindberg, Ronald L. Numbers, John Hedley Brooke, and Lawrence M. Principe. Steven Weinberg, a Noble Prize–winning physicist, is another of the few New Atheists to give considerable attention to the historical literature. For Weinberg's discussion of the historiography, see Steven Weinberg, *To Explain the World: The Discovery of Modern Science* (New York: HarperCollins, 2015), esp. 100, 155–56, 376–77, 378.

16. Richard Carrier, "Christianity Was Not Responsible for Modern Science," in John W. Loftus, ed., *The Christian Delusion* (Amherst, NY: Prometheus Books, 2009), 396–419, long quotation on 413–14. On 419, Carrier cites Peter Harrison, *The Bible, Protestantism, and the Rise of Natural Science* (Cambridge: Cambridge University Press, 1998); and Harrison, "Curiosity, Forbidden Knowledge, and the Reformation of Natural Philosophy in Early-Modern England," *Isis* 92 (2001): 265–90. On early Christianity's supposed hostility to science, Carrier also cites Lorraine Daston, William Eamon, G.E.R. Lloyd, Marshall Clagett, and Neil Kenny. For personal information, see http://www.richardcarrier.com. Published too late to be used in this chapter is Richard Carrier, *The Scientist in the Early Roman Empire* (Durham, NC: Pitchstone Publishing, 2017).

17. Stephen Jay Gould, "Darwinian Fundamentalism," *New York Review of Books*, June 12, 1997; Gould, "Evolution: The Pleasures of Pluralism," *New York Review of Books*, June 26, 1997. See also Daniel C. Dennett and Stephen Jay Gould, "'Darwinian Fundamentalism': An Exchange," *New York Review of Books*, August 14, 1997.

18. Stephen Jay Gould, *Rocks of Ages: Science and Religion in the Fullness of Life* (New York: Ballantine, 1999), 4 (goodwill, science, religion), 6 (magisterial), 8 (agnostic), 71–74 (Galileo), 97–170 (historical reasons), 210 (anathema), 214 (syncretist). See also Numbers, "Aggressors, Victims, Peacemakers," 47–48, on which this paragraph is partially based.

19. Richard Dawkins, *Unweaving the Rainbow: Science, Delusion and the Appetite for Wonder* (Boston: Houghton Mifflin, 1998), 193 (poetry); Dawkins, *God Delusion*, 54–60, quotation on 55.

20. Brandon Vogt, "Atheism, Philosophy, and Science: An Interview with Dr. Michael Ruse," Strange Notions, December 2013, http://www.strangenotions.com/interview-with-atheist-philosopher-dr-michael-ruse (skeptic); Michael Ruse to Daniel C. Dennett, February 19, 2006, quoted in Andrew Brown, "When Evolutionists Attack," *Guardian*, March 6, 2006, http://www.guardian.co.uk; Madeleine Bunting, "Why the Intelligent Design Lobby Thanks God for Richard Dawkins," *Guardian*, March 27, 2006, http://www.guardian.co.uk.

21. Jerry Coyne, "Michael Ruse Advises the Faithful on How to See the Evolution of Humans as Inevitable," *Why Evolution Is True* (blog), September 12, 2012, https://whyevolutionistrue.wordpress.com/2012/09/12; Dawkins, *God Delusion*, 66–69 (Chamberlain); William Demkski, "Remarkable Exchange between Michael Ruse and Daniel Dennett," UncommonDescent.com, February 21, 2006, http://www.uncommondescent.com.

22. David Sloan Wilson, "Beyond Demonic Memes: Why Richard Dawkins Is Wrong about Religion," *eSkeptic*, July 4, 2007, http://www.skeptic.com/eskeptic/07-07-04; "Richard Dawkins Replies to David Sloan Wilson," *eSkeptic*, July 11, 2007, http://www.skeptic.com/eskeptic/07-07-11.

23. See, e.g., Borden W. Painter Jr., *The New Atheist Denial of History* (New York: Palgrave Macmillan, 2014).

24. Mark Noll, "Science, Religion, and A. D. White: Seeking Peace in the 'Warfare between Science and Theology'" BioLogos, http://biologos.org/uploads/projects/noll_scholarly_essay2.pdf.

25. John F. Haught, *God after Darwin: A Theology of Evolution* (Boulder, CO: Westview, 2000), 6–7 (no conflict); John F. Haught, *Deeper Than Darwin: The Prospect for Religion in the Age of Evolution* (Boulder, CO: Westview, 2003), 18 (skirmishes), 19 (information-hungry), 73 (engagement). See also Haught, *God and the New Atheism: A Critical Response to Dawkins, Harris, and Hitchens* (Louisville, KY: Westminster John Knox Press, 2008).

26. Dawkins, *Devil's Chaplain*, 151; Alister McGrath, *Dawkins' God: Genes, Memes, and the Meaning of Life* (Oxford: Blackwell, 2005), 139 (propaganda), 142 (stereotype). In further support of his claim about a revolution in the writing of history (139–59), McGrath also mentions works by Mario Biagoli, David N. Livingstone, Colin Russell, Frank M. Turner, Geoffrey Cantor, and John Hedley Brooke.

27. Rabbi Jonathan Sacks, *The Great Partnership: Science, Religion, and the Search for Meaning* (New York: Schocken Books, 2011), 11 (stop worrying), 211 (Prometheus), 216 (gardener), 62 (universe, left- and right-brain), 2 (take things apart). Michael Shulson, "Jonathan Sacks on Richard Dawkins: 'New Atheists Lack a Sense of Humor,'" *Salon*, September 27, 2014, http://www.salon.com/2014/09/27/jonathan_sacks_on_richard_dawkins_new_atheists_lack_a_sense_of_humor/ (fundamentalism).

28. Isaac Chotiner, "Interview: Richard Dawkins Keeps Making New Enemies," *New Republic*, October 28, 2013, https://newrepublic.com/article/115339/richard-dawkins-interview-archbishop-atheism (wet sponge).

29. Kenneth R. Miller, *Finding Darwin's God: A Scientist's Search for Common Ground between God and Evolution* (New York: HarperCollins, 1999), 185; Miller, "Science and Religion: Incompatible?," *The Blog*, March 7, 2012, *Huffington Post*, https://www.huffingtonpost.com/kenneth-r-miller/post_3059_b_1310826.html.

30. Denis Alexander, *Rebuilding the Matrix: Science and Faith in the 21st Century* (Grand Rapids, MI: Zondervan, 2001), 209–19. See also http://www.bethinking.org/atheism/a-clash-of-fundamentalisms (clash of orthodoxies). For Coyne's description of Alexander, see "A Christian Scientist Tells Us Why Evolution and Religion Are Compatible, but Gets It All Wrong," *Why Evolution Is True* (blog), July 6, 2014, https://whyevolutionistrue.wordpress.com/2014/07/06/a-christian-scientist-tells-us-why-evolution-and-religion-are-compatible-but-gets-it-all-wrong.

31. Jerry Coyne, "Francis Collins Pollutes Science with Religion," *Why Evolution Is True* (blog), July 27, 2009, https://whyevolutionistrue.wordpress.com/2009/07/27/francis-collins-pollutes-science-with-religion (wackaloon); Coyne, "Francis Collins as NIH Director," *Why Evolution Is True* (blog), July 10, 2009, https://whyevolutionistrue.wordpress

.com/2009/07/10/francis-collins-as-nih-director (superstitious); Sam Harris, "Science Is in the Details," *New York Times*, July 27, 2009, A21 (epidemic of ignorance).

32. Karl W. Giberson and Francis S. Collins, *The Language of Science and Faith* (Downers Grove, IL: InterVarsity Press, 2011), 84. To support their claim, they reference John Hedley Brooke, *Science and Religion: Some Historical Perspectives* (Cambridge: Cambridge University Press, 1991).

33. Phillip E. Johnson, *Darwin on Trial* (Downers Grove, IL: InterVarsity Press, 1991); Johnson, *Reason in the Balance: The Case against Naturalism in Science, Law and Education* (Downers Grove, IL: InterVarsity Press, 1995), 15, 26. See also Johnson, "Atheist Crusaders," *Touchstone* 23 (March–April 2010). For a concise history of ID, see Ronald L. Numbers, *The Creationists: From Scientific Creationism to Intelligent Design*, expanded ed. (Cambridge, MA: Harvard University Press, 2006), chap. 17.

34. Owen Gingerich, *God's Universe* (Cambridge, MA: Harvard University Press, 2006), 74 (single-handedly); quoting Dembski's thank you note, Madeleine Bunting, "Why the Intelligent Design Lobby Thanks God for Richard Dawkins," *Guardian*, March 27, 2006, http://www.guardian.co.uk. For Dawkins's opinion of ID, see, e.g., *God Delusion*, 138, 144–61.

35. David Berlinski, *The Devil's Delusion: Atheism and Its Scientific Pretensions* (New York: Crown Forum, 2008), xii (jihad); Alvin Plantinga, *Where the Conflict Really Lies: Science, Religion, and Naturalism* (New York: Oxford University Press, 2011), ix (overall claim; italics in the original), 6 (simplistic), 63 (evolution), 75 (lunatic fringe), 144 (universal flood), 310 (cogent).

36. "Evolution's Evangelists," *Acts & Facts* 37 (May 2008): 10; Brian Thomas, "Empiricism: A Glaring Flaw of New Atheism," *Acts & Facts* 37 (September 2008): 15; Christine Dao, "Dawkins Supports 'No God' Ads," *ICR News*, October 23, 2008, http://icr.org; Dao, "Atheists' Christmas Campaign: 'Yes' to Goodness, 'No' to God," *ICR News*, November 18, 2008, http://icr.org; Dao, "Anti-God Ads Hit Dallas," *ICR News*, April 7, 2009, http://icr.org; Brian Thomas, "Dawkins' Latest Book: The Greatest Lie on Earth," *ICR News*, September 23, 2009, http://icr.org. On the war between what creationists liked to call the naturalist and supernaturalist worldviews, see Henry Morris III, "The Holy War," *Acts & Facts* 39 (July 2010): 22. On the history of "creation science," see Numbers, *The Creationists*.

37. Ken Ham, "The 'New Atheists,'" *Ken Ham* (blog), January 16, 2007, Answers in Genesis, https://answersingenesis.org/world-religions/atheism/the-new-atheists; Ham, "A New Atheist Billboard?," *Ken Ham* (blog), October 12, 2013, Answers in Genesis, https://answersingenesis.org/blogs/ken-ham/2013/10/12/a-new-atheist-billboard.

38. Coyne has helpfully collected many relevant posts at *Why Evolution Is True* (blog), https://whyevolutionistrue.wordpress.com/2009/06/12/the-big-accommodatinism-debate-all-relevant-posts.

39. Colin A. Russell, "The Conflict of Science and Religion," in Gary B. Ferngren, ed., *Science and Religion: A Historical Introduction* (Baltimore, MD: Johns Hopkins University Press, 2002), 9.

40. Mary Midgley, *Evolution as a Religion: Strange Hopes and Stranger Fears* (London: Methuen, 1985). See also Karl Giberson, *Saving Darwin: How to Be a Christian and Believe in Evolution* (New York: HarperOne, 2008), 175; Sacks, *Great Partnership*, 211.

CHAPTER THIRTEEN

Neo-Harmonists

PETER HARRISON

In a 2009 interview for the Pew Forum, the renowned geneticist and evangelical Christian Francis Collins set out a direct challenge to new atheist claims of the inevitability of conflict between science and religion: "[T]hose of us who are interested in seeking harmony here have to make it clear that the current crowd of seemingly angry atheists, who are using science as part of their argument that faith is irrelevant, do not speak for us. Richard Dawkins, Sam Harris and Christopher Hitchens do not necessarily represent the consensus of science."[1] Collins is himself representative of a significant group of individuals, many of them scientists, who see no cause for ongoing conflict between science and religion. Religiously inclined counterparts to the New Atheists (see chapter 12 in this volume), this group may be designated new or "neo-harmonists." This chapter begins with some reflections on the term "neo-harmonist," which has a somewhat complicated provenance. It develops a working definition of that category—partly by identifying that group of thinkers who might be regarded as the original harmonizers—and then moves to a discussion of the positions of three key contemporary figures regarded as emblematic of this position: Rodney Stark, Denis Alexander, and Francis Collins.

Will the True Neo-Harmonists Step Forward?

The term "neo-harmonist" is not in wide usage, even in the extensive literature on the relations between science and religion. The original source of the expression seems to have been historian David Hollinger's essay "Justification by Verification," which first appeared in 1989 and was subsequently reprinted in 2013. In that piece, Hollinger advances the claim that recent science-and-religion historiography is subtly "neo-harmonist" in its dual reluctance, first, to acknowledge that beneath all the newly discovered complexities in the relations of science and Christianity there has persisted an authentic struggle over the epistemic principles that shape modern culture and, second, to confront in relation to that struggle the gradual and historic de-Christianization of intellectual discourse in the United States.[2] He identifies the volume *God and Nature* (1986), edited by David C. Lindberg and Ronald L. Numbers, as the chief exemplar of the neo-harmonist historical literature. Hollinger also singles out James R. Moore's *Post-Darwinian Controversies* (1979), which he describes as "the most ambitious and polemical contribution to 'neo-harmonist' historiography."

The fact that Hollinger's essay was reprinted in 2013 without emendation suggests that he continues to regard this characterization as an apt one. For our present purposes, however, applying the label to this group is not particularly helpful. The target of the Lindberg and Numbers volume was neither the epistemic principles that shape modern culture, nor the historical de-Christianization of the intellectual discourse of the United States, but simply the false historical claim that since antiquity "science" and "religion" in the West have been engaged in a perpetual warfare. Indeed, the bulk of the contributions in the collection deal with the complexities of science-religion interactions before the nineteenth century. One of the main burdens of Moore's book is to disabuse readers of the idea that military metaphors and talk of science-religion conflict are a helpful way to think about the nineteenth-century post-Darwinian controversies. But nothing follows from this important revisionist work about the significance of other topics not pursued. What is shared by the eighteen authors of the Lindberg and Numbers volume (and James Moore is one of them) is not some implicit conviction about modern American intellectual discourse, but a desire to set the historical record straight. *Pace* Hollinger, it seems implausible to suggest that beneath their diverse historical interests lies some uniform reluctance to address a putatively authentic struggle and a common, if covert, commitment to harmony. What unites these authors is what they deny, not what they affirm or for good reason choose to ignore. Few if any of the historians in question would contest the fact that there were past historical episodes that might attract the label "science-religion conflict." Their point is rather that Western history has not been characterized by an inevitable and enduring battle between two reified entities "science" and "religion."

Since the 1980s, a general consensus about the conflict myth has developed among historians of science that more or less reflects the position pioneered by Moore, Lindberg, and Numbers. The most significant contribution from that time has been John Hedley Brooke's classic, *Science and Religion: Some Historical Perspectives* (1991). Brooke is critical of the conflict model, but his targets include both "warriors" *and* "harmonizers." The real lesson from history, he insists, is complexity.[3] This model of complex historical relations between science and religion, along with its negative implications for any simple, overarching story, has become the default position among historians of science.[4] Crucially, they are not united by any *normative* claim about what the proper relationship between science and religion ought to be (although if this general consensus is correct, it will disqualify certain arguments for conflict that rest on false historical premises). "Neo-harmonizer" seems to be the wrong label for members of this group. A more fitting descriptor would be "revisionists" or "myth busters"—historians of science whose goal is the negative one of demonstrating the falsity of the conflict myth.[5]

That said, Hollinger's misidentification at least provides a clue as to those who were the *original* harmonizers. These would be the subjects of Moore's analysis who, in a post-Darwinian world, continued to insist that a long-standing harmony between science and religion could and should still be upheld. These original harmonizers would include such individuals as James McCosh, Henry Ward Beecher, John Fiske, Joseph Le Conte, Lymann Abbott, Mino Judson Savage, Asa Gray, and George Frederick Wright. To these names might also be added the names of British figures such as James Iverach, Aubrey L. Moore, and Charles Kingsley. (We might also include figures such as Alexander Winchell and Newman Smyth who explicitly take up the conflict motif and reject it.) This is not the place for an extended commentary on the views of these figures, and in any case there are excellent treatments of their thought and its significance.[6] But, in short, these individuals explicitly maintained that there ought to be harmony between science and religion and that conflict could only arise when there is some kind of fundamental misunderstanding. Geologist Alexander Winchell (1824–91) thus spoke on a number of occasions of the "harmony between science and religion." Congregationalist minister Newman Smyth maintained that "there is and can be no conflict between religion and Science."[7] Philosopher and principal of the College of New Jersey (now Princeton University) James McCosh (1811–94), insisted on "the harmony of Genesis and geology."[8] Harvard philosopher and historian John Fiske (1842–1901) maintained that the conflict between religion and science arises out of "a lamentable confusion of thought."[9] A common motif in the writings of these thinkers, rehearsed by our neo-harmonists, was that of "the two books"—the book of nature and the book of scripture, which on account of their common divine authorship were necessarily in harmony.[10]

It is also true that long before the nineteenth century, the conviction that truths about the natural world were in harmony with the central truths of Christianity was commonplace. But earlier advocates of compatibility did not express their views in terms of a relationship between science and religion, since our conception of "science" did not emerge until the nineteenth century. Consequently, their convictions about compatibility were not expressly directed against a clearly articulated conflict model (which again, emerged most powerfully in the nineteenth century). The original harmonizers, then, we will take to be those who took up their stance in response to the growing currency of the nineteenth-century conflict model. Darwinism is part of this picture, for while it was by no means the most important driver of the conflict model, it was rarely far from discussions about the compatibility of science and religion and since the late nineteenth century has often been taken as a proxy for "science."

Granting all this, "*neo*-harmonizers" will be those who, in the context of controversies about the relations between science and religion, assert against contemporary advocates of conflict that science and religion are in fact compatible. While for some this may entail a corrective descriptive account of the past, it will also include a normative stance about what the relations between science and religion *ought to be*. Their targets include the New Atheists, whose views have been considered in the previous chapter, and like their nineteenth-century counterparts, evolution will often be at the forefront of their concerns. Three representative figures will be considered—Stark, Alexander, and Collins.

Stark Alternatives

Rodney Stark is a US sociologist of religion and presently distinguished professor of social sciences at Baylor University, Texas. He originally made his name in the 1970s and 1980s when, with William Sims Bainbridge, he advanced a model of religion that drew upon rational choice theory and treated religious affiliation as if it operated on market principles. Against a prevailing orthodoxy that associated religious pluralism with a decline in religious adherence and in the plausibility of religious belief, Stark maintained that a pluralistic and competitive religious market would lead to higher levels of religious participation. This, he argued, explained the robust religiosity of the United States, while also accounting for the highly secularized condition of northern Europe. In the former, religious choice correlates with high participation, whereas in the latter, dismal participation is the result of state churches maintaining a highly regulated religious monopoly. Religious decline in Europe was a problem of supply, not demand.[11]

In the 1990s, Stark focused this theoretical lens on American church history and, with Roger Finke, advanced the intriguing theory that religiosity during the colonial period was low and increased only when the US Constitution reduced the regulation of the religious market, allowing for a degree of denominational pluralism.[12] Fatefully, Stark then turned his attention to the rise of Western Christianity, again seeking to account for its growth in uncompromising economistic terms. Christianity succeeded, he suggests, because it represented by far "the best religious 'bargain' around" (the inconveniences of martyrdom notwithstanding).[13] It is fair to say that while this treatment offers some fascinating new insights into the social demography of early Christianity, its reception by the guild of church historians has been cool.

This foray into history seems to have emboldened Stark to embark on even more ambitious historical excursions, and he has subsequently produced a number of more expansive pieces on the role of Christianity in the rise of the West. Two of these titles deal with the relations between Christianity and science: *For the Glory of God: How Monotheism Led to Reformations, Science, Witch-Hunts, and the End of Slavery* (2003) and *The Victory of Reason: How Christianity Led to Freedom, Capitalism, and Western Success* (2005). In the first of these books, Stark argues that religious factors played a significant and positive role in the scientific revolution of the seventeenth century. As for the nineteenth century, Stark maintains that Darwinism was not necessarily in tension with Christianity but was hijacked by a "social movement" that used it as an instrument in an ideological battle against religion.[14] In *The Victory of Reason* he reasserts these claims, again highlighting religious contributions to the rise of science: "The so-called Scientific Revolution of the sixteenth century has been misinterpreted by those wishing to assert an inherent conflict between religion and science. Some wonderful things were achieved in this era, but they were not produced by an eruption of secular thinking. Rather, these achievements were the culmination of many centuries of systematic progress by medieval Scholastics, sustained by that uniquely Christian twelfth-century invention, the university.

Not only were science and religion compatible, they were inseparable—the rise of science was achieved by deeply religious Christian scholars."[15]

The general conclusion about science and religion is this: "It is *not* my claim that scientists should include God within their cosmologies, or, indeed, that nonbelievers can't do good science—at least not once the system is in place. I *do* argue that religion and science are compatible, and that the *origins* of science lay in theology."[16] These statements provide a good prima facie basis for granting Stark neo-harmonist credentials.

Insofar as Stark's arguments in these recent books are historical, it would seem that he shares with the historian myth busters the goal of demolishing the conflict myth. Indeed, a number of his general points about the religious origins of modern science and ideological appropriations of Darwin have also been made by others, including our myth-busting historians of science, albeit couched in less bracing rhetoric.[17] But there is a significant difference in strategy. Stark sets about his task not so much by exposing the complexities of historical science-religion relations as by constructing a countermyth. Unlike the myth busters, he offers his own alternative narrative of Western history and, in the process, seeks to restore Christianity to what he considers to be its rightful place as a major agent of progressive historical change. Stark also seems to have a broader target in mind than the myth busters, setting himself not only against past fabricators of history such as John William Draper and Andrew Dickson White, but also against an American intellectual establishment that, in his eyes, is infected with an animus against religion and perpetuates in various ways the Draper-White historiography that proposed an enduring warfare between science and religion.

A further difference is that, like the harmonists of the nineteenth century, Stark seeks directly to address contemporary debates about the status of evolution. He has characteristically strong views about this. Linking the present debate to his revisionist historical narrative, he writes, "The battle over evolution is not an example of how heroic scientists have withstood the relentless persecution of religious fanatics. Rather, from the very start it primarily has been an attack on religion by militant atheists who wrap themselves in the mantle of science."[18] Richard Dawkins is here singled out for special mention. Again, like a number of nineteenth-century harmonists, Stark is ambivalent about evolution, maintaining the plague-on-both-your-houses line that he is "neither a creationist nor a Darwinist."[19] This admission of ambivalence has gained him both critics and admirers, particularly in the blogsphere, and confirms the prejudices of the New Atheist camp on the one hand, while giving some minimal encouragement to creationists and Intelligent Design theorists on the other.[20] This attitude distinguishes him from Alexander and Collins, both of whom enthusiastically endorse evolution.

Stark's disavowal of creationism brings us to the question of his religious commitments, which have been something of a moving target. In the 1980s, during his sociologist days, Stark described himself as "incapable of religious faith." When he penned *The Rise of Christianity* in 1996, he admitted only to being "an admirer, but not a believer." Subsequently, in an interview conducted in 2004—the year he accepted a post at Baylor University—he stated that while he was not a man of faith, neither could he be called an atheist. Most recently, in 2007, he has described himself as "an independent Christian."[21] Aligning these biographical details with his publication history, *For the Glory of God* would have appeared just before his "independent Christian" phase. On balance, this would suggest that the book is not intended, explicitly or otherwise, to be a work of Christian apologetics. Stark himself says something like this in his introduction: the book "is a work of social science" and consequently "my personal religious beliefs are of concern only to me."[22] This suggests that he is not writing for an exclusively Christian audience nor is engaged purely in religious apologetic.

This is not the occasion for a comprehensive assessment of Stark's views and whether they represent good history. But it is worth offering some brief evaluative comments, since these will put some distance between him and the myth busters. To start with, there may be something to be said for the strategy of constructing a counternarrative. Complexity has proven to be a hard sell, and the combined efforts of historians of science over the past forty years have failed to lay the conflict myth to rest. Part of the explanation for this is that simple stories, with a few emblematic heroes or villains, are difficult to dislodge from popular consciousness; and all the more so when they appear to be confirmed by contemporary events—in this case the clash between doctrinaire neo-Darwinists and religious fundamentalists. But whatever can be said for Stark's countermyth construction and his crusade against perceived secularist bias, his work makes for problematic history. My own view is that Stark is correct to assert that the role of Christianity in contributing to key institutions in the modern West has often been underplayed or misunderstood. But his polemical approach has its own limitations.

Specifically, his "victory of reason" unhappily conflates "reason" with science. Here again, we encounter an ahistorical understanding of "reason." Moreover, "modern science," from the seventeenth century to the present, has involved networks of different epistemic practices, some of which were actually premised on a deep suspicion of human reason (as conceived at the time) and its deliverances. This was especially true for the dominant early modern tradition of experimental science, which expressly opposed itself to the "speculative" reason-based approaches of the so-called Continental rationalists.[23] Second, his mission requires that to some extent he accept the premises of his opponents. This leads him to operate with abstract or reified notions of "science," "Christianity," and "religion" of the kind that most historians of science and religion have long abandoned.[24] These lapses are owing to Stark's unfamiliarity with relevant primary sources and lack of knowledge of the specific detail of much recent work in the history of science. This leads to the third and most serious deficiency: in spite of his apparent opposition to triumphalist historical narratives, Stark happily accepts the basic plot lines of the Whiggish histories he seems to be opposing. His goal is not to denounce this long-discredited historiography but simply to insert religion back into it, and ensure that Christianity gets its due as the major agent behind the progress of Western civilization—and as the unique selling proposition that gives the West its cultural superiority.

Is "neo-harmonist" an apt description of Stark's position? At first glance it may seem that he is better classified as an unwelcome gate-crasher at the party of the historian myth busters. He shares their frustration with the constant rehearsal of historical untruths and oversimplifications, but in his impatience to demolish the conflict myth and his flouting of the canons of historical scholarship, he comes close to fathering a new set of equally problematic oversimplifications of his own. Overall, I think he is right to argue that Christianity has not been accorded its proper place in the history of modernity, but he is right for all the wrong reasons. That said, though, he is a far better candidate for the neo-harmonist label than the historians originally singled out by Hollinger. Like the original harmonists, he takes the normative stance that there should be agreement between science and religion. He also believes that attempts to manufacture conflict are motivated by nonscientific factors. Like many of the original harmonists, he too, is inclined to think that there is some wriggle room in both evolutionary theory and Christianity that might need to be deployed in order for harmony to break out. In this respect, he differs from my next candidate for the neo-harmonist label, Denis Alexander.

Rebuilding the Matrix of Science and Religion

Denis Alexander was a practicing natural scientist who for many years chaired the Molecular Immunology Programme at the Babraham Institute, Cambridge. In 2008, he moved from his lab to assume full-time directorship of Cambridge's Faraday Institute for Science and Religion, which he helped found in 2006. Under his leadership, the Faraday Institute grew to become a key center for research and public engagement on science-religion issues. A long-time editor of the journal *Science and Christian Belief*, his activities in the science-and-religion space have included books, public lectures, broadcasts, and opinion pieces. He is one of the United Kingdom's most influential commentators on science and religion. Unlike Stark, his religious commitments have always been relatively clear: he identifies as an evangelical Christian. Throughout his career, he has consistently argued that science and mainstream Christianity are compatible: "The 'Book of God's Word' and the 'Book of God's Works' can be held firmly together in harmony."[25] Author or editor of over ten books, Alexander's key publications are *Rebuilding the Matrix: Science and Faith in the 21st Century* (2001), *Creation and Evolution: Do We Have to Choose?* (2009, 2014), *Rescuing Darwin: God and Evolution in Britain Today* (with Nick Spencer, 2009), and *Biology and Ideology from Descartes to Dawkins* (2010), coedited with Ronald L. Numbers.

Rebuilding the Matrix offers a helpful guide to Alexander's overall position. He notes at the outset that in popular culture, science is often associated with hostility to Christian faith. This is because, on the one hand, "a small but vocal group of scientists have insisted on using science as a weapon for attacking religious belief," and on the other, "creationists have carried out a vigorous campaign to ban the teaching of evolution in American schools."[26] Alexander aims to show that both sides are wrong, that science and religion can coexist, and that science need not be the dehumanizing force that it is often thought to be. The approach taken is multidisciplinary and draws on history of science, sociology, philosophy, theology, and, of course, science. The argument cannot be rehearsed in detail, but here are some of the main topics. Much of the book concerns the historical interactions between science and religion, comprising the origins of modern science, the Galileo affair, and Darwin and his religious supporters.

Generally, Alexander shows himself to be well informed by the history of science literature. While not unsympathetic to some of the claims advanced by Stark, he tends to endorse the myth buster's consensus that neither the conflict myth nor the model of an enduring harmony between science and religion are sustainable.[27] There are also topical treatments of evolution and big bang theory. On the first issue, Alexander suggests that evolution has no religious significance; on the second, he provides an outline of anthropic reasoning and fine-tuning arguments. The conclusion is not merely that science and religion are compatible, but that resituating contemporary science within the theistic frame in which it was originally born in the seventeenth century offers the best prospect of rehumanizing it in the present. This would constitute a rebuilding of the theological matrix within which Western science had its birth.

The later work, *Creation or Evolution*, offers some insights into how the public perceptions of science and religion changed in the first decade of the twenty-first century. This period witnessed the growing influence of scientific creationism and Intelligent Design in the United Kingdom and Europe. Once widely regarded as a parochial North American movement, religiously motivated opposition to evolution has now become a global phenomenon.[28] Research sponsored by the London-based Theos think tank, and released in the report *Rescuing Darwin*, revealed the surprising statistic that only 37 percent of the UK population believe that Darwin's theory of evolution has been established "beyond reasonable doubt." Moreover, 32 percent think that young-earth creationism is either definitely or partly true, while 51 percent believe that evolution alone is incapable of explaining the complexities of living things.[29] And this is the situation 150 years after the publication of Darwin's *On the Origin of Species*. These figures can be attributed to the growing international reach of the home-grown US product, but also, paradoxically, to the vocal insistence of the New Atheists that evolution represents some kind of refutation of religion, and that belief in evolution is incompatible with religious belief. Faced with this stark choice, many religionists have opted to reject evolution. Both camps, in other words, seem to promote the idea of science and religion as rival explanations of human origins.

While *Rebuilding the Matrix* was aimed at a broad, general audience, Alexander addresses *Creation or Evolution* primarily to evangelical Christians who believe, like him, "that the Bible is the inspired Word of God from cover to cover," and seeks to persuade them that biblical Christianity is compatible with acceptance of evolution by natural selection.[30] To this end, the book offers concise accounts both of the biblical doctrine of creation, and of the scientific consensus about evolution. It also includes an (approving) account of the views of the original nineteenth-century harmonists. Alexander also points out that a number of the authors of the *Fundamentals*, those pamphlets produced in 1910–15 from which the term "fundamentalism" derives, were committed to some form of evolutionary thinking. A reason for the subsequent parting of the ways, he suggests, was the hijacking of evolutionary theory for a variety of nefarious purposes, one of which was its use to promote an atheistic view of the world.[31]

In proposing an alternative to the conflict model, the book also goes well beyond simply gesturing to the existence of a Deity who lies behind the evolutionary process. It tackles a number of specific and taxing issues. Was there an original historical pair, Adam and Eve, who bore the image of God? Are human beings the accidental end products of evolution, or can evolutionary processes guarantee the appearance of human beings? Was there a historical Fall from original perfection, and if not, how does this square with Christian teaching concerning the necessity for redemption? Alexander does not seek to present *the* definitive set of answers to these questions but offers a number of models that show how traditional Christian beliefs about creation might be preserved while still allowing for evolutionary commitments.

Another objective of this book is to question one recent initiative that seeks to offer a third way between young-earth creationism and full-blown atheistic evolutionism: Intelligent Design. Advocates of ID, unlike scientific creationists, argue that there are good scientific reasons for asserting that living things have certain features that are so complex that they could not have arisen from the unguided processes of evolutionary change. Such features as the bacterial flagellum or the vertebrate blood-clotting system are said to be examples of irreducible complexity or specified complexity that are inexplicable without recourse to the idea of a designer. Alexander gives these arguments short shrift, suggesting that they are neither good science nor good theology. His conclusion is again an advocacy of harmony between evolution and science: "Holding to evolution as a biological theory should not affect one whit the Christian's belief in the uniqueness of humankind made in God's image, the Fall, the reality of Sin and our need for redemption through the atoning work of Christ on the cross for our sins."[32]

Francis Collins and *The Language of God*

Francis Collins, like Denis Alexander, has had a long career as a practicing scientist working in the life sciences. After undertaking a PhD in physical chemistry at Yale, Collins studied medicine at the University of North Carolina, eventually moving into genetics research. During the 1980s, Collins and his collaborators were involved in the identification of genes related to a number of medical conditions, including cystic fibrosis and Huntington's disease. In 1993, he succeeded James Watson (codiscoverer of the structure of DNA) as leader of the Human Genome Project, which in the early years of this century successfully produced a reference version of the human genome sequence. In 2009, President Barack Obama nominated him as director of the National Institutes of Health, the parent body of some twenty-seven institutes and centers that undertake research into biomedical science, which makes it the largest biomedical research organization in the world.

Collins has described how he began his graduate studies as an agnostic who gradually drifted toward atheism. At that time, he believed that "no thinking scientist could seriously entertain the possibility of God without committing some kind of intellectual suicide."[33] During his medical studies, however, his encounters with dying patients, along with a gradual acceptance of the case for Christianity set out in C. S. Lewis's *Mere Christianity*, led him to become an evangelical Christian.[34] His firm belief in the compatibility of his scientific vocation and his religious beliefs prompted him to publish *The Language of God: A Scientist Presents Evidence for Belief* (2006), which endeavors to show, among other things, "that belief in God can be an entirely rational choice, and that the principles of faith are, in fact, complementary with the principles of science."[35] The book soon found its way onto the *New York Times* best-seller list and remained there for sixteen weeks.

Owing to the popularity of *The Language of God*, Collins found himself inundated with queries from readers seeking clarification on various points made in the book. In response to these requests, Collins decided that he would develop a website devoted to responding to "frequently asked questions" on science and religion, and more specifically on evolution and the Christian faith. To this end, he also established the BioLogos Foundation in 2007, with the website of the same name being launched in 2009. (The main FAQ's are also available in book form in *The Language of Science and Faith*, co-authored with Karl W. Giberson.)[36] In addition to overseeing the website, the foundation sponsors events, disburses grants, and produces educational resources—all devoted to presenting a Christian perspective on science-religion issues.

The Language of God provides a comprehensive account of Collins's ideas about science and religion. It begins with an introductory biographical section, makes reference to cosmology, the fine-tuning argument, and briefly touches on the historical relations between science and religion. But like Alexander, Collins focuses primarily on the major site of contemporary science-religion conflict—evolution. Indeed, of our three exemplars, it is Collins who most consistently uses the terminology of "harmony" when speaking of the science-religion relation. In *The Language of God*, he enumerates four possible responses to the contentious interactions between evolution and faith in God: first, atheism and agnosticism (science trumps faith); second, creationism (faith trumps science); third, Intelligent Design (science needs some divine help); and fourth, Collins's preferred option, what he calls "BioLogos" (science and faith in harmony). This fourth option is really another label for theistic evolution—the idea that "God chose the elegant mechanism of evolution to create microbes, plants, and animals of all sorts." This view, Collins maintains, "is the dominant position of biologists who are also serious believers," and he notes that it is the position that many Christians adopt. (Like Alexander, he is dismissive of Intelligent Design on both religious and scientific grounds.) The neologism "BioLogos" is intended as a synonym for theistic evolution: "'BioLogos' expresses the belief that God is the source of all life and that life expresses the will of God."[37]

Collins's advocacy of harmony goes some way beyond simply asserting the compatibility of evolution and Christianity. Thus one might think of other compatibilist options such as "uneasy coexistence," "mutual indifference," or simply "not incompatible." But Collins's version of "harmony" is stronger than these options. In his view, science actually provides positive support for religious belief. Evolution, he suggests, provides "elegant evidence of the relatedness of all living things"; it is "the master plan of the same Almighty who caused the universe to come into being," and it provides "an occasion for awe." "The tools of science," he maintains, "uncover some of the awesome mysteries of His creation." This leads Collins to the venerable seventeenth-century idea that the formal study of nature is a form of religious worship. God, he writes, "can be worshipped in the cathedral or in the laboratory," and "science can be a form of worship."[38] Specifically in relation to evolution, the BioLogos website states that acceptance of evolutionary creation "can aid the church's mission: to worship our Creator

God, raise Christian young people, and bring people to Christ."[39] Claims such as these represent as strong a harmonist perspective as one is likely to find.

As significant as the arguments Collins presents for the harmony between science and religion is his status as a high-profile scientist. Collins's scientific credentials and research record are impeccable; he is the leader of what can reasonably be regarded as the largest scientific institution in the world; and his area of expertise, genetics, is contiguous with the theory that generates most tension with religious faith—evolution. His biography, then, gives the lie to the notion that one cannot be a respectable scientist and orthodox evolutionist and still cherish conventional Christian beliefs. His appointment to the NIH was seen by some as an explicit statement by the Obama administration of its view that the war between science and religion was a phony one.[40] Not unexpectedly, there were rumbles about his appointment from Sam Harris, Jerry Coyne, and Steven Pinker—critics of religion who argue for the incompatibility of evolution and Christianity—perhaps prompted in part by the realization that this high-profile scientific role, combined with outspoken profession of religious belief, represents a powerful argument for the compatibility of evolution and Christian belief.[41]

Finally, it should be pointed out that like Alexander, Collins promotes harmony not only because he thinks that Christianity will be damaged by ongoing conflict with science. He also believes that the social standing of science is seriously harmed by the advocates of conflict. Thus, the hostility toward evolution that persists in the United States, and indeed has spread globally, can be partly attributed to the view now expressed most vociferously by the New Atheists, who insist that evolution is not compatible with religious belief: evolution or religion, but not both. It is certainly possible that pursuing this line will turn some away from religion, which is presumably the goal of the New Atheists. But for the many who value their religious beliefs over commitment to scientific orthodoxy, the choice will be made for religion and against evolution. This, insist our neo-harmonists, is a false choice, and is to the detriment of both science and religion.

Conclusion

In this chapter, I have used the term "neo-harmonist" to pick out a group of contemporary thinkers who argue for the compatibility of science and religion. This version of compatibility is not just a matter of suggesting that science and religion had largely harmonious relations in the past but that, properly understood, science and religion *ought to* have a harmonious coexistence *in principle*. The three figures taken to exemplify this position are by no means the only possible candidates, and many other names from the present and recent past could be added: historians Reijer Hooykaas and Eugene Klaaren; theologians Keith Ward, Nancey Murphy, John Haught, Alister McGrath; scientists Stanley Jaki, Francisco Ayala, John Polkinghorne, Arthur Peacocke, Simon Conway Morris, George Ellis, and Robert J. Russell. Nor is this list exhaustive; it could be supplemented with the names of many of the members of the International Society for Science and Religion, which exists "to facilitate dialogue between the two academic disciplines of science and religion."[42]

It is worth asking, then, whether this group is sufficiently coherent to warrant its own label and whether that label ought to be "neo-harmonist." Like any "type" or unit of formal classification, the users of the label run into the dangers of oversimplification, misidentification, and essentializing. Nonetheless, analytical categories like this one can be helpful and sometimes offer a useful starting point for productive discussions.[43] The important thing is to be clear about how the label is to be used. My suggestion has been that part of the logic of this particular term lies in the fact that neo-harmonists can be thought of as those who, in the present, project attitudes similar in many respects to those religious figures in the nineteenth century who sought a rapprochement between evolution and Christian theology. Admittedly, there is a complication here, because the original harmonists included two subgroups that are not analogous to our contemporary neo-harmonists. First were those who believed that Christian theology needed significant modification in order to be consistent with Darwinism, while a second faction embraced a version of evolution but remained doubtful about the mechanism of natural selection. These two subdivisions have their twenty-first-century counterparts—the first in those religious thinkers who, partly because of pressure from evolutionary ideas, have moved in the direction of process theology or panentheism (the view that the world in some sense exists within God) in order to better accommodate theology to modern science. These would include individuals such as A. N. Whitehead, Charles Hartshorne, John B. Cobb, David Ray Griffin, Joseph Bracken, and, to some extent, Ian Barbour (process theology), along with figures such as Arthur Peacocke and Phillip Clayton (panentheism). The second group would correspond to advocates of Intelligent Design, who accept some version of evolution but not the whole package. Of our three neo-harmonists, Stark is closer to these latter groups in his identification as an "independent Christian" and in his reluctance to commit to a fully fledged neo-Darwinism. Alexander and Collins, however, are unequivocal in their acceptance of evolution and their profession of religious orthodoxy (assuming that evangelical Protestantism can be so classified).

A further issue is raised by the prefix "neo-," with its implication that only after a significant hiatus did a new breed of advocates of harmony between science and religion arise in the late twentieth century, and that there was something noteworthy about their reemergence. But there is nothing particularly exceptional about the view that science and religion are compatible, and indeed it has been a standard position since the discussion began in the nineteenth century. Is there, then, a more adequate term that might encapsulate the positions set out above, which involve a dual commitment to the truth of traditional Christianity and reliability of contemporary science? As noted at the outset, the historians who have criticized the conflict myth do not fit here, since their historical endeavors require neither Christian commitment nor the normative view that science and religion *should not* be in conflict. So the appellation "historian myth buster" remains a more adequate label for them. "Accommodationist" is an alternative. It is often intended to be a pejorative term, redolent of the kind of craven compromise that characterized the actions of Neville Chamberlain, and it has been applied by atheist apologists Jerry Coyne and PZ Myers to traitorous scientists (and historians, too) who have refused to acknowledge that religion is the real enemy.[44] But while "accommodationist" has certainly been applied to religious figures such as Collins and the Roman Catholic molecular biologist Kenneth Miller, it has also been routinely used to describe atheist philosopher Michael Ruse and representatives of the National Center for Science Education such as Eugenie C. Scott. These latter individuals might believe that there is no conflict between science and religion, but they do not speak explicitly of harmony and do not necessarily endorse any form of religion. So "accommodationist" is too broad a term for the group in question.

In sum, perhaps it is best to have in mind three overlapping groups in this discussion—"myth busters," who simply point out that *history* is not characterized by an enduring conflict between science and religion; "accommodationists" who argue that there is no need, *in the present*, for science-religion conflict; and our third group, who typically belong in each of the previous categories, but who are also motivated by overt religious concerns. "Neoharmonist" could work for this last group, but the "neo-" brings with it misleading connotations. "Christian accommodationist" would work just as well, or perhaps simply "harmonist."

NOTES

1. "'Evidence for Belief': An Interview with Francis Collins," Pew Forum, April 17, 2008, http://www.pewforum.org/2008/04/17/the-evidence-for-belief-an-interview-with-francis-collins.

2. David A. Hollinger, *After Cloven Tongues of Fire: Protestant Liberalism in Modern American History* (Princeton, NJ: Princeton University Press, 2013), 84. A similar line is pursued in Hollinger's *Science, Jews, and Secular Culture: Studies in Mid-Twentieth-Century American Intellectual History* (Princeton, NJ: Princeton University Press, 1996).

3. John Hedley Brooke, *Science and Religion: Some Historical Perspectives* (Cambridge: Cambridge University Press, 1991), 2–5, 42–51.

4. See, e.g., Steven Shapin, *The Scientific Revolution* (Chicago: University of Chicago Press, 1996), 195; David C. Lindberg and Ronald L. Numbers, eds., *When Science and Christianity Meet* (Chicago: University of Chicago Press, 2003), 2–3; Numbers, ed., *Galileo Goes to Jail and Other Myths about Science and Religion* (Cambridge, MA: Harvard University Press, 2009), 6, 79–89; Thomas Dixon, *Science and Religion: A Very Short History* (Oxford: Oxford University Press, 2008), 15; Thomas Dixon, Geoffrey Cantor, and Stephen Pumfrey, eds., *Science and Religion: New Historical Perspectives* (Cambridge: Cambridge University Press, 2010); Richard G. Olsen, *Science and Religion, 1450–1900: From Copernicus to Darwin* (Baltimore, MD: Johns Hopkins University Press, 2004), 4–7; Colin A. Russell, "The Conflict of Science and Religion," and David B. Wilson, "The Historiography of Science and Religion," both in Gary Ferngren, ed., *Science and Religion: A Historical Introduction* (Baltimore, MD: Johns Hopkins University Press, 2002), 3–12, 13–29; Matthew Stanley, *Huxley's Church and Maxwell's Demon* (Chicago: University of Chicago Press, 2015), 264–70.

5. For the revisionist label, see William Durbin, "Science," in Philip Goff, *The Blackwell Companion to Religion in America* (Oxford: Blackwell, 2010), chap. 22.

6. In addition to Moore's book, see Ronald L. Numbers, *Darwinism Comes to America* (Cambridge, MA: Harvard University Press, 1998); Jon H. Roberts, *Darwinism and the Divine in America: Protestant Intellectuals and Organic Evolution* (Madison: University of Wisconsin Press, 1988; repr., Notre Dame, IN: University of Notre Dame Press, 2001); David Livingstone, *Darwin's Forgotten Defenders: The Encounter between Evangelical Theology and Evolutionary Thought* (Vancouver: Regent College Publishing, 1984).

7. Alexander Winchell, *Reconciliation of Science and Religion* (New York: Harper and Brothers, 1877), 305, cf. iv, 222, 358, 379; Newman Smyth, *Old Faiths in a New Light* (New York: Charles Scribners's Sons, 1879), 28.

8. James McCosh, *The Religious Aspect of Evolution* (New York: G. P. Putnam's Sons, 1888), 69.

9. John Fiske, *The Idea of God as Affected by Modern Knowledge* (Boston: Houghton, Mifflin, 1885), 96. Cf. Fiske, *Darwinism and Other Essays* (London: Macmillan, 1879), 7.

10. On the modern uses of the metaphor, see K. van Berkel and Arjo Vanderjagt, eds., *The Book of Nature in Early Modern and Modern History* (Leuven: Peeters, 2006), esp. the essay by Ronald L. Numbers, "Reading the Book of Nature through American Lenses," 261–74.

11. Rodney Stark and William S. Bainbridge, *A Theory of Religion* (Toronto: Peter Lang 1987); Rodney Stark, "Rational Choice Theories of Religion," *Agora* 2 (1994): 1–5; Rodney

Stark and Laurence R. Iannaccone, "A Supply-Side Reinterpretation of the 'Secularization' of Europe," *Journal for the Scientific Study of Religion* 33 (1994): 230–52.

12. Rodney Stark and Roger Finke, *The Churching of America, 1776–1990: Winners and Losers in Our Religious Economy* (New Brunswick, NJ: Rutgers University Press 1992).

13. Rodney Stark, *The Rise of Christianity: A Sociologist Reconsiders History* (Princeton,, NJ: Princeton University Press, 1996), 167.

14. Rodney Stark, *For the Glory of God: How Monotheism Led to Reformations, Science, Witch-Hunts, and the End of Slavery* (Princeton,, NJ: Princeton University Press, 2003), 185.

15. Rodney Stark, *The Victory of Reason: How Christianity Led to Freedom, Capitalism, and Western Success* (New York: Random House, 2005), 12.

16. Stark, *For the Glory of God*, 197.

17. My own work, for example, argues strongly for the Christian origins of certain key features of modern science. See Peter Harrison, *The Bible, Protestantism and the Rise of Natural Science* (Cambridge: Cambridge University Press, 1998); Harrison, *The Fall of Man and the Foundations of Science* (Cambridge: Cambridge University Press, 2007), Harrison, *The Territories of Science and Religion* (Chicago: University of Chicago Press, 2015).

18. Rodney Stark, "Fact, Fable, and Darwin," *Meridian Magazine*, May 2, 2010, http://web.archive.org/web/20050404045331/http://www.meridianmagazine.com/ideas/050210darwin.html. And elsewhere in this article: "[T]he theory of evolution is regarded as the invincible challenge to all religious claims."

19. Stark, "Fact, Fable, and Darwin."

20. John S. Wilkins, "Rodney Stark's Idiotic History," ScienceBlogs, September 6, 2008, http://scienceblogs.com/evolvingthoughts/2008/09/06/rodney-starks-idiotic-history; Alex Williams, "The Biblical Origins of Science," review of *For the Glory of God* by Rodney Stark, Creation.com, n.d., http://creation.com/the-biblical-origins-of-science-review-of-stark-for-the-glory-of-god; Andrew S. Kulikovsky, "A Revealing Look at the World's Religious Belief Systems," review of *Discovering God: The Origins of the Great Religions and the Evolution of Belief* by Rodney Stark, Creation.com, http://creation.com/review-stark-discovering-god.

21. For discussions of Stark's religious commitments, see David Lehmann, "Rational Choice and the Sociology of Religion," in Bryan D. Turner, ed., *The New Blackwell Companion to the Sociology of Religion* (Oxford: Blackwell, 2010), 181–200, esp. 183–84, and more recently the 2004 interview by Michael Aquilina, National Institute for the Renewal of the Priesthood, http://www.jknirp.com/stark.htm; and the 2007 interview by Massimo Introvigne, Centro Studi sulle Nuove Religion, http://www.cesnur.org/2007/mi_stark.htm.

22. Stark, *For the Glory of God*, 13.

23. See Harrison, *Fall of Man*; Peter Anstey, "Experimental versus Speculative Natural Philosophy," in P. Anstey and J. Schuster, eds., *The Science of Nature in the Seventeenth Century: Patterns of Changes in Early Modern Natural Philosophy* (Dordrecht: Springer, 2005), 215–42.

24. Stark, *For the Glory of God*, 4, 124–25; Stark, *Victory of Reason*, 12–16.

25. Denis Alexander, *Creation or Evolution: Do We Have to Choose?* (Oxford: Monarch, 2008), 13.

26. Alexander, *Creation or Evolution*, 8.

27. Denis Alexander, *Rebuilding the Matrix: Science and Faith in the 21st Century* (Grand Rapids, MI: Zondervan, 2001) 62–64, 478n3.

28. Simon Rogers, "God or Darwin: The World in Evolution Beliefs," *Guardian*, July 1, 2009, http://www.theguardian.com/news/datablog/2009/jul/01/evolution; Ronald L. Numbers, *The Creationists: From Scientific Creationism to Intelligent Design*, rev. ed. (Cambridge, MA: Harvard University Press, 2006), chap. 18; R. Koenig, "Creationism Takes

Root Where Europe, Asia Meet," *Science* 315 (2007): 579a; G. S. Levit, U. Hoszfeld, L. Olsson, "Creationists Attack Secular Education in Russia," *Nature* 444 (2006): 265; S. Halmeed, "Science and Religion: Bracing for Islamic Creationism," *Science* 322 (2006): 1637–38.

29. Nick Spencer and Denis Alexander, *Rescuing Darwin: God and Evolution in Britain Today* (London: Theos, 2009), 9.

30. Alexander, *Creation or Evolution*, 11.

31. Alexander, *Creation or Evolution*, 179–80.

32. Alexander, *Creation or Evolution*, 351.

33. Francis Collins, *The Language of God: A Scientist Presents Evidence for Belief* (New York: Free Press, 2007), 16.

34. Collins, *Language of God*, 16–31.

35. Collins, *Language of God*, 3.

36. Karl W. Giberson and Francis S. Collins, *The Language of Science and Faith* (Downers Grove, IL: InterVarsity Press, 2011).

37. Collins, *Language of God*, 159–212, quotes on 201, 199, 203.

38. Collins, *Language of God*, 199, 211, 230.

39. "Why Should Christians Consider Evolutionary Creation?," BioLogos, https://biologos.org/common-questions/christianity-and-science/why-should-christians-consider-evolutionary-creation.

40. Steve Waldman, "Francis Collins as a Culture War Statement," *Wall Street Journal*, July 10, 2009, http://www.wsj.com/articles/SB124718874130520769.

41. Sam Harris, "Science Is in the Details," *New York Times*, July 26, 2009, http://www.nytimes.com/2009/07/27/opinion/27harris.html?pagewanted=all&_r=0. For criticisms surrounding the appointment, see Amanda Gefter, "New Détente in Science-Religion War?," *New Scientist*, July 14, 2009, http://www.newscientist.com/blogs/shortsharpscience/2009/07/does-nih-pick-signal-detente-i.html.

42. International Society for Science and Religion, http://www.issr.org.uk/about issr.

43. For standard typologies of science-religion relations, see Ian G. Barbour, "Ways of Relating Science and Theology," in Robert J. Russell, William R. Stoeger, S.J., and George V. Coyne, S.J., eds., *Physics, Philosophy, and Theology: A Common Quest for Understanding* (The Vatican: Vatican Observatory Publications, 1988), 21–48; Geoffrey Cantor and Chris Kenny, "Barbour's Four-Fold Way: Problems with Its Taxonomy of Science-Religion Relationships," *Zygon* 36 (2001): 765–81; John Haught, *Science and Religion: From Conflict to Conversion* (New York: Paulist Press, 1995), 9; Mikael Stenmark, *How to Relate Science and Religion: A Multidimensional Model* (Grand Rapids, MI: Eerdmans, 2004).

44. See, e.g., Michael Ruse, "Accommodationism in the Religion-Science Debate: Why It's Incomplete," *Huffington Post*, May 25, 2011, http://www.huffingtonpost.com/michael-ruse/accommodationism-and-why-_b_715915.html.

CHAPTER FOURTEEN

Historians

JOHN HEDLEY BROOKE

My aim in this chapter is to sample a few of the historical studies that, during the past fifty years, have exposed defects in the classic narratives of a conflict between science and religion, often using them as didactic devices to introduce more sophisticated historiographical principles. The examples are discussed chronologically and have been chosen for their distinctiveness, direct engagement with a conflict historiography, and cumulative impact. They illustrate a range of strategies that historians have employed to qualify, reconfigure, or undermine the resilient idea of a necessary and therefore perennial antagonism between religious belief and the culture of science. Critically, this idea can be challenged by the disclosure of enormous diversity in science-religion discourse, in many cases not involving oppositional stands. Historicizing the conflict theses, explaining why they came to prominence when they did, has been another strategy in stripping them of timeless pretensions. Revisionist historiographies had already turned the tables during the 1930s, as in claims for a beneficial impress of the Christian doctrine of creation on the metaphysical foundations of early modern science[1] and in Robert Merton's thesis of a catalytic stimulus to the practical sciences from Puritan values that sanctioned diligent study of the creation as a means of glorifying its Creator.[2] Andrew Dickson White's nineteenth-century warfare model had also been challenged by Perry Miller, who in his classic account of New England Puritans argued that such indifference as they showed to scientific developments was not to be confused with hostility.[3] A plea for greater discrimination, as in distinctions between liberal and conservative members of a faith tradition, or between the theological implications of different sciences, or between the scientific implications of different religious doctrines, has been another strategy. In a telling observation, Frank Turner has noted that "a man of science who was thoroughly naturalistic in his own area of research might embrace a broad theism or ethical idealism for his understanding of the scientific enterprise as a whole."[4] Philological critiques have also played their part, as in Peter Harrison's contention that there could not have been conflict between *science* and *religion* before the seventeenth century, and arguably until the nineteenth, because the two words "science" and "religion" did not yet have the meanings they have in popular discourse today.[5]

A different solvent has been a reconfiguring of the conflict thesis by transferring the locus of conflict to somewhere other than a straightforward clash between scientific and religious propositions. When contextualizing conflicts assumed to have been between science and religion, historians have favored "thicker" descriptions, often involving social and political issues as the mainsprings of controversy. Relocation to a clash between different modes of science, where theology might stake a claim in one rather than another, has been one reconfiguration, as in accounts of Galileo (1564–1642) and his trial (1633), where the damage is inflicted by theologians too ready to sanctify an Aristotelian, to the detriment of a heliocentric, cosmology. Or the relocation might be to a conflict between religion and religion, as when religious reformers appeal to contemporary science when seeking to purge their religious tradition of irrationality. An example would be Joseph Priestley's (1733–1804) insistence that science and religion were fighting on the same side against infectious superstition.[6] Seemingly high-profile science-religion disputes have also been reconfigured as "proxies for more deep-seated ideological or, in its broadest sense, 'theological' battles," as with religiously motivated antievolutionists "who fear not the 'science' as such, but the secularist package of values concealed in what they perceive to be the Trojan horse of evolutionary theory."[7] Or again, the conflict ostensibly between science and religion in Victorian Britain has been reconfigured as dissonance between competing methods of harmonization.[8]

I begin my survey in the mid-1960s, when the historian Bruce Mazlish could still claim that White, cofounder of Cornell University and author of the widely read *History of the Warfare of Science with Theology in Christendom* (1896), had established his warfare thesis beyond reasonable doubt.[9] This was a decade during which "science and religion" emerged as a comprehensive field of study[10] and when the positivism that had frequently underpinned the conflict narratives was challenged by new models of scientific methodology and epistemology.[11] Thomas Kuhn, among others, was arguing that "facts" were theory-laden and that, at times of revolutionary change in science, what counted as verification depended on the theoretical "paradigms" in contention.[12] The rhetoric of a single, omni-competent "scientific method," so often contrasted with and marshaled against religious belief, was also under attack. Possibilities for a greater flexibility in the interpretation of the classic "conflicts between science and religion" arose as a consequence, as they did from new visions of how the history of science as a discipline should be practiced.

The question of best historical practice arose in the course of a lively, if overheated, debate between so-called internalists and externalists—the former broadly content to analyze the internal, technical development of scientific theories; the latter increasingly drawn to the external cultural factors that made possible, and shaped, scientific research in specific historical contexts.[13] There was room for the integration of the two approaches, but as the discipline became increasingly professionalized, as it embraced the methods and techniques of social, economic, and political history, one consequence was a greater willingness to investigate how religious beliefs and institutions had affected scientific activity. By the mid-1970s, sophisticated scholarship was appearing that drew attention to subtle correlations between scientific and religious sensibilities. A striking example was Arnold Thackray's biographical analysis of early members of the Manchester Literary and Philosophical Society, revealing a preponderance of Unitarians among proponents of science and industry.[14] Within their circle, worshipping at Manchester's Crosse Street Chapel, there was no conflict between science and their dissenting religion.

The year in which Thackray published his analysis (1974) also saw the inauguration of an Open University (OU) distance-learning course, "Science and Belief from Copernicus to Darwin," in which the teaching units were designed to introduce students to serious historical scholarship on controversies caricatured in popular literature. A team of authors, including the OU's first professor of the history of science, Colin Russell, encouraged students to grasp the shortcomings of a Whiggish historiography and the dependence of the conflict narratives on it. The course materials emphasised the political complexity of the Galileo affair, the interplay between physics and metaphysics in the mechanization of nature, the sometimes intimate interpenetration of natural philosophy and religion in British natural theology, the putative links between science, technology, and religious dissent, and the diversity of religious responses to Darwinian evolution. Students were steered away from glib generalizations, as when Russell pointed to the "paradox" that organized Christianity, as represented by Catholic and Protestant churches, provided the bitterest opposition to, and an ultimate inspiration for, the new sciences of the seventeenth century.[15] I include this OU course, with its affiliated radio and TV broadcasts, in my sample of distinctive revisionist literature because student registration for the class, and for its successor, "Science and Belief from Darwin to Einstein," extended to thousands. It was a successful vehicle for propelling a critical assessment of the conflict thesis into the public domain.

One writer for the "Darwin to Einstein" course was James Moore, whose interests included the role of Calvinist theology where it facilitated rather than blocked the assimilation of Darwin's science. Moore's analysis in *The Post-Darwinian Controversies* (1979) was distinctive because it included a relentless, one hundred–page attack on the conflict thesis, its military metaphors, and triumphalist rhetoric. Moore's claim was bold: Darwin's theory of evolution by natural selection could be accepted in substance *only* by those Christians whose theology was distinctly orthodox.[16] Whereas liberal theologians blithely placed an optimistic gloss on evolutionary progress, Calvinists who had struggled with the problem of reconciling free will and predestination were better able to cope with the conjunction of chance and necessity in Darwin's natural selection.

One of Moore's strategies exemplified the technique of shifting the locus of conflict away from "religion versus science" to a different polarity. In this case, it was to philosophy versus science, or more precisely to a Baconian, inductivist philosophy of science, fortified with a Platonist conception of the fixity of species, versus the more speculative hypothetico-deductive science of Darwin. As Moore put it, Christian anti-Darwinism emerged from a conflict between Darwinian doctrines and "certain fundamental philosophical, rather than specifically Christian beliefs."[17] This thesis was not without problems, since there was no shortage of anti-Darwinian Presbyterians. Critics might also ask: Why the "rather than" in the above sentence? Why not "both . . . and"? When philosophical critiques of Darwin's *On the Origin of Species* (1859) were launched by Christians, might not their motivation have been primarily theological? Samuel Wilberforce (1805–73), bishop of Oxford, self-consciously devoted the majority of his review to philosophical shortcomings in Darwin's argument, yet he also maintained that the theory was "utterly irreconcilable" with Christian teaching on creation and redemption.[18]

Nevertheless, to argue that conflicts about science and religion were, deep down, conflicts about something else, was a seductive way of rewriting the narratives of both White and John William Draper, whose *History of the Conflict between Religion and Science* (1874) was equally influential. Another was to challenge simplistic categories through case studies that exposed richness and diversity in the ways science and religion had been mutually relevant. The landmark text here was *God and Nature* (1986), edited by David Lindberg and Ronald Numbers. With its eighteen chapters, ranging from science in the early Christian church to the appraisal of science in twentieth-century Protestant theologies, a rich tapestry was produced of the ways in which relations between science and religion had been constructed.[19] Diversity among the early church fathers in their attitudes to the study of nature was underlined by Lindberg. Exposure of a variety of opinion on the latest science, even within the same branch of Christianity, featured prominently, notably in William Ashworth's account of seventeenth-century Catholic scholars (Marin Mersenne, Pierre Gassendi, René Descartes, Blaise Pascal, and Nicolaus Steno) who, as did their Protestant counterparts, differed markedly in their theologies of nature. By 1986, the store of quality historical literature on the interrelations between science and religion had been significantly increased. On the subject of Puritanism and science, for example, Charles Webster had stressed the role of millenarian theologies in utopian visions of a society improved by better agriculture and medicine.[20] Webster contributed to *God and Nature*, highlighting principles that he believed were paramount: "It would be totally unrealistic to write the history of the emergence of English science in the period 1560–1660 in terms of monocausal derivation with respect to Protestantism, Anglicanism, or Puritanism, however they are defined." But, he added, "to equate this history with the story of the great discoveries, or to construe science as an entirely autonomous development, unrelated to the Reformation and the Puritan Revolution or to the socioeconomic framework of which Puritanism was a constituent element, is to eliminate vital factors in the explanatory mosaic." Importantly, "any truly historical account of the Scientific Revolution must pay due attention to the deep interpenetration of scientific and religious ideas." It would seem perverse, he surmised, "to deny religious motivation in the numerous cases where this was made explicit by the scientists themselves, often with painful emphasis."[21]

Historiographical precepts sensitive to past interpenetration of science and religion have become increasingly visible. There was no suggestion in *God and Nature* that scientific innovation had not sometimes threatened conservative religious beliefs. Indeed, by comparing Johannes Kepler

(1571–1630), René Descartes (1596–1650), and Isaac Newton (1642–1727), Richard Westfall correlated the decline of orthodox Christianity with the rise of science. The editors also expressed unease about overblown revisionist history written by religious apologists. The Calvinist Reijer Hooykaas and the Catholic Stanley Jaki were both accused of having "sacrificed careful history for scarcely concealed apologetics."[22] Hooykaas, for example, had stressed the role of Reformed theology in a de-deification of nature that he considered a prerequisite of modern science. This raises a delicate issue: Do definitive criteria exist for discriminating between apologetic and nonapologetic history when a suspected apologia is not overt? Critics of both Hooykaas and Jaki certainly did complain of their reluctance to acknowledge scientific achievement in theological cultures other than their own.[23] *God and Nature* therefore raised the absorbing question of whether knowledge of an author's biography is indispensable, desirable, or irrelevant when evaluating a thesis that casts a particular tradition of religious belief in a favorable light.

The extent to which different facets of Christianity were auspicious or inauspicious for a viable science of nature was far from settled in the 1970s and 1980s, as critics of the conflict thesis were themselves subjected to criticism. Of these critiques of critiques, that of Rolf Gruner was one of the more telling. Gruner failed to see how, from a religious value system in which nature was highly esteemed as a divine creation but knowledge of nature was not, it was possible to arrive at a scientific value system in which knowledge of nature was prized but nature itself was not.[24] This was, however, more a philosophical than a historical critique, and it was impossible to read *God and Nature* without concluding that the conflict theses, in their doctrinaire forms, were no longer tenable. They had been thoroughly disaggregated, reconfigured, and denounced. For example, in a chapter on the professionalization of geology, Martin Rudwick insisted that specific episodes of conflict should be regarded as episodes in which people on both sides appealed to some aspect of nature to support and justify their own view of the *meaning* of personal and social life. The historical relationship between science and religion had involved a process of "gradual differentiation and divergence, rather than the replacement of one by the other, as the older positivist tradition maintained."[25]

Historiographical refinement when discussing the post-Darwinian debates was achieved by Jon Roberts, whose *Darwinism and the Divine in America* was published in 1988. Roberts eschewed the triumphalist conflict narratives but was equally critical of another "fiction," in which genuine tensions between Christianity and a science of evolution were minimized. These tensions had contributed to the privatization of religion in North America by 1900. In fact, they were more "insidious" in their consequences than straightforward warfare might have been. This was because *indifference* to a theophanic view of nature was ultimately more subversive than an oppositional stance, which, in the very act of confronting religious positions, tacitly conferred importance on them. There was also irony in Roberts's reconstruction, one of many in which ideas once safely embedded in Christian literature (such as mechanical models of nature in seventeenth-century Europe) have subsequently backfired on the very theologies that nurtured them. It is, Roberts concluded, "one of the ironies of history that after freeing themselves from the Catholic church's authority, Protestants increasingly found their own understanding of the relationship between nature and nature's God limited, even determined, by their affiliation with science."[26] In a much older book, John Dillenberger's *Protestant Thought and Natural Science* (1959), the irony was that a science-inspired natural theology dug its own grave by exaggerating evidence for design. It had also damaged Christian faith by deflecting attention from a theology of redemption to one of creation.[27] From a Catholic perspective, Michael Buckley would argue a comparable case in 1987.[28]

The extensive bibliographic essay in my own *Science and Religion: Some Historical Perspectives* (1991) shows how a burgeoning historical literature could benefit from synthesis and critical appraisal.[29] This monograph has been seen as facilitating a new consensus about the "the right way" to approach the history of engagement between science and religion.[30] Striving for a balanced critique, historical examples were enlisted not only against the conflict thesis, but also against all meta-narratives that streamlined a rich, diverse, and complex history for polemical or apologetic purposes. A distinctive feature was the refusal to treat past science-religion controversies as if two forces were battling for the same territory, occupying the same flat plane. To capture the many different levels on which there could be mutual relevance, it was imperative to be thinking in three dimensions. Accordingly, I offered a conceptual analysis, using historical examples to show how religious and metaphysical beliefs had penetrated the study of nature as presuppositions, motives, and sanctions, regulating methodology and playing a selective role when deciding between competing scientific theories. This typology, which also included the explanatory role of religious belief when it functioned as primitive science, allowed one to say that there could simultaneously be conflict on certain levels and harmony or separation on others. The book constituted a plea for sharper differentiation on other matters: on distinctions between the different sciences as they had impinged on religious sensibilities in different ways, and between different doctrines within the same religion, where some might be constructive and others obstructive to scientific inquiry. My assault on uncritical harmonization narratives showed that strenuous attempts to demonstrate compatibility were often in response to already existing tensions or allegations of conflict—as with the compatibilist strategies in Galileo's *Letter to the Grand Duchess Christina* (1615), when Galileo already knew that the Copernican system was eliciting clerical disapproval.[31] It is an intriguing question why my monograph appears to have been appreciated more for its critique of the conflict thesis than for its admonition concerning claims for an underlying harmony. Perhaps antireligious polemicists do not generally feel the need of help from historians, whereas critiques of perennial conflict can ingratiate themselves to religious apologists.[32]

Soon to be dignified as the author of the "complexity thesis," I concluded my anti-essentialist critique with the assertion that "there is no such thing as *the* relationship between science and religion. It is what different individuals and communities have made of it in a plethora of different contexts."[33] This was a protest against dehumanized philosophical approaches to the subject, preoccupied with attempts to determine the "best account" of the relationship between science and religion—as if some single magic formula could meet an imagined all-encompassing need. In this connection, Noah Efron has observed that those seeking from history comforting guidance on how to relate a personal faith to the scientific enterprise might find my quasi-nominalist approach, with its multiplication of non-superimposable stories, demoralizing.[34]

The year 1991 also saw the publication of Geoffrey Cantor's biography of Michael Faraday (1791–1867), which challenged the common view that Faraday kept his science and religious faith completely separate. A lifelong member of a minority Christian sect, the Sandemanians, Faraday took his Bible seriously in a life of stern moral discipline, regular worship, and pastoral duty. Cantor suggested that Faraday's religious practice found expression in a disciplined scientific life, in which Sandemanian social philosophy was transferred to the organization of science.[35] Faraday's vision was of a scientific community without avarice, partisan interests, or personal disputes. In the practice of science, as in the practice of religion, he mistrusted earthly rewards, called into question the entrepreneurial spirit, and disliked interventionist patronage. The scientist was a moral agent whose knowledge was for sharing and edification. Cantor's biography was a subtle assault on essentialist, ideological positions that proclaimed either the mutual irrelevance of science and religion or a necessary conflict between them. That Faraday, and his illustrious successor in electrodynamics, James Clerk Maxwell (1831–79), combined a life of science with a conservative, biblically regulated faith, was a clear embarrassment to interpretations premised on inevitable conflict. It had always been a weakness of the conflict narratives that they struggled with the category of the deeply religious scientist. Cantor even proposed that features of Faraday's physics were almost certainly shaped by his deepest theological principles. Belief that the universe was the product of a single mind could encourage the conviction that the forces of nature were ultimately one and therefore interconvertible. Only the gravitational force eluded Faraday's unification, which had succeeded with the magnetic, electrical, and chemical.

My collaboration with Cantor in our 1995 Gifford Lectures reinforced a critique of the common master-narratives with the insistence that whenever questions are raised concerning relations between science and religion, the first response must be "whose science and whose religion?" In our book *Reconstructing Nature* (1998), the Galileo affair was analyzed to expose the many different parameters (scientific, epistemological, personal, political, and theological) that must be considered when examining the contingencies that led to the tragic trial. A distinctive feature of the book was a determination to show how different styles of historical scholarship could be illuminating in different ways. Thus a conventional "history of ideas" approach was adopted to show how information from natural history and natural philosophy was incorporated into a genre of natural theology that survived well into the nineteenth century. The importance of aesthetic judgment and rhetorical technique to the language of natural theology was also explored. A social history approach was used to illuminate the role of a religious minority—the Quakers—in science, an exploration that led to Cantor's important comparative analysis in *Quakers, Jews and Science*.[36] The particular value of biographical studies was underlined in *Reconstructing Nature* by comparing four Victorian scientists, an Anglican, Adam Sedgwick (1785–1873); a Roman Catholic, St. George Mivart (1827–1900); a Unitarian dissenter, William Carpenter (1813–85); and an agnostic, John Tyndall (1820–93). Bringing unique individuals back to life was also a means of transcending simplistic categorization. If one asked, for example, which of Ian Barbour's four categories of conflict, integration, independence, or dialogue best fitted these Victorian figures, the answer in the case of Mivart would have to be: all four *simultaneously*, depending on which issue he was addressing. Mivart did perceive conflict between what he saw as the Darwinians' overconfident allegiance to natural selection and his own grasp of the human condition in which mental and moral attributes could not be explained by it. Mivart used an independence strategy in self-defense when arguing that the Galileo affair proved that theologians should not interfere in scientific matters. At the same time, his own work was driven by integrationist strategies that presupposed a divine architect ultimately responsible for the archetypal structures in living things.[37] *Reconstructing Nature* added another ironic thesis. Given that natural theology had provided a value system through which the natural sciences received part of their justification, its eventual loss meant that the very agent of corrosion, newly professionalized science, lost one of its pillars of cultural support. How the moral vacuum left by the exclusion of natural theology from a scientific culture was filled, if it ever has been, was left as a tantalizing question.

The quest for moral meanings in the natural world featured prominently in another book published in 1998: Peter Harrison's *The Bible, Protestantism, and the Rise of Natural Science*. This was an important reexamination of the case made by Merton, Hooykaas, and others for direct connections between the Protestant Reformation and openness to reform in the study of nature. Merton had not argued that Protestant religion constituted a primary, independent variable on which science depended but that, through reciprocal interaction, a higher value was gradually placed on the practical sciences that promised to alleviate suffering. The influence of Protestantism for which Harrison argued was less direct and more diffuse.[38] Protestant exegetes, he argued, had taken the lead in reforming the multilayered hermeneutics of Christian scholars who looked for several levels of meaning in each biblical text. Minimally, there was the literal sense, the moral, and the allegorical. In reactions against elaborate allegorical, typological, and symbolic readings, preeminence was gradually given to the literal. Paradoxically, given the obstructive role biblical literalism would later play in popular resistance to innovative science, this hermeneutic shift to the literal in the sixteenth and seventeenth centuries arguably created space for a new science of nature. This was because, when the book of nature had been read by analogy with the book of God's word, objects in nature had been studied for their symbolic meanings (animals serving as moral tutors, for example) rather than as objects to be studied in relation to each other, in patterned phenomena subject to nature's (ultimately God's) laws. The book of God's works could now be read as an anthology of God-given resources for human appropriation and, in seventeenth-century physico-theology, as a repository of evidence for divine craftsmanship, wisdom, and design. Harrison's sophisticated analysis of the subtle and indirect role played by Protestant biblical theology in shifting perceptions of the natural world transcended discussion structured by metaphors of conflict and harmony. Anticipating the subject of a later book, Harrison gave a nuanced account of how belief in a historic Fall, and the desire to regain as much as possible of the knowledge and dominion originally possessed by Adam, could provide a theological justification for the necessity of empirical over rationalist methods in the scrutiny of nature.[39]

The beginning of the new millennium heralded a large encyclopedia devoted entirely to these topics, *The History of Science and Religion in the Western Tradition*. This volume contained many articles touching on the conflict thesis. Two in particular, by Colin Russell and David Wilson, addressed it directly. Its defects were summarized by Russell: its obscuring of relations other than conflict; its ignoring examples of close alliance; its outmoded tradition of positivist, Whiggish historiography; its neglect of diversity within both scientific and religious movements; and its exalting minor squabbles into major conflicts.[40] And yet, despite the best efforts of historians, the idea was still influential. Why? Wilson suggested three interesting reasons. It had probably been more successfully dispelled for the seventeenth century than for the nineteenth; revisionist historians were prone to statements that could be construed as more supportive of the thesis than they intended; and among contemporary scientists it was still prevalent, despite the efforts of some, such as Stephen Jay Gould, to correct the mythology that science and religion were "natural antagonists."[41]

Gould later contended that there need be no conflict between them, as long as their respective magisteria were kept separate, the sciences having jurisdiction over facts and theories, religion over values and morals. For most historians this was too neat a solution, given the permeability of many scientific and religious beliefs. But how could permeation of science by religion be detected? Had it always been detrimental to both? A systematic engagement of these questions surfaced in 2001 in a volume of collected essays, *Science in Theistic Contexts*.[42] Evidence from several case studies yielded no comforting generalities for those believing that such permeation might still be beneficial to science; but they did provide examples where it arguably had been in the past. Richard England argued that, during the "eclipse of Darwinism" in the decades around 1900, two Oxford biologists, Edward Poulton (1856–1943) and Frederick Dixey (1855–1935) defended the primacy of natural selection in a university where the theologian Aubrey Moore (1848–90) was praising Darwin as a friend to Christianity for having purged it of a *deus ex machina*. As Anglican entomologists, they were encouraged in their Darwinism by personal acquaintance with Moore and the knowledge that their defense of natural selection had a theological imprimatur.[43] Coincidentally, Peter Bowler's *Reconciling Science and Religion* appeared the same year, with its demonstration of a "concerted effort" in early-twentieth-century Britain to bring about "reconciliation between science and religion after their alienation in the Victorian era." Bowler could declare that "the old model of inevitable conflict (still visible in the writings of extremists on either side) has been qualified, if not abandoned."[44] Qualifications abounded in *Science in Theistic Contexts*. This was where Bernard Lightman urged that much of what passed for conflict between science and religion in the Victorian period was actually discord between different ways of harmonizing them.[45] In Lightman's interpretation, even T. H. Huxley (1825–95), for all his vehement attack on theology, had an understanding of "religion" in terms of an inner spirituality that made talk of conflict with science inappropriate.[46]

In a generous review, Frank Turner suggested that *Science in Theistic Contexts* was a suitable text for university teaching purposes.[47] It was, however, scarcely a typical undergraduate text. That lacuna was filled, again by Lindberg and Numbers, with their anthology *When Science and Christianity Meet*.[48] The title indicated a limitation in that faith traditions other than Christianity were not discussed. Draper and White still haunted the introduction; but the editors could now refer to studies such as John Heilbron's *The Sun in the Church* with its oft-to-be-repeated assertion that "[t]he Roman Catholic church gave more financial and social support to the study of astronomy for over six centuries, from the recovery of ancient learning during the late Middle Ages into the Enlightenment, than any other, and, probably, all other institutions."[49] Heilbron's study was more about the use of cathedrals as scientific laboratories than about a special synergy between Catholic theology and scientific exploration; but (*pace* Draper) he reaffirmed the key role of Jesuit educators in mathematics, astronomy, and the physical sciences.

From within a largely consensual historiography, Lindberg and Numbers addressed the common hypostatization of terms: "We often speak of what science states or what Christianity claims, as though science and Christianity were existing things, capable of speaking and claiming." But they would not let their student audiences forget that "it is people who do the believing, the speaking, the teaching, and the battling. . . . And when human beings are involved, so are human agendas and interests."[50] A chapter by David Livingstone provided the perfect illustration. Livingstone observed Darwin's very human capacity to "rig up diverse models of himself to suit his politics, his family, his friends, his public." Livingstone particularly echoed a warning from the historian Frank Burch Brown not to underestimate "the degree to which a human being—and especially a Victorian—can hold apparently incompatible beliefs and can vacillate time and again between them."[51] By Darwin's own admission, he was among those who experienced such fluctuation of belief.[52]

Livingstone's chapter introduced to a wider audience his distinctive emphasis on the importance of local context when examining religious reactions to innovative science. Comparing receptivity to Darwin among Presbyterians in different countries and regions, he noted that high-profile local events could make a big difference to perceptions. Edinburgh and Belfast were two locations where reactions to Darwin could be broadly contrasted. In the Edinburgh of the 1870s, opposition to Darwin among Presbyterians was relatively inconspicuous, whereas in Belfast, at least after 1874, there was growing public hostility. The local circumstances that made the difference were, in Scotland, the distraction of what to many was a more notorious threat to the faith from the Old Testament scholar William Robertson Smith (1846–94), whose adoption of the methods of biblical criticism led to a trial for heresy; and, in Ireland, the Belfast address of the physicist John Tyndall who, speaking in 1874 as president of the British Association for the Advancement of Science, went on the offensive, declaring that scientists would wrest from theology the entire domain of cosmological theory. In that address, Tyndall constructed a history of conflict between science and theology as the scope of naturalistic explanation had expanded. His references to Darwin in a polemic against the churches and the poor state of science education in Irish schools meant that, in hostile reactions from Catholics and Protestants alike, Darwin's science would henceforward be the more readily associated in Belfast with materialism and atheism. The model of inherent, systemic conflict was "not a fine enough tool for slicing history."[53]

Such calls for greater finesse have regularly featured in biographical studies where subjects with both scientific and religious commitments often cannot easily be sliced at all. Nicolaas Rupke's anthology of *Eminent Lives in Twentieth-Century Science and Religion* (2007) adds salient examples by including several evolutionary biologists such as Ronald Fisher (1890–1962), an Anglican; Theodosius Dobzhansky (1900–1975), a member of the Eastern Orthodox Church; Julian Huxley (1887–1975), humanist and agnostic; and E. O Wilson (1929-), described as an "evolutionist with the soul of a Baptist."[54] Its aim is to "uncover the presence and significance of religiosity" in the scientific careers under scrutiny.[55] The value of biographies sensitive to the religiosity of their subjects was demonstrated in the same year in Matthew Stanley's portrait of the Quaker physicist Arthur Eddington (1882–1944) and his leadership of the 1919 expedition to test Einstein's general theory of relativity. In *Practical Mystic* (2007), Stanley expands the taxonomy of ways in which religious dispositions could have a bearing on science when he argues for the moral significance of Eddington's Quaker pacifism. Crucially, Eddington strives to restore the international character of scientific research when, during World War 1, relations had soured between British and German scientists. In Stanley's account of the 1919 expedition there is something deeply symbolic about a British physicist leading the effort to confirm the theory of a German scientist.[56]

An ever-widening gap between popular subscription to the conflict narratives and the exposure by historians of their inadequacies was addressed in 2009 in a concerted display of myth busting, when the fruits of historical scholarship were placed before a nonspecialist readership in the volume *Galileo Goes to Jail and Other Myths about Science and Religion.*[57] Among the twenty-five myths exploded are some that had featured in extravagant ripostes to the conflict models, such as the claims that Christianity gave birth to modern science and that Einstein believed in a personal God. But most of the falsehoods are derivatives of the conflict narratives, such as the claim that the medieval Christian church suppressed the growth of science; that medieval Christians taught that the earth was flat; that Newton's mechanistic cosmology eliminated the need for God; and that Thomas Huxley defeated Samuel Wilberforce in their 1860 Oxford debate over evolution.

At a less popular level, new guidelines for research appeared in the proceedings of an international conference held at Lancaster University in 2007.[58] Minimally, these included reexamination of themes already present in the literature. For example, Frank Turner gives the best account to date of the circumstances that explain why the conflict thesis gained currency when it did.[59] In another chapter, Adam Shapiro revisits the notorious Scopes trial by taking another look at the textbook industry in 1920s Tennessee. He argues that a "complex network of concerns over economy, culture, industry, demography and education was aligned by competing groups . . . and then replaced by the universalizing rhetoric of science-and-religion conflict." In short, "the claim of science-religion-conflict became self-fulfilling in Dayton when conflicting groups recast their battles in terms of science and religion."[60]

In his introduction, Thomas Dixon argues for wider global perspectives. These are represented by B. Harun Küçük's exploration of how extreme and conflicting views on the history of Islamic science have served to accentuate claims about attitudes toward science within Christianity. Attitudes toward evolution and creationism in the Islamic world are explored by Salman Hameed, who notes the paradox that, whereas Draper has been made responsible for popularizing the conflict thesis in the West, his sympathetic reference to a "Mohammedan theory of evolution," which he attributed to al-Khazini in the twelfth century, has resulted in his book's frequent citations on websites supporting compatibility between evolution and Islam.[61] Further global dimensions are developed by Sujit Sivarsundaram in an innovative account of the role of missionaries in science globalization. In an observation with contemporary political relevance, he describes how Western science has not infrequently been resisted as symbol and instrument of colonial subjugation.[62]

Two contributions to the Lancaster University conference proceedings deserve special mention because they raise the question that gives Geoffrey Cantor's chapter its title: "What Shall We Do with the 'Conflict Thesis'?" Both Cantor and Peter Harrison express concern that the facility with which historians have exposed the crudities of a tired conflict *thesis* risks the neglect of genuine conflicts that have stimulated original thinking and played a legitimate critical role when religious thinkers have sought to restrain extravagant cultural ambitions voiced by scientific parties. Cantor notes the possibility of numerous potential tensions between scientific propositions and religious faith, at least some of which will be actualized in informed individuals. From the spiritual diary of John Rutty, a Dublin apothecary of the eighteenth century, he displays Rutty's inner conflict and deep ambivalence about his pursuit of medicine and science. The satisfaction he gleaned from the study of nature was tempered by a conscience that spoke to him of a neglect of spiritual purity and the medicine of the soul. For Cantor, it is incumbent on the historian to investigate how dissonance and ambivalence are handled and the constructive thinking that may ensue.[63]

The new perspective in Harrison's chapter derives from a philological critique of the conflict thesis, later refined and expanded in *The Territories of Science and Religion* (2015). It is often noted how the words "science" and "religion" changed their meanings over time, and how abstracting modern meanings from the practices and presuppositions of earlier periods can result in artificiality and anachronism. Harrison's achievement has been to trace with unprecedented sensitivity not only the stages in which these transformations of meaning occurred, but also their implications for any general thesis that assumes stable meanings. Thus, with reference to the Middle Ages, he warns that "to speak of the relationship between science and theology in this period is to ignore the categories that the historical actors themselves were operating with." There could still be fruitful historical exploration of the relationship between natural philosophy and theology in this period; but the fact that both these disciplines were speculative sciences "makes an important difference."[64] The point is thrown into relief by John Locke's (1632–1704) remark that he could not see how natural philosophy could ever become a "science" in the sense of embodying demonstrable knowledge. Theories about the workings of nature might be accorded different degrees of probability, but that was not the same as deductive certainty. A close reading of natural history and natural philosophy in the eighteenth century reveals how anachronistic it would be to try to identify various kinds of relationship between "science" and "religion" in that century when these were not yet independent entities that might bear a positive or negative relation to each other. Not until the nineteenth century does it become possible to discuss a "relationship between science and religion," either of conflict or conciliation, when several specialized scientific disciplines were finally placed under the single banner of "science." In common with Cantor, and with an eye on the present, Harrison observes that too close a rapprochement between scientific and religious goals, such as indiscriminate religious and moral approval of every biotechnological advance, can prevent theology from properly distancing itself from dominant cultural forces. For the fulfilment of its prophetic role, there has to be room for "legitimate conflict."[65]

Whether one sees a particular protest movement as legitimate invariably depends on where one is coming from. Noah Efron has recently described the encounter of a group of young rabbis with the Kentucky Creation

Museum. Their anguish at finding the Torah enlisted to oppose the scientific culture they saw as integral to the America they loved, the America which by its separation of church and state had protected science as a secular profession open to Jews, is brilliantly described in an indispensable study of science and secularization.[66] The felt need for more multifaith and global perspectives is gradually being met, as in *Science and Religion around the World* (2011), where each chapter examines the changing place of science in a different religious culture.[67] An illuminating chapter on sub-Saharan Africa describes societies in which local medical understanding, for example, long defied conventional Western categories of science and religion.

Conclusion

A growing sensitivity to the problems in relating those categories, especially where those relations are reduced to conflictual modes, has been the main theme of this chapter. The intention has not been to imply a teleological process, culminating in an undisputed consensus. But serious historical scholarship on science and religion does illustrate the revisionist trends I have identified. A contemporary example, with which I complete this fifty-year review, is Matthew Stanley's *Huxley's Church and Maxwell's Demon* (2015). This text highlights conflict, but the author insists that his story is not about "science versus religion."[68] Sensitive to Harrison's concern about the meanings of words, Stanley recognizes that "Victorian science saw many dramatic shifts in what counted as 'science.'"[69] Without denying there were scientific naturalists of Huxley's generation seeking to attack established religion, Stanley's analysis exemplifies a critique in which the primary locus of conflict shifts from generalities about "science" and "religion" to conflict between two forms of science: *naturalistic* science as championed by Huxley and *theistic* science as defended by Maxwell. The difference did not reside in prescriptions for how scientific research should proceed. Virtually identical methodological values were inscribed in both. The crucial difference lay in the justification each would give for precepts fundamental to scientific inquiry: a belief in the order, uniformity, and unity of nature. For Maxwell, as for Faraday, these principles were grounded in the monotheistic belief that the universe had been created by a single, intelligent Mind, the constancy of whose will guaranteed the uniformity. For Huxley, this theological foundation was simply otiose. A Humean empiricist understanding of laws of nature was sufficient. Strikingly, representatives of the competing metaphysics vehemently claimed that their presuppositions, and theirs alone, could provide science with a rational basis. For advocates of theistic science, the uniformity of nature would be imperiled by a lack of belief in a sustaining deity; for Huxley and his allies, belief in an active deity (particularly an intervening God) was precisely that which imperiled belief in nature's uniformity. Because Stanley's story returns us to the time when the original conflict thesis gained its purchase, it marks a fitting conclusion to this chapter. It comes with a reminder that at the core of the scientific naturalists' strategy was the reconstruction of a history of science in which science's theistic past was erased. The task of restoration will doubtless continue. Whether that will ever convince those who believe in "conflict between science and religion" because they *want* to believe it, or because it has political utility, is a very different question.

NOTES

1. M. B. Foster, "The Christian Doctrine of Creation and the Rise of Modern Science," *Mind* 43 (1934): 446–68; Foster, "Christian Theology and Modern Science," *Mind* 44 (1935): 439–66 / 45 (1936): 1–27.
2. Robert K. Merton, *Science, Technology and Society in Seventeenth-Century England* (1938; New York: Harper, 1970).
3. Ronald L. Numbers, "Science and Religion," in Sally Gregory Kohlstedt and Margaret W. Rossiter, eds., *Historical Writing on American Science: Perspectives and Prospects* (Baltimore, MD: Johns Hopkins University Press, 1985), 59–80, esp. 63.
4. Frank M. Turner, "The Late Victorian Conflict of Science and Religion as an Event in Nineteenth-Century Intellectual and Cultural History," in Thomas Dixon, Geoffrey Cantor, and Stephen Pumfrey, eds., *Science and Religion: New Historical Perspectives* (Cambridge: Cambridge University Press, 2010), 87–110, 88.
5. Peter Harrison, *The Territories of Science and Religion* (Cambridge: Cambridge University Press, 2015).
6. John Hedley Brooke, "Joining Natural Philosophy to Christianity," in John Hedley Brooke and Ian Maclean, eds., *Heterodoxy in Early Modern Science and Religion* (Oxford: Oxford University Press, 2005), 319–36.
7. Harrison, *Territories*, 197.
8. Bernard Lightman, "Victorian Sciences and Religions: Discordant Harmonies," in John Hedley Brooke, Margaret J. Osler, and Jitse M. van der Meer, eds., *Science in Theistic Contexts: Cognitive Dimensions* (Chicago: University of Chicago Press, 2001), 343–66.
9. Bruce Mazlish, preface to Andrew Dickson White, *A History of the Warfare of Science with Theology in Christendom*, abridged ed. (1896; New York: Free Press, 1965), 18.
10. Ian G. Barbour, *Issues in Science and Religion* (London: SCM Press, 1966).
11. Imre Lakatos and Alan Musgrave, eds., *Criticism and the Growth of Knowledge* (Cambridge: Cambridge University Press, 1970).
12. Thomas S. Kuhn, *The Structure of Scientific Revolutions* (Chicago: University of Chicago Press, 1962).
13. George Basalla, ed., *The Rise of Modern Science: Internal of External Factors?* (Lexington, MA: D. C. Heath, 1968).

14. Arnold Thackray, "Natural Knowledge in Cultural Context: The Manchester Model," *American Historical Review* 79 (1974): 672–709; Paul Wood, ed., *Science and Dissent in England, 1688–1945* (Aldershot, UK: Ashgate, 2004), 1–18.

15. Colin A. Russell, *The Conflict Thesis and Cosmology* (Milton Keynes, UK: Open University Press, 1974), 9.

16. James R. Moore, *The Post-Darwinian Controversies: A Study of the Protestant Struggle to Come to Terms with Darwin in Great Britain and America, 1870–1900* (Cambridge: Cambridge University Press, 1979), ix, 15.

17. Moore, *Post-Darwinian Controversies*, 14–15.

18. Samuel Wilberforce, *Essays Contributed to the Quarterly Review* (London: Murray, 1874), 1:52–103, 94.

19. David C. Lindberg and Ronald L. Numbers, eds., *God and Nature: Historical Essays on the Encounter between Christianity and Science* (Berkeley: University of California Press, 1986).

20. Charles Webster, *The Great Instauration: Science, Medicine and Reform 1626–1660* (London: Duckworth, 1975).

21. Charles Webster, "Puritanism, Separatism, and Science," in Lindberg and Numbers, *God and Nature*, 192–217, quotations on 213.

22. Lindberg and Numbers, introduction to *God and Nature*, 5 (quotation), 12; Richard S. Westfall, "The Rise of Science and the Decline of Orthodox Christianity: A Study of Kepler, Descartes, and Newton," in Lindberg and Numbers, *God and Nature*, 218–37; Lindberg, review of *Religion and the Rise of Modern Science* by R. Hooykaas, *Journal of the American Scientific Affiliation* 26 (1974): 176–78; Numbers, review of *The Road of Science and the Ways to God* by Stanley L. Jaki, *Church History* 50 (1981): 356–57.

23. David B. Wilson, "The Historiography of Science and Religion," in Gary B. Ferngren, ed., *The History of Science and Religion in the Western Tradition* (New York: Garland, 2000), 3–11.

24. Rolf Gruner, "Science, Nature, and Christianity," *Journal of Theological Studies* 26 (1975): 55–81.

25. Lindberg and Numbers, introduction to *God and Nature*, 9 (quoting Martin J. S. Rudwick, "Senses of the Natural World and Senses of God: Another Look at the Historical Relation of Science and Religion," in Arthur R. Peacocke, ed., *The Sciences and Theology in the Twentieth Century* [Notre Dame, IN: University of Notre Dame Press, 1981], 241–61); and Martin Rudwick, "The Shape and Meaning of Earth History," in Lindberg and Numbers, *God and Nature*, 297.

26. Jon H. Roberts, *Darwinism and the Divine in America* (Madison: University of Wisconsin Press, 1988), 242.

27. John Dillenberger, *Protestant Thought and Natural Science* (London: Collins, 1959).

28. Michael J. Buckley, *At the Origins of Modern Atheism* (New Haven, CT: Yale University Press, 1987).

29. John Hedley Brooke, *Science and Religion: Some Historical Perspectives* (Cambridge: Cambridge University Press, 1991), 348–403 (Canto Classics ed. [Cambridge University Press, 2014], 475–552).

30. Dixon et al., introduction to *Science and Religion: New Historical Perspectives*, 15.

31. Brooke, *Science and Religion: Some Historical Perspectives*, 45–46 (Canto Classics ed., 61–62).

32. For a recent refined example, see Joshua M. Moritz, *Science and Religion: Beyond Warfare and toward Understanding* (Winona, MN: Anselm Academic, 2016).

33. Brooke, *Science and Religion: Some Historical Perspectives*, 321 (Canto Classics ed., 438).

34. Noah Efron, "Sciences and Religions: What It Means to Take Historical Perspectives Seriously," in Dixon et al., *Science and Religion: New Historical Perspectives*, 247–62, reference on 253.

35. Geoffrey N. Cantor, *Michael Faraday: Sandemanian and Scientist* (London: Macmillan, 1991), 201–5.

36. Geoffrey N. Cantor, *Quakers, Jews and Science* (Oxford: Oxford University Press, 2005).

37. John Hedley Brooke and Geoffrey N. Cantor, *Reconstructing Nature: The Engagement of Science and Religion* (Edinburgh: T & T Clark, 1998), 276.

38. Peter Harrison, *The Bible, Protestantism, and the Rise of Natural Science* (Cambridge: Cambridge University Press, 1998), 8.

39. Harrison, *The Bible*, 226–35; Harrison, *The Fall of Man and the Foundations of Science* (Cambridge: Cambridge University Press, 2007).

40. Colin A. Russell, "The Conflict of Science and Religion," in Ferngren, *History of Science and Religion*, 14–15.

41. Wilson, "Historiography of Science and Religion," in Ferngren, *History of Science and Religion*, 7–8.

42. Brooke, Osler, and van der Meer, *Science in Theistic Contexts*.

43. Richard England, "Natural Selection, Teleology, and the Logos," in Brooke, Osler and van der Meer, *Science in Theistic Contexts*, 270–87.

44. Peter J. Bowler, *Reconciling Science and Religion: The Debate in Early Twentieth-Century Britain* (Chicago: University of Chicago Press, 2001), 3–5.

45. Lightman, "Victorian Sciences and Religions."

46. Cited by Lightman, "Victorian Sciences and Religions," 348.

47. Frank Turner, review of *Science in Theistic Contexts*, *Isis* 94 (2003): 684.

48. David C. Lindberg and Ronald L. Numbers, eds., *When Science and Christianity Meet* (Chicago: University of Chicago Press, 2003).

49. John L. Heilbron, *The Sun in the Church: Cathedrals as Solar Observatories* (Cambridge, MA: Harvard University Press, 1999), 3.

50. Lindberg and Numbers, introduction to *When Science and Christianity Meet*, 3.

51. David Livingstone, "Re-placing Darwinism and Christianity," in Lindberg and Numbers, *When Science and Christianity Meet*, 186.

52. Francis Darwin, ed., *The Life and Letters of Charles Darwin* (London: Murray, 1887), 1:304.

53. Livingstone, "Re-placing Darwinism," in Lindberg and Numbers, *When Science and Christianity Meet*, 195. Livingstone's major investigation into Presbyterian responses to Darwin in different localities was later published in his *Dealing with Darwin: Place, Politics, and Rhetoric in Religious Engagements with Evolution* (Baltimore, MD: Johns Hopkins University Press, 2014).

54. Nicolaas A. Rupke, ed., *Eminent Lives in Twentieth-Century Science and Religion* (Frankfurt am Main: Peter Lang, 2007), 230.

55. Rupke, *Eminent Lives*, 7.

56. Matthew Stanley, *Practical Mystic: Religion, Science, and A. S. Eddington* (Chicago: University of Chicago Press, 2007).

57. Ronald L. Numbers, ed., *Galileo Goes to Jail and Other Myths about Science and Religion* (Cambridge, MA: Harvard University Press, 2009).

58. Dixon et al., *Science and Religion: New Historical Perspectives*.

59. Turner, "Late-Victorian Conflict."

60. Adam R. Shapiro, "The Scopes Trial beyond Science and Religion," in Dixon et al., *Science and Religion: New Historical Perspectives*, 212.

61. Salman Hameed, "Evolutionism and Creationism in the Islamic World," in Dixon et al., *Science and Religion: New Historical Perspectives*, 143.

62. Sujit Sivarsundaram, "A Global History of Science and Religion," in Dixon et al., *Science and Religion: New Historical Perspectives*, 177–97.

63. Geoffrey Cantor, "What Shall We Do with the 'Conflict Thesis'?," in Dixon et al., *Science and Religion: New Historical Perspectives*, 289–90.

64. Peter Harrison, "'Science' and 'Religion': Constructing the Boundaries," in Dixon et al., *Science and Religion: New Historical Perspectives*, 25.

65. Harrison, "'Science' and 'Religion,'" in Dixon et al., *Science and Religion: New Historical Perspectives*, 41.

66. Noah J. Efron, *A Chosen Calling: Jews in Science in the Twentieth Century* (Baltimore, MD: Johns Hopkins University Press, 2014).

67. John Hedley Brooke and Ronald L. Numbers, eds., *Science and Religion around the World* (New York: Oxford University Press, 2011).

68. Matthew Stanley, *Huxley's Church and Maxwell's Demon* (Chicago: University of Chicago Press, 2015), 4.

69. Stanley, *Huxley's Church*, 80.

CHAPTER FIFTEEN

Scientists

ELAINE HOWARD ECKLUND AND
CHRISTOPHER P. SCHEITLE

> Now Dr. [Francis] Collins says, "Well, God did it. And God needs no explanation because God is outside all this." Well, what an incredible evasion of the responsibility to explain. Scientists don't do that. Scientists say, "We're working on it. We're struggling to understand."
>
> —*Richard Dawkins*

> Religion is based on dogma and belief, whereas science is based on doubt and questioning. In religion, faith is a virtue. In science, faith is a vice.
>
> —*Jerry Coyne*

The relationship between science and religion has once again become a contentious modern-day topic in the United States. And the conflict model for framing that relationship has taken center stage in debates about topics as varied as human embryonic stem cell research and the teaching of evolution in public schools.[1] A conflict model also seems to have a global reach in modern-day media. The European Union has witnessed a resurgence in religious opposition to scientific research, with public leaders in the United Kingdom and France worrying that a recent influx of Muslim immigrants may pose unique faith-based challenges to science.[2] And scientists in Italy are concerned that the influence of Catholicism on the culture will unnecessarily harm a developing science infrastructure.

Scientists as Purveyors of Conflict between Science and Religion

Vocal scientists like UK evolutionary biologist Richard Dawkins and US biologist Jerry Coyne shape a public perception of scientists as not only atheists, but as *antireligious* atheists. This sentiment contributes to the larger idea that science is intricately linked to secularization and that scientists themselves are secularization's mouthpieces. As the nineteenth-century sociologist Max Weber put it, "the fate of our times is characterized by rationalization and intellectualization and, above all, by the 'disenchantment of the world.'"[3]

But how much do we really know about American scientists' religious views, much less how they compare with scientists in other countries and with general publics in their own nations? How do the religious views of scientists in particular national contexts shape their practice, dissemination, and interpretation of science? And how does their scientific practice shape their religious interpretation and understanding? And what does all of this have to say about the role of scientists in contributing to the perpetuation of the conflict model?

Finding answers to such questions has enormous implications for religious and scientific communities, for scholars of all disciplines that are concerned with the interface between science and religion, and for members of different national general publics. To begin to answer these questions, we analyze both survey and interview data from the first comprehensive global study of scientists. We include biologists and physicists at all career stages from France, Hong Kong, India, Italy, Taiwan, Turkey, the United Kingdom, and the United States.

Scientists and Secularization

This research emerges from an underlying desire to explore a theoretical question, namely, how are the religious views of scientists connected to the particular piece of secularization that is about science and scientists? The general assumption is that if science is connected to the decline of religion, then scientists are secularization's mouthpieces. Indeed, much of the research on the religious beliefs and practices of scientists in the United States has appeared to reinforce the idea that science plays a role in destroying religious belief and authority in the lives of individual scientists and—by extension—in the groups of people that scientists interact with most. For example, James Leuba's 1916 and 1934 surveys on the attitudes of American scientists toward Christian belief—defined as participation in Christian worship and acceptance of the Christian theology of life after death—discovered that scientists were less likely than the general public to be religious.[4] In particular, the most successful scientists were the least likely to be involved in religion. One implication of Leuba's findings was that religion ought to completely assent to science in order to remain an influence on American society.

However, other scholars argue that rather than discovering conflict between scientific and religious ways of knowing, these studies simply assumed the conflict.[5] Additional scholars have offered empirical findings and theories that counter those of Leuba's studies. They found that individuals with an active religious faith are no less likely to pursue academic careers than those without an active religious faith,[6] and historically, scientists did not view religion as being in opposition to their work.[7] C. Mackenzie Brown also contends that Leuba, and later scholars who supported Leuba's work, used a narrow conception of religion that is no longer relevant for much of the American public (and we would add that this conception is especially irrelevant for a global public).[8] Such findings indicate that it would be irresponsibly limiting for future research to assume a conflict between religion and science among scientists without empirically investigating the veracity of this framing.

Moreover, there is good reason to believe that even the relationship US scientists have to religion is more complex than a previous generation of scholarship has indicated. For example, our own earlier work reveals that nearly 50 percent of scientists who work at the top US research universities have some form of a religious identity (though such identities are often very different in character from the religious identities found among the general public).[9] In addition, Ecklund finds that more than 20 percent of *atheist* scientists still consider themselves spiritual to some extent; we call this group "spiritual atheists." We have also shown that US scientists are looking for members of their own scientific ranks to serve as "boundary pioneers," crossing the picket lines between science and religion, and helping their colleagues do a better job of translating scientific discoveries to a largely religious American public.[10]

Studying the Religiosity of Scientists in International Context

Yet, if we know that the picture of the relationship US scientists have to religion is more complicated than commonly believed, then why examine scientists' identities, beliefs, and practices on a global stage? In short, science is a global enterprise, and the relationship between religion and science has global consequences. We need to study scientists across different national contexts to understand how the ecology of these contexts might have an impact on the beliefs and practices of scientists. For example, Europe, on the whole, is more secular than the United States across several dimensions, including public religion as well as personal religious beliefs and identities of the public.[11] And many European scholars are quite outspoken when it comes to their views on the relationship between religion and science.[12] Could European scientists be purveyors of conflict while US or Indian scientists are actually more friendly to religion? Researchers have never addressed these questions because no work thoroughly examines the religious beliefs, identities, and practices of scientists in different global contexts.[13]

Nations vary in ways that might matter for how scientists view religion. For example, they vary greatly in the level of their science infrastructures and cultures. In the United States there is an emphasis on commercialization coming from the federal government and universities (top-down), while some in the scientific community and many in the general public push individual scientists to tie their research to commercially viable applications (bottom-up). Italy (to some extent) provides a clear opportunity to see what happens during a massive buildup of science capacity. For its part, Italy—home to some of the most important figures in modern science, including Leonardo da Vinci and Galileo Galilei—has a long history of participating at the forefront of scientific development. Italy also makes an interesting case in that the Catholic Church finds its seat in Rome. Today, Italy continues to maintain a strong output in physics, with a large presence at CERN (European Council for Nuclear Research), the world's foremost particle physics experiment.[14] India is in the institution-building stage, working to establish a strong science-and-technology infrastructure; as a result, it often looks to the United States and other countries with a more developed scientific infrastructure for guidance on how to progress.[15] Further, the great extent to which India exports scientists to other countries of the world, including the United States, is very important to our understanding of the country's own science infrastructure, as well as its relation to the rest of the world. The United Kingdom has a broad and well-developed science infrastructure, spending 1.88 percent of its gross domestic product (GDP) on research and design (R&D). Of the eight national contexts, the United States has the best-developed science infrastructure (in terms of number of scientists, science funding, and universities and institutes devoted to scientific research). The United States spends more than any other country on R&D—$389 billion in 2008, which is 2.64 percent of its GDP.

These nations are also important because of the demographics of the scientists within them—particularly their migration streams. This within-country diversity among scientists could lead to different perspectives on religion and ethics among them, based on the scientists' backgrounds and countries of origin.[16] For example, previous research reveals that US scientists (particularly when compared with those in Italy and, to a lesser extent, the United Kingdom) are much more globally diverse than the general population.[17] The nations and regions where we study scientists' religiosity also have very different levels and types of religiosity among their general population, and thus—if the religious beliefs of scientists are at all local—their scientists likely have different ways of viewing the relationship between religion or spirituality and science.

This chapter draws on data from the Religion among Scientists in International Context (RASIC) study. RASIC is a mixed-methods study comprising both nationally representative surveys (a total of 9,422 respondents) and 609 qualitative, in-depth interviews with physicists and biologists in eight countries and regions: France, Hong Kong, India, Italy, Taiwan, Turkey, the United Kingdom, and the United States. To create the survey sampling frames (the universe of scientists eligible to participate in the study) we employed what sociologists call a two-stage sampling procedure. In the first stage we drew a random sample of organizations stratified by discipline (physics or biology) and academic reputation (elite or non-elite). We determined the elite status of each institution through a multidimensional process that involved assessing research productivity as measured by the Web of Science database, factoring in existing published rankings, and consulting scientists in each national or regional context for insider perspectives on appropriate university ranking. In the second stage, we drew a random sample of scientists from each of the eligible organizations depending on the number and characteristics of scientists. We compiled lists of active scientists engaged in research at these organizations through organizational and departmental websites. We obtained, when possible, roughly equal sample sizes in each stratum, resulting in oversampling of both women and higher-ranking scientists.[18] The survey sample was stratified by discipline (biology or physics), status (elite or non-elite), and, in some nations, across regions. In order to allow for national comparisons in each region, we designed the survey instrument by utilizing some questions from broader general population surveys such as the World Values Survey (WVS) and the International Social Survey Programme (ISSP).

We drew the list of people we wanted to interview from the pool of respondents who completed the survey and granted permission to be contacted again about further research, which we asked as the last question on the survey. Our strategy in picking scientists to interview took into account geographic location as well as a desire to achieve a rough balance of the interview sample across gender, discipline, rank, and personal religiosity. Interviews were semi-structured in format, allowing the respondents some margin of freedom in directing the course of the response and conducted both in person and over the phone or Skype.

We first present results from our survey data that are relevant to the "conflict narrative," and then we add detail and nuance to these statistics with our interview data.

Religiosity of Scientists

Religious Identities and Practices

We begin by comparing the religiosity of scientists across contexts. Table 1 shows several measures of religiosity for the scientists in each of the eight national contexts. Turkey stands out as having the highest levels of religiosity among scientists, although India is a close second. About half of the scientists in these two nations report being a religious person and praying once a day or more, and about one-quarter to one-third report attending religious services once a week or more. On the other hand, scientists in France report the lowest levels of religiosity across all of the measures. Only 5 percent of French scientists, for instance, state that they believe in God with no doubts. Scientists in the United Kingdom and the United States also show relatively low levels of religiosity in comparison with the other nations. Hong Kong, Italy, and Taiwan fall between these extremes. Italian and Taiwanese scientists, for example, identify as religious at a rate similar to those in India and Turkey, but their frequency of prayer and religious service attendance are substantially lower.

While comparing scientists across nations and regions to each other is interesting, it might provide more insight to compare the scientists to the general populations in their respective nations. Because the RASIC surveys included some questions from cross-national surveys, we have the ability

TABLE 1

Select Measures of Religiosity among Scientists across Eight Nations (in percentages)

	France	Hong Kong	India	Italy	Taiwan	Turkey	United Kingdom	United States
Claims to be at least slightly a religious person	16	39	59	52	54	57	27	30
Reports praying once a day or more	3	11	48	17	13	54	9	11
Reports attending religious services once a week or more	3	13	26	17	12	33	8	11
I know God exists, no doubts	5	17	26	16	20	61	9	10
N	645	276	1,606	1,262	776	431	1,531	1,779

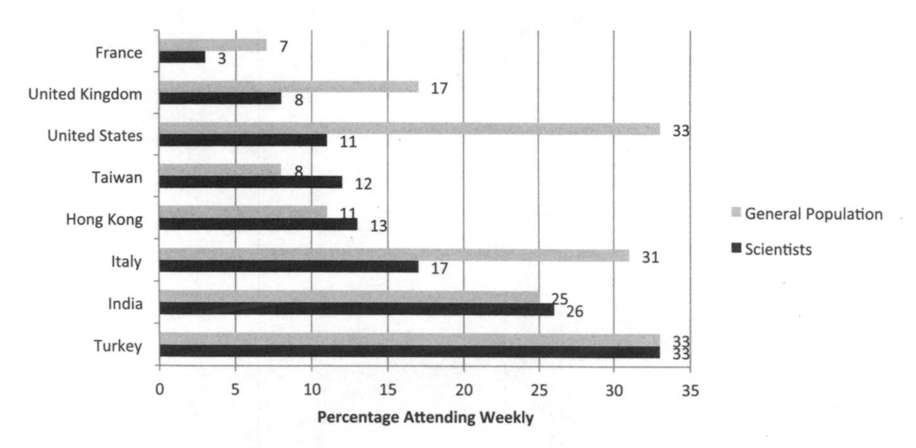

Fig. 1. Biologists and Physicists by Region Attending Religious Services Weekly Compared with General Population (in percentages). *Note*: The general population percentages for the United Kingdom, France, and Italy come from the 2005–2009 World Values Survey. The remaining nations' percentages come from the 2010–2014 World Values Survey.

to conduct such comparisons. Figure 1 compares the frequency of religious service attendance of a nation's general population to its scientist population. On this measure the scientists in India, Turkey, Hong Kong, and Taiwan are quite similar to the general population. Scientists in Taiwan and Hong Kong actually appear to attend religious services slightly more than the general populations of those locations. In France, the United Kingdom, the United States, and in Italy, however, scientists are one-half to one-third less likely than the general population to say they attend religious services weekly. Figure 2 shows that scientists' own perceptions of their religious commitment also often differ from that expressed by the general populations of their nations. Except for Hong Kong and Taiwan, in all the other regions surveyed, scientists self-identified as "religious" at lower rates than the general population. It appears that the United States and western Europe have more of a scientist–general population gap in religiosity than those nations outside of these regions, meaning that the scientists in those nations are significantly *less* religious then members of the general population in the nations. We expect that if there are big religious differences between scientists and the general population, this could create very negative implications for public dialogue about religion and science.

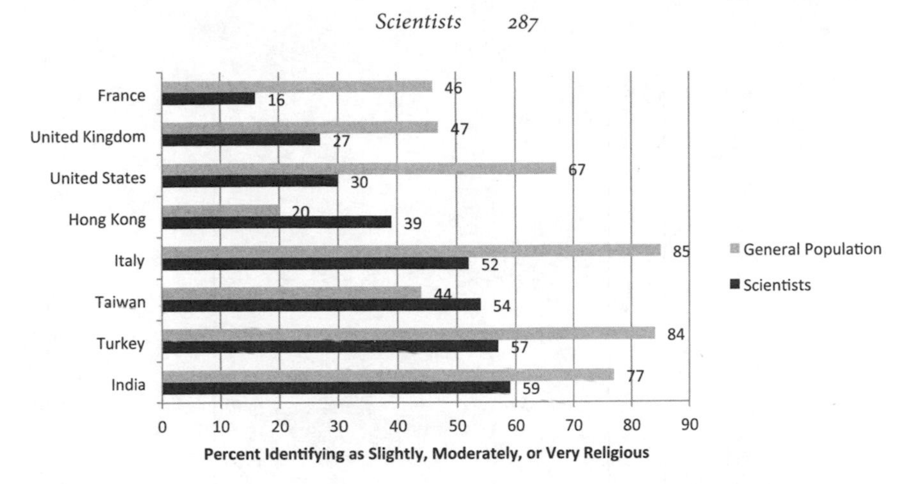

Fig. 2. Scientists by Region Identifying as Religious Compared with General Population (in percentages). *Note*: United Kingdom, France, and Italy general population percentages come from the 2005–2009 World Values Survey. The remaining nations' percentages come from the 2010–2014 World Values Survey.

Changes in Scientists' Religiosity

One argument concerning scientists' relatively low religiosity, at least in some nations, is that learning more about science caused them to become less religious.[19] It is hard to know whether this is the result of causation or selection. That is, do religious individuals in such nations become scientists and then become less religious, or is it that the less religious population in these nations are disproportionately entering the sciences? The RASIC survey did ask questions that provide some information on this issue; scientists were asked what their religious affiliation was at age sixteen, which we can compare with a similar question about their current religious affiliation. And they were asked whether they felt their scientific knowledge and training had made them more or less religious. Table 2 shows the responses to these questions.

Scientists in India and Turkey do not show much change in affiliation with a religion between age sixteen and the time of the survey. Of course, there could have been substantial changes in the specific affiliations that are masked by this overall affiliation measure. When we look at France, Italy, the United Kingdom, and the United States, however, we find that scientists' current rate of religious affiliation is about 20 percent lower than the rate of affiliation at age sixteen. Scientists in Hong Kong and Taiwan are actually

TABLE 2

Measures of Change in Religiosity among Scientists across Eight Nations (in percentages)

Change in religious affiliation	France	Hong Kong	India	Italy	Taiwan	Turkey	United Kingdom	United States
Identified with some religious affiliation at age sixteen	53	20	98	84	48	90	55	60
Identifies with some religious affiliation now	30	31	94	65	58	84	37	39
Would you say that your scientific knowledge and training have . . .								
Made you much more religious?	1	10	6	4	3	11	3	4
Made you slightly more religious?	3	7	7	8	10	16	3	4
Had no effect on how religious you are?	69	64	57	61	73	50	59	57
Made you slightly less religious?	9	7	10	9	8	7	16	13
Made you much less religious?	18	12	20	18	6	16	19	22
Total	100	100	100	100	100	100	100	100
N	645	276	1,606	1,262	776	431	1,531	1,779

somewhat more likely to say they have a current religious affiliation compared with when they were sixteen. For instance, 31 percent of scientists in Hong Kong report having a religious affiliation now, while 20 percent said they had one when they were sixteen.

If science and religion were incompatible, thus requiring a choice between the two in the lives of individual scientists, then we imagine that an individual entering science would let go of religion. Yet, our data show different results. We tested this assumption by asking scientists to reflect on whether their scientific training and knowledge has made them more or less religious in their own sense of things. A majority of scientists in all eight countries and regional contexts reported that it has had no effect. Of scientists who state that some effect has occurred, the majority in each nation report that science has made them less religious. UK and US scientists are the most likely to say that science has made them less religious. On the one hand, only 4 percent of French scientists, 6 percent of scientists in the United Kingdom, and 8 percent of scientists in the United States report that their exposure to science has made them more religious. On the other hand, 27 percent of Turkish scientists reported becoming more religious as a result of their scientific training and knowledge.

Perceptions of the Religion and Science Interface

We now turn to examining scientists' views and experiences with the religion-science relationship. Despite the pervasive nature of the conflict thesis, the scientists we surveyed did not overwhelmingly perceive a conflict. As shown in table 3, when asked how they perceive the relationship between science and religion, a plurality of scientists in all eight national contexts said they perceive a relationship of "independence" between the two. These scientists do place religion and science in separate spheres, but do not consider them to be oppositional. In France, Italy, the United States, and the United Kingdom, the second-most-common view is that religion and science are in conflict and that scientists view themselves as being on the side of science. This view is particularly salient in the United Kingdom, where 35 percent of scientists said that religion and science are in conflict and that they are on the side of science. In Hong Kong, India, Taiwan, and Turkey, however, the second-most-popular view is that religion and science have a relationship of "collaboration; each can be used to help support the other." Almost no scientist in any nation stated that she views religion and science as in conflict and that she sides with religion.

The RASIC survey also asked whether our scientist respondents perceived their scientist colleagues as having a negative attitude toward religion, which could be seen as yet another way to assess potential hostility or conflicts between the two spheres. As seen in table 4, scientists in the United Kingdom and the United States are the most likely to agree to some extent that their peers have negative attitudes toward religion. In both of these nations, over 50 percent of scientists agreed with this statement. This compares with 13 percent in Hong Kong, 24 percent in India, 18 percent in Taiwan, and 26 percent in Turkey.

Narratives of Conflict

The narrative interviews add nuance to these statistics. Interview respondents provided a variety of conflict stories, again disproving the notion that all scientists, if they do perceive a conflict, perceive it in the same way. Some identified a clash between science and religion in terms of competing identities. A UK scientist described how he interpreted these competing realms: "And if you can go to work one day and be a scientist and you do a good job, I'm sure you can still be a good scientist, but then on a Sunday morning you get up and go to church and take that completely out. You're living two lives, in my opinion. There's a Jekyll/Hyde about it."[20]

A US biology professor explained the conflict as a mental issue or even disorder, saying, "Religious scientists have to be a little schizophrenic and not question some of their beliefs, not ask for evidence for them."[21] Another scientist specifically referenced schizophrenia when explaining why he believes there is a conflict between religion and science: "It is a conflict between the two. . . . They're incompatible. You cannot on the one hand believe in a kind of all-powerful spirit which is capable of let's say creating matter from nothing and then on the other hand believe in the scientific process. [There is] some kind of schizophrenia in the brains of those people."[22]

Others find conflict because of the difficulty in defining religion. A biology lecturer in the United Kingdom provided an explanation for why some might experience tension, stating, "In my experience there is not a conflict and I know of many good scientists who are also very religious and they don't see any conflicts and I think sometimes misunderstanding of religion or misinterpretation of some of the religious precepts or scriptures could lead to that conflict, but I don't relate to that. Maybe there could be some conflicts when the different religions are taken in an orthodox [manner]."[23]

TABLE 3

Scientists Perceptions' of the Religion-Science Relationship across Eight Nations (in percentages)

For me personally, my understanding of science and religion can be described as a relationship of . . .	France	Hong Kong	India	Italy	Taiwan	Turkey	United Kingdom	United States
Conflict: Side of religion	0	0	1	0	0	2	0	0
Conflict: Side of science	27	17	18	21	9	24	35	29
Conflict: Not sure which side	0	1	1	0	1	0	0	0
Independence	58	44	44	58	62	35	47	51
Collaboration	7	24	29	15	21	33	12	12
Don't know	8	14	7	5	7	7	7	7
Total	100	100	100	100	100	100	100	100
N	645	276	1,606	1,262	776	431	1,531	1,779

TABLE 4

Perceptions of Peers' Attitudes about Religion among Scientists across Eight Nations (in percentages)

In general, I feel that scholars in my discipline have a negative attitude toward religion.	France	Hong Kong	India	Italy	Taiwan	Turkey	United Kingdom	United States
Strongly agree	8	3	4	7	2	5	13	14
Somewhat agree	28	10	20	34	16	21	42	41
Have no opinion	21	34	19	14	22	10	15	14
Somewhat disagree	13	18	16	16	18	23	8	10
Strongly disagree	12	14	23	11	23	34	9	6
Don't know	18	21	18	18	19	7	13	15
Total	100	100	100	100	100	100	100	100
N	645	276	1,606	1,262	776	431	1,531	1,779

In some ways these scientists are responding to ongoing scholarship about and debate over the definition of religion, for example, regarding literal versus more interpretive readings of religious texts. Related to the definitional aspects of conflict, others emphasize the distinction between religion and spirituality. Finding a conflict then depends on how a scientist defines what it means to be religious or spiritual. When scientists talked with us about doing science, they often alluded to the ability to think rationally or critically. When asked if it is possible to be rational and religious, a nonreligious Turkish scientist replied, "I guess. I mean for me it's hard to understand but still you can believe that there is some form that we don't understand, whatever. [*Laughs*] It's hard to explain. Energy, harmony, you know something that's . . . hard to grab, so and all that can . . . somehow explains in the basic formula of the world or something. So maybe—this can be a kind of spirituality. Or I wouldn't say it's a religion, but I mean this is different in any way for me—religious people say this is a sect. And they say 'this is bad but we are good'; it's all just, from my opinion [made] up in one's brain."[24] For this respondent, traditional religion is definitely in conflict with science, but spirituality is a completely different story.

Independence

Despite the success of the conflict paradigm in Western culture, which is at times validated by scientists themselves, other models of thinking exist. In particular, scientists outside the West may consider the relationship between religion and science as one of independence. For example, a biology graduate student, who is in fact from the United States, recognized that others around the world can place religion and science in two spheres that do not overlap and therefore do not conflict: "You can see very clearly when I talk to them [US scientists] about religion that they are very much stuck in [a framework of] 'both of these things are incompatible with one another, that they actually—they actually are two different realms of human thought,' where I don't think they're necessarily incompatible. It's a false dichotomy that has been created in America and Europe especially I think about it, because definitely when I go overseas and talk to other people in Southeast Asia, people who have religions as a big part of culture, they really don't see that as being a bad thing for their own science, for instance."[25]

Our scientist respondents found it striking that both religious and nonreligious scientists across the national contexts considered the possibility of religion and science as independent entities. A UK professor of physics who identifies as a Christian said, "I try and keep them separate. It sometimes—because I'm a Christian so I attend the church about every other week. And I like to keep them separate because there's science and science is something that you can test, and then there's religion, which you can't really test, so I just see them as very separate."[26]

And a female Indian biology professor explained why the two can in fact coexist: "They must be taking it as two different things, but I think that religion and science, most of the time they are complementary because they both can go together, it's not that only you have to have one, no it's not so, for me, but actually religion makes a person refined, so any type. That may be a refined scientist also, that may be a refined person, because if you follow it, in the true sense, then it makes a person refined."[27] For this biologist, religion and science do not need to be in conflict and can coexist well, especially when the highest form of each is followed.

Collaboration

As mentioned earlier, a significant minority in several of the national contexts views the relationship between religion and science as one of collaboration. Scientists employed a variety of themes in their narratives to explain this relationship. The relationship was usually framed in one of two ways: narratives either emphasized a bi-directional influence (e.g., science and religion enrich one another; science and religion have principles that reinforce one another; and science and religion share the same goal) or they emphasized a one-way influence (e.g., religion motivates scientific inquiry, science enriches belief, and science is a tool for religious exploration). The complexity of a complementary relationship is reflected in the numerous and nuanced types of explanations offered by scientists.

A Hong Kong professor of biology saw religion and science as complementary because they both share the same goal: discovery of truth. He said, "I think that Buddhism is more like a way of looking at things differently. . . . [I]n terms of principles, it places every emphasis on finding out the truth, so in some ways, it has similarities with science, so I do like Buddhism more."[28] Another Hong Kong scientist explained that religion and science ask complementary questions: "It is really hard to say. I think . . . religion is a higher stage of science. I think we can ask the first question, the second question, and the third question in science. But when it comes to the fourth question, we need to use religion to explain and answer it. Because . . . what people can learn and observe [is really limited]. Even though a person is conducting research for his whole lifetime, he could not know how this world is operating. So . . . if we keep asking questions, there will not be an ending point."[29]

Other scientists, while seeming to reference a hierarchical relationship, still emphasized the collaborative nature of the relationship between religion and science. For example, science is used as a tool for religious exploration, explained a Hong Kong biology professor: "Because from the perspective of a Christian, I believe science is part of God's revelation. . . . Naturally, I believe that the principles and mechanisms that I am studying are all revelations of God. In other words, they are creation of God. . . . So, I believe what my faith and my science have in common is the search for truth."[30]

Others expressed that their science was enriched by their religious beliefs. Religion was a reason for doing science. Unusual given her national context, a French physicist spoke about her experiences as a religious scientist this way: "Because I believe in God, then when I discover something, I know . . . my experiment for me is reflective of this. . . . [I have] an obligation to understand the universe, and this is not going to be [a] contradiction. . . . I think my belief enriches my scientific discoveries, and then my scientific discoveries also enrich my faith, at least in my case."[31]

Even from an institutional standpoint, science and religion can be seen as working together. An Italian associate professor of physics referenced the Vatican and its relationship to science: "I see an interplay which can be considerable at some times. . . . Even the Vatican has a conservatory. . . . Conflict is another subject . . . there can be some influences that way, here it doesn't, but also I don't know, also a kind of somehow cooperation, it's not just conflict, it's more complex."[32] When speaking about the relationship between religion and science, these scientists brought up the notion of collaboration, thus indicating that other narratives exist other than conflict and even independence.

Religion in the Scientific Workplace

With regard to their colleagues, scientists generally *perceive* that others yield to a conflict perspective, even if they do not. They say in our survey that religion rarely if ever comes up in the scientific workplace and they think that many scientists have a negative attitude toward religion. The question in table 4 asked whether a scientist agreed or disagreed that "scholars in my discipline have a negative attitude toward religion." Scientists in the United States and United Kingdom lean toward agreeing with this statement. Italian scientists also tend to agree, although less strongly than their US and UK counterparts. Indian scientists, on the other hand, appear to lean more toward disagreeing with this statement. It is important to point out that sizable proportions of the scientist populations in all eight national contexts either had no opinion or stated that they "don't know" whether other scholars in their discipline have a negative attitude toward religion. This suggests that religion might not even be a salient issue or topic of discussion for many scientists, and thus there is no need for the conflict model to exist within the workplace. It is also worth noting that when asked "Have there been conflicts about religion in your department?," there is not much variation across the eight national contexts in response to this question, and most scientists overwhelmingly report that they have not seen conflict related to religion in their departments. Only about one in ten scientists in each context reported any such conflict. But when we sat with scientists in their offices and laboratories we found that religion does indeed sometimes enter the scientific workplace and—even for those who are not personally religious—when it does, they must respond.

Avoidance

Even if they are not personally religious, scientists must often acknowledge the presence of religion and react to it. Not surprisingly, however, these scientists lack tools to navigate religion in the workplace. The secularity of the United Kingdom, in particular, and the accepted norm that religion is a private affair, in some cases seems to amplify scientists' awareness of students' religiosity.

The most common pattern of dealing with religion at work tends to be avoidance, meaning that scientists are aware of religious symbols and practices sometimes entering work environments but construct boundaries out of respect for religious tradition or to segment science and religion from one another. A UK associate professor of biology touched on the theme of avoidance, saying, "You have to watch anything you say here. It's almost like Northern Ireland. And say there was [a] huge population of Muslim students. So we are advised not to put in any reference to religion. That goes as far now as not to be able to talk about Easter break any more . . . or Christmas break. 'Take this word out of your slides—just call it the break.' [*Chuckle*] . . . Management—and management follows obviously the directives of government."[33]

Like faculty members, science students we interviewed also did not see religion coming up as a topic of discussion with colleagues or teachers very often. Yet even if initially they reported that religion did not come up, some scientists did recall small instances as they began to speak more on the topic. A female biology PhD student in India could recall select instances when religion might come up in her lab: "No, not really. I mean, yeah one instance I can cite is that when—especially when people are in sixth year they are applying for postdocs, so one obvious thing is that to apply for visas, there definitely it's a problem if you belong to the Muslim section. It's the name that becomes a problem. So that kind of discussion sometimes [comes up] in the lab. It happens that that particular name will create a problem for you when you're applying for postdocs."[34]

In terms of religion in the classroom, one Indian associate professor in biology cited religious diversity in the classroom as a reason not to bring up a discussion of religion with students. He reasoned, "Because in the classroom, there are different students and they come from different religions, so if you are discussing any religion, there may be someone uncomfortable there, so I don't think there should be any discussion about religion in the class."[35]

Perceived Discrimination

We found that perceived discrimination can also create a boundary that obstructs the observation of religious beliefs and practices in the workplace (we stress here that through the interview and survey data that we have collected we cannot tell whether religious scientists are actually being discriminated against, *but* we can report their perceptions of discrimination). Although overt religious discrimination is rare, some scientists did provide examples of students being prevented from observing their religious beliefs and practices, motivated by the goal of protecting the autonomy of practice in science. For example, when asked whether he encounters religion in his scientific work, a UK senior lecturer in biology admitted misleading Muslim students about the source of animal models for experiments, remarking, "I actually did say that it was lamb rather than pig. And that satisfied everybody, and they carried on with the experiment. That perhaps was wrong of me."[36] This obstruction of practice was motivated by his desire to carry out a course without having too many students opt out.

Discrimination can also surface as a result of scientists' attempts to protect the legitimacy of science. One example of this is a reluctance to hire individuals with creationist beliefs for research positions. Despite the official structures of the university emphasizing neutrality toward religion, another UK biology professor recounted a time when he deterred a student who was technically skilled, but a creationist, from joining the university: "I said, 'look you know you're coming into a research environment where we all accept evolution as a fact.' . . . And she kind of took the hint . . . and went away."[37]

Accommodation

We also found evidence for the accommodation of religion in the workplace that permits the practice of both science and religion to occur alongside one another. Common examples of this include accommodating the absence of Muslim students for Friday prayers and changing schedules to work around fasting during the month of Ramadan for Muslim students. A UK biology professor spoke of a time when he offered to accommodate a student's religious practices, explaining, "I once found the prayer mats in a tissue culture room. . . . She's not supposed to have prayer mats in [the] tissue culture [room]. I told her, if she wanted to pray, we could find somewhere else for her to pray."[38] Accommodation was a much more prevalent phenomenon than discrimination, although less frequent than avoidance. Thus, when scientists accommodate religion in the workplace, whether in research or teaching, they are allowing science and religion to coexist without a problem. This behavior eliminates the need for conflict.

Conclusion

The conflict thesis, used as a way to characterize the relationship between religion and science, is popularized in the media and utilized by various "celebrity" scientists.[39] Despite how the relationship is portrayed, there has not been any widespread, international investigation into the opinions of scientists themselves. Here we have begun this investigation.

Although the survey results presented in the tables and figures above have many nuances, they can be broadly summarized with a few conclusions. It is notable that, first, a minority of scientists in all eight national contexts view science and religion as being in conflict in the abstract. Similarly, very few scientists credit scientific training and knowledge for any declines in their own personal religiosity. When differences can be drawn between the eight contexts, it tends to be with nations like France, the United Kingdom, and the United States on one side and nations like India and Turkey on the other. Specifically, scientists in the latter tend to be more religious, more similar to their general populations' religiosity, and more likely to see potential collaboration between religion and science. Scientists in the former group of nations tend to be less religious overall and less religious relative to their respective national populations. They also tend to perceive more conflict between religion and science in the abstract and more negativity toward religion among their colleagues. There is a third group of contexts that includes Italy, Hong Kong, and Taiwan and is more difficult to characterize. Scientists in Italy appear to be more religious than those in the France–United States–United Kingdom group, but they also share some characteristics of that group. Scientists in Hong Kong and Taiwan are unique in that they tend to show higher levels of religiosity than their national populations and report increases in their religiosity from childhood.

In our interviews, most scientists in all of the national and regional contexts noted that religion does not often come up in the workplace. Yet, in the instances when it did arise, there were several themes to describe scientists' reactions: avoidance, discrimination, or accommodation. Avoidance was the most common pattern of behavior, in which scientists tried to establish or maintain a respectful boundary between the two domains. This boundary can lead to discrimination, as was raised in a few interviews, specifically with regard to hiring decisions or obstructing certain religious practices. Finally, accommodation also occurs and allows for religion to exist in the workplace.

In sum, the perception of abstract religion-science conflict does seem to be somewhat nation specific, with Western scientists being less religious overall and more oriented toward the idea of conflict, and Indian scientists almost universally thinking the conflict model does not apply to their approach to the relationship between religion and science. And the bottom line is that the majority of scientists in these eight national contexts do not see religion and science as in conflict.

ACKNOWLEDGMENTS

Research for this chapter was funded by the Templeton World Charity Foundation, Religion among Scientists in International Context Study TWCF0033/AB14, Elaine Howard Ecklund PI, Kirstin R. W. Matthews and Steven Lewis, Co-PIs.

NOTES

Epigraphs: Richard Dawkins in debate with Francis Collins, interview by David Van Biema, "God vs. Science," *Time*, November 5, 2006, http://content.time.com/time/magazine/article/0,9171,1555132,00.html. Jerry Coyne quoted in Steve Connor, "For the

Love of God . . . Scientists in Uproar at £1M Religion Prize," *Independent*, April 6, 2011, http://www.independent.co.uk/news/science/for-the-love-of-god-scientists-in-uproar-at-1631m-religion-prize-2264181.html.

1. A. Binder, "Gathering Intelligence on Intelligent Design: Where Did It Come from, Where Is It Going, and How Should Progressives Manage It?," *American Journal of Education* 113, no. 4 (2007): 549–76; Francis Collins, *The Language of God: A Scientist Presents Evidence for Belief* (New York: Simon and Schuster, 2006).

2. Karina Piser, "French Secuarism Is in Crisis: What Does That Mean for Muslim Youth?," *Nation*, January 8, 2018, https://www.thenation.com/article/french-secularism-is-in-crisis-what-does-that-mean-for-muslim-youth.

3. Max Weber, *The Vocation Lectures: "Science as a Vocation" "Politics as a Vocation,"* ed. David Owen and Tracy B. Strong (Indianapolis, IN: Hackett, 2004).

4. James Henry Leuba, *The Belief in God and Immortality: A Psychological, Anthropological, and Statistical Study* (Boston: Sherman, French, 1916); Leuba, "Religious Beliefs of American Scientists," *Harper's Magazine* 169 (1934): 291–300.

5. C. M. Brown, "The Conflict between Religion and Science in Light of the Patterns of Religious Belief among Scientists," *Zygon: Journal of Religion and Science* 38, no. 3 (2003): 603–32; J. H. Evans and M. S. Evans, "Religion and Science: Beyond the Epistemological Conflict Narrative," *Annual Review of Sociology* 34 (2008): 87–105; R. K. Merton, *Science, Technology, and Society in Seventeenth-Century England* (New York: H. Fertig, 1970); R. L. Numbers, ed., *Galileo Goes to Jail and Other Myths about Science and Religion* (Cambridge, MA: Harvard University Press, 2009); John Polkinghorne, *Science and Theology* (Minneapolis, MN: Fortress Press, 1998); Robert Wuthnow, "Science and the Sacred," in P. E. Hammond, ed., *The Sacred in a Secular Age* (Berkeley: University of California Press, 1985), 187–203.

6. A. M. Greeley and P. G. Rossi, *The Education of Catholic Americans* (New York: John Wiley and Sons, 1967); A. M. Greeley, "The 'Religious Factor' and Academic Careers: Another Communication," *American Journal of Sociology* 73, no. 5 (1973): 1247–55; Wuthnow, "Science and the Sacred," in Hammond, *The Sacred in a Secular Age.*

7. Merton, *Science, Technology, and Society*; Numbers, *Galileo Goes to Jail*; R. Stark, *For the Glory of God: How Monotheism Led to Reformations, Science, Witch-Hunts, and the End of Slavery* (Princeton, NJ: Princeton University Press, 2003).

8. Brown, "The Conflict between Religion and Science."

9. Elaine Howard Ecklund and Christopher Scheitle, "Religion among Academic Scientists: Distinctions, Disciplines, and Demographics," *Social Problems* 54 (2007): 289–307; Elaine Howard Ecklund, *Science vs. Religion: What Scientists Really Think* (New York: Oxford University Press, 2010).

10. Ecklund and Scheitle, "Religion among Academic Scientists"; Ecklund, *Science vs. Religion*; Elaine Howard Ecklund and Elizabeth Long, "Scientists and Spirituality," *Sociology of Religion* 72, no. 3 (2011): 253–74.

11. P. L. Berger, ed., 1999. *The Desecularization of the World: Resurgent Religion and World Politics* (Washington, DC: Ethics and Public Policy Center; Grand Rapids, MI: Wm. B. Eerdmans, 1999); R. Inglehart and P. Norris, *Sacred and Secular: Reexamining the Secularization Thesis* (Cambridge: Cambridge University Press, 2004); J. Stolz and O. Favre, "The Evangelical Milieu: Defining Criteria and Reproduction across the Generations," *Social Compass* 52 (2005): 169–83.

12. Richard Dawkins, *The God Delusion* (New York: Houghton Mifflin, 2006); A. McGrath and J. C. McGrath, *The Dawkins Delusion? Atheist Fundamentalism and the Denial of the Divine* (Downers Grove, IL: InterVarsity Press, 2007); Polkinghorne, *Science and Theology.*

13. Two exceptions include one early-twentieth-century case, when Drawbridge questioned British Royal Society Fellows on their religious beliefs, revealing that 60 percent of respondents believed in a spiritual domain; and a more recent study of the Royal Society claims that most reject the supernatural. For these studies of the Royal Society, see C. L. Drawbridge, *The Religion of Scientists: Being Recent Opinions Expressed by Two Hundred Fellows of the Royal Society of Religion and Theology* (London: Benn, 1932); and M. Stirrat and R. E. Cornwell, "Eminent Scientists Reject the Supernatural: A Survey of the Fellows of the Royal Society," *Evolution: Education and Outreach* 6 (2013): 33.

14. See *CERN Courier*, November 2, 1999, http://cerncourier.com/cws/article/cern/28116.

15. David Baltimore, "A Global Perspective on Science and Technology," *Science* 332 (2008): 544–51.

16. New migration streams generally increase religious diversity (Fenggang Yang and Helen Rose Ebaugh, "Transformations in New Immigrant Religions and Their Global Implications," *American Sociological Review* 66, no. 2 [2001]: 269–88), and religion often has an impact on gender relations and gender diversity within occupational structures (Jeremy Freese, "Risk Preference and Gender Difference in Religiousness: Evidence from the World Values Survey," *Review of Religious Research* 46, no. 1 (2004): 88–91; Jonathan Kelley and Nan Dirk De Graaf, "National Context, Parental Socialization, and Religious Belief: Results from 15 Nations," *American Sociological Review* 62, no. 4 [1997]: 639–59).

17. Ecklund and Scheitle, "Religion among Academic Scientists."

18. The sample contained only those scientists with valid contact information. Although still considered part of the broader population, we removed from the sample scientists whom we had no means of contacting for participation in the survey.

19. Leuba, *Belief in God and Immortality.*

20. RASIC_UK22, lecturer, biology, male, nonreligious, December 4, 2013.

21. RASIC_US24, professor, biology, male, nonreligious, April 1, 2015.

22. RASIC_FR29, researcher, biology, male, nonreligious, July 7, 2015.

23. RASIC_UK71, professor, biology, male, slightly religious, May 8, 2014.

24. RASIC_TK27, postdoctoral researcher, biology, male, nonreligious, May 29, 2015.

25. RASIC_US03, graduate student, biology, female, nonreligious, March 2, 2015.

26. RASIC_UK43, associate professor, physics, female, religious, December 6, 2013.

27. RASIC_IND34, professor, biology, female, slightly religious, May 22, 2014.

28. RASIC_HK02, assistant professor, biology, male, religious, December 11, 2014.

29. RASIC_HK21, postdoctoral researcher, physics, male, religious, November 20, 2012.

30. RASIC_HK22, professor, biology, male, religious, November 21, 2012.

31. RASIC_FR76, researcher, physics, female, religious, August 27, 2015.

32. RASIC_ITA06, associate professor, physics, male, religious, July 23, 2014.

33. RASIC_UK68, associate professor, biology, male, slightly religious, April 23, 2014.

34. RASIC_IND11, PhD student, biology, female, slightly religious, March 18, 2014.

35. RASIC_IND32, associate professor, biology, male, religious, May 22, 2014.

36. RASIC_UK60, senior lecturer, biology, male, slightly religious, April 1, 2014.

37. RASIC_UK34, professor, biology, male, nonreligious, December 5, 2013.

38. RASIC_UK60, senior lecturer, biology, male, slightly religious, April 1, 2014.

39. Dawkins, in *Time* interview by Van Biema; Dawkins, *God Delusion*.

CHAPTER SIXTEEN

Social Scientists

THOMAS H. AECHTNER

The conflict model of science-religion interactions is a narrative that was frequently articulated by many of the social sciences' founding figures. In fact, the original architects of these fields recurrently expressed religion-science discord via deterministic evolutionary models of human culture and societies, which anticipated the inevitable victory of science over religion. Such developmentalist theories depicted religion and science as competing worldviews, offering divergent means of interpreting and manipulating nature. A quintessential case of this early trend can be found in the work of Auguste Comte, often described as the founder of sociology. Comte maintained that, like other sciences, sociology should include specific laws, and he thus proposed the law of three stages. Underlying this law was the eventual substitution of supernatural, primitive knowledge with facts derived from observation, arranged "according to the general ideas or laws of an entirely positive order."[1] In like manner, Edward Burnett Tylor, who has been labeled "the father of anthropology," assumed that science would eventually displace religion.[2] Conflict between religion and science was inevitable because of their irreconcilable descriptions of the physical world. Though contemporary social scientists tend not to endorse these models of cultural evolutionary progression overtly, it is apparent that adherence to the conflict model still persists within anthropology and sociology. In fact, as this chapter demonstrates, contemporary postsecondary textbooks and reference materials from both fields still present the conflict model narrative as *the* historical account of religion-and-science interactions. Hence, while many nineteenth-century notions of religion, science, and cultural advancement are eschewed by social scientists, the conflict model perseveres as a conspicuous historical narrative in modern university-level pedagogical materials.

This chapter shows the ways in which modern introductory anthropology and sociology publications continue to endorse religion-science myths, while they advance a chronicle of inevitable and enduring science-religion discord. In order to do so, contemporary materials produced only within the twenty-first century are analyzed, including works printed by the foremost publishing houses retailing social scientific textbooks and reference materials. These presses include AltaMira, Blackwell, Macmillan, McGraw-Hill, Pearson, Polity, Routledge, Sage, and Thomson Wadsworth; all of which offer many of the most widely used university-level teaching materials employed on postsecondary campuses around the world. Additionally, several of the texts analyzed here have undergone numerous editions, are still being used, and are intended to introduce undergraduates and novices to central concepts in both fields. What is essential to note regarding the introductory materials identified here are the ways in which they portray the history of religion-and-science interactions. Conspicuously, these anthropology and sociology texts perpetuate widespread religion-science myths while discussing the Scientific Revolution and the Enlightenment, as well as religious responses to heliocentrism and Darwin's theory of evolution. Such tales, which have been readily dismissed by historians of science, include the conflict model's central supposition that science and religion have always been, and will continue to be, rancorous enemies.

The Scientific Revolution, the Enlightenment, and Perennial Conflict

In his notorious text *History of the Conflict between Religion and Science*, the nineteenth-century polemicist and professor of chemistry John William Draper asserted, "The history of Science is not a mere record of isolated discoveries; it is a narrative of the conflict of two contending powers, the expansive force of the human intellect on one side, and compression arising from traditionary faith and human interests on the other."[3] Importantly, in the same way that contemporary anthropologists and sociologists reject many of the deterministic theories of cultural progress that initially characterized both fields, so too have modern historians of science discarded Draper's infamous contentions. Nonetheless, allegations resembling those found in the work of Draper, as well as other nineteenth-century proponents of the conflict model, can still be identified throughout modern anthological and sociological introductory materials. In 1869, Andrew Dickson White, the first president of Cornell University and staunch proponent of secular education, voiced his opinion that the conflict between religion and science has been an invariable historical "struggle which has been going on for so many centuries."[4] Remarkably, within anthropology and sociology introductory materials one can locate numerous similar offhanded allusions to a chronic religion-science history of conflict, as well as remarks involving deep-seated irreconcilabilities between science and religion.

In *Sociology: A Global Introduction*, a contemporary textbook assigned to undergraduates at the very university for which I work, John J. Macionis and Ken Plummer discuss religion in the twenty-first century. While addressing the topic of secularization and atheism as a social movement, the authors explain to students that from a historical perspective, the "modern world can indeed be seen as a long battle between 'science' and religion.'"[5] Elsewhere, in yet another sociology textbook, Macionis again engages with the subject matter of secularization. This time considering modern religious fundamentalist opposition to secular humanism, he emphasizes to students that there "is nothing new in this tension between science and religion; it has existed for centuries."[6] Somewhat similarly, while remarking on sociological issues associated with fundamentalist religious groups and American religiously motivated antievolutionism in *Sociology in a Changing World*, William Kornblum notes casually that there remains an "ancient tension between science and religion."[7] *The Wiley-Blackwell Companion to Sociology* includes similar conflict model references, which also appear within brief discussions of religious fundamentalism and creationism in the United States. In this context, the sociologists Mark Erickson and Frank Webster first mention the "ongoing clash of science and religion," before articulating to readers that quarrels between the two "are, of course, not new."[8] In like manner, anthropological introductory texts make references to perennial religion-and-science discord, albeit often through relatively subtle remarks. For example, in the *Encyclopedia of Anthropology*, edited by H. James Birx, an article addressing religion and the environment alludes to such conflict in association with the topic of spiritual ecology. The author explains with apparent surprise, "Likewise, spiritual ecology is conducive to such cooperative initiatives between representatives from science and religion," even after "several centuries of antagonism."[9] Though indirect, this comment incorporates a perceptible opinion regarding the overall historical condition of religion-science interactions.

More direct cases can be found in *21st Century Anthropology: A Reference Handbook*, in passages serving to explain secularization and associated theses to students. The author of an article titled "Enlightenment and Secularism," for instance, first states, "The Copernican revolution generally marks the advent of the scientific revolution because of the tremendous implications of heliocentrism. In replacing the earth with the sun as the center of the universe, Copernicus revolutionized humankind's understanding of science and religion." He then makes clear, "Indeed, the Copernican revolution marks the beginning of the disagreements between science and religion that would quickly become a persistent theme of history thereafter," and consequently, "science-versus-religion debates have increasingly developed as a struggle between conflicting worldviews, of enlightenment versus orthodoxy." Underscoring the point further, a subsection within the same chapter, named "The Controversy of Science versus Religion," begins with the following unambiguous proclamations: "The relationship between science and religion has been uneasy throughout much of history. Revolutionary scientific developments such as the Copernican, Newtonian, and Darwinian revolutions all significantly strained the relationship."[10]

Noticeably, several statements about enduring past and present religion-science conflict appear along aside references to the Scientific Revolution and the Enlightenment. Within both anthropology and sociology introductory materials, these historical periods serve as conflict model motifs and are identified as specific junctures in human history that sparked science-and-religion conflict. "Around 1400, philosophers started to use what would be called the 'scientific method,' systematic, experimental studies that uncover the facts of the natural world," explain Michael Kimmel and Amy Aronson, authors of *Sociology Now*, in a chapter devoted to analyzing both religion and science as institutions.[11] "Unfortunately, the facts they uncovered," students are then told, "often disagreed with religious doctrine," such that: "The sun doesn't revolve around Earth. The equatorial regions are not too hot to support life. Earth is much more than 6,000 years old." The authors then conclude, "The Church conceded some points, but not others, and the competition between religion and science began." Accordingly, the Scientific Revolution and the Enlightenment are often framed as historical stages in which humanity finally surmounted the cultural paralysis, or medieval mediocrity, of religion, religious cosmologies, and religiously entrenched ideas. These phases are thus narratively categorized as key points of history in which humanity successfully fashioned a split between religion and science, leading to a societal move toward secular, sensible, scientific thinking. For it was "the rise of science, with its claims for rationality" that served as a "serious challenge to Christianity during the Enlightenment and the Industrial Revolution."[12] Hence, in an encyclopedia entry on the Enlightenment, Emil Visnovsky unabashedly writes that this period "was the age of reason, which could not eliminate faith as such, despite all its efforts; rather, it replaced the religious faith with the secular faith in reason itself."[13] Accordingly, it was during this era that "Enlightenment thinkers replaced the universalism of theology with the universalism of scientific conceptions."[14]

Perhaps unsurprisingly, the historical formation of the social sciences, which includes the work of Comte and Tylor described above, is also narratively linked with the genealogy of this Enlightenment story.[15] As a result, textbook accounts relating the genesis of the social sciences connect its disciplines to the historical rise of scientific rationalism over-and-against religion. Correspondingly, in *Social and Cultural Anthropology: The Key Concepts*, Nigel Rapport and Joanna Overing provide the following report: "In succeeding centuries, the humanism of the Renaissance gave on to the Enlightenment and the rise of science, with its belief in rationality, as opposed to (religious) revelation, as an adequate source of human knowledge; also on to liberalism, and a belief in the inherent dignity of individuals and their right to freedom and self-determination; and also on to social science, and its belief in the possibility and necessity of applying knowledge about human affairs and individual relations to an improvement of the sociocultural conditions of human life."[16]

What is essential to note regarding these treatments of the Scientific Revolution and the Enlightenment is that they disclose no sense of the complex relationships that were thought to have existed between religion and what today might be called science. This is particularly remarkable, considering that even a cursory glance at scholarly accounts of religion and science during the Enlightenment provided by modern historians disclose far more intricate portrayals of events. Accordingly, John Brooke has maintained that to "reduce the relations between science and religion to a polarity between reason and superstition is inadmissible, even for that period when it had such rhetorical force."[17] Indeed, Dan Edelstein has gone so far as to insist that histories depicting the Enlightenment as an age that moved humanity from ignorance to scientific reason are very much just that; narratives primarily molded and delivered to us by the *philosophes*: "The key contribution made by these French scholars, writing between 1680 and 1720, was less epistemological than narratological. In other words, they did not propose a new method of reasoning or advocate a new philosophical understanding of the world. Rather, they offered a seductive account of the events and discoveries of the past century, in conjunction with a more overarching history of human civilization."[18] As Edelstein goes on to say, "Calling this narrative a myth underscores its constructed and partial nature, reminding us that this story should not be mistaken for an accurate history."[19] In spite of these academic assessments, however, introductory anthropology and sociology publications simply pass on to students the *philosophes*' Enlightenment narrative *as* accurate history. "During the 18th-century Age of Enlightenment," explains Elisa Ruhl in the *Encyclopedia of Anthropology*, "the emphasis shifted toward shedding superstitions and recognizing the human capacity for reason."[20]

Heliocentrism and Religion-Science Discord

Within anthropology and sociology introductory texts, references to the Scientific Revolution and the Enlightenment often include some mention of Copernicanism, Galileo, and the Roman Catholic Church's resistance to heliocentrism. Hence, sociologist Jon Shepard initially explains to students in a chapter on religion and society, "As scientific knowledge began to increase with the Enlightenment, religious beliefs in the West also began to change." He then communicates that this revolution brought with it a new "understanding of the universe," which "slowly emptied the prisons of scientists who disputed the theological truism of an earth-centered solar system."[21] Curiously, Shepard seems to be implying that during the Enlightenment prisons were teeming with supporters of heliocentrism, while simultaneously overlooking the nontheological contestations at play during the Galileo affair. Similar misconstruals regarding heliocentrism, and the myth of Galileo's imprisonment pepper other sociology reference materials, particularly when providing instruction on historical science-religion interactions. For instance, in this context Steve Bruce explains to readers that the church "imprisoned Galileo for continuing to promote the Copernican view that the earth moved around the sun," while the author of *Sociology: A Global Perspective* notes that in 1633 "powerful church inquisitors threatened to torture and kill Galileo, who had embraced Copernicus's theory, if he did not renounce it; upon renouncing it, Galileo was imprisoned for life."[22]

In a subsection of *Sociology* titled "Does Science Threaten Religion?," Macionis also mentions Galileo, whom he asserts not only "observed the stars and found that Earth orbited the sun, not the other way around," but also "discovered some of the laws of gravity" by dropping objects from the Leaning Tower of Pisa. "For his trouble," recounts Macionis, "Galileo was challenged by the Roman Catholic Church, which had preached for centuries that Earth stood motionless at the center of the universe." Moreover, "Galileo only made matters worse by responding that religious leaders had no business talking about matters of science." He then concludes with what might be taken as an archetypical conflict model statement: "As Galileo's treatment shows, right from the start, science has had an uneasy relationship with religion."[23] Statements resembling Macionis's Galileo storyline, along with other flawed notions regarding religion and heliocentrism, can be identified throughout other sociology and anthropology texts when authors make some mention of religion-science relationships. For example, in an *Encyclopedia of Sociology* article describing materialist philosophies, Robert J. Antonio claims that it was in fact the threat of "materialism as a subversive force" that moved the church to convict Galileo.[24]

Each and every case involving this subject matter within these introductory texts presents the church as a religiously zealous oppressor of scientific truth, which blindly persecuted scientific pioneers. Under a subsection titled "Religion, Science, and Intelligent Design," in *The Tapestry of Culture: An Introduction to Cultural Anthropology*, the book's authors identify how many "aspects of the natural world that were formerly explained by religious ideology are now explained by means of science." The Catholic Church's refusal to accept heliocentrism is then subsequently appealed to as representative of science's defiant and progressive victory over failing religious dogmatism. "In the seventeenth century," summarize the textbook's writers, "the Catholic Church insisted that the earth was at the center of the solar system, and persecuted Galileo for his scientific research, which demonstrated that Copernicus's earlier conclusion was correct: that the sun, not the earth, was at the center of our solar system." Additionally this oppression occurred in spite of "conclusive evidence supporting this scientific view of the solar system," and unfortunately, "it took the Catholic Church hundreds of years to officially accept the scientific explanation."[25] Consequently, in much the same perfunctory manner that such introductory texts address the Enlightenment, anthropology and sociology materials also fail to appropriately communicate the many intricacies that helped engender Rome's reaction to heliocentrism.

An unmistakable conflict model legend associated with the Galileo affair, and which appears throughout anthropology and sociology materials, is the allegation that a primary intellectual triumph of the Copernican revolution was its displacement of humanity from the center of the universe. For instance, in an encyclopedia entry that provides a brief overview of the premodern era, sociologist Julie M. Albright relates that religious opposition to heliocentrism arose because it demoted humans from their esteemed place within the cosmos. "Copernicus challenged the Church's contention that the sun revolves around the Earth," Albright explains, "and thus that man is the center of the universe."[26] In this way, these introductory texts recapitulate a popular myth associated with Galileo, once promulgated by Bertolt Brecht's play *Leben des Galilei*. This tale features the church construing Copernican's astral displacement as a nonbiblical downgrading of humanity and the earth's divine status. "In Europe during the Renaissance (after c. 1450 A.D.)," explain the authors of *Anthropology: A Global Perspective* while describing the Scientific Revolution to students, "scientific discoveries began to challenge conceptions about both the age of the Earth and humanity's relationship to the rest of the universe." As a result "humans could no longer view themselves and their planet as the center of the universe, which had been the traditional belief."[27] This "traditional belief" was the church's ossified view of the universe, supported by scripture and a sort of theological literalism. Similarly, while outlining the history of naturalism, Bill Cooke insists, "Since the Renaissance, naturalism has returned to the center stage of all intellectual and scientific life. Three great revolutions in naturalism have been instrumental in achieving this." The first of these, he affirms, "was what is now called the Copernican revolution," which ceased to make it "credible to see Jerusalem, or even the entire planet Earth, as the center of the universe."[28]

Cooke's reference to Jerusalem appears to be a deliberate allusion to religious cosmologies, and the notion that Copernicanism endangered the theologically superlative location of the earth, humanity, and the holy city. However, even a brief analysis of contemporary history of science scholarship demonstrates otherwise. As Ernan McMullin has concluded, the Galileo affair's primary issue was not whether Copernicanism displaced humanity from the center of the cosmos, but *who* had the authority to interpret scripture in a post-Reformation world. "Galileo had the misfortune to bring the Copernican claims to public notice at just the wrong time," McMullin clarifies, "a time when sensitivities in regard to questions involving scriptural interpretation and Church authority were at their most intense." In fact, he postulates that had "Galileo made his case for Copernicanism a century earlier or a century later, it seems unlikely that it would have evoked the strong response it did on the part of the Roman theologians." It is for this reason that McMullin states decisively, "The Galileo affair ought not then be construed, as it so often has been, as primarily a clash between rival cosmologies, with the resistance of the Church authorities to the new cosmology to be explained by their stubborn adherence to an outmoded Earth-centered cosmos."[29]

David C. Lindberg has further noted that even within the church itself some clergy acted as Galileo's most vocal supporters, and in many ways the entire affair was essentially an interscientific dispute.[30] Dennis R. Danielson has made analogous claims, while acknowledging that in terms of the church and geocentrism versus heliocentrism, "*anthropocentrism* is a figurative term only" that does not theologically demand a physically central location within the universe. Consequently, it "is true of course that in the seventeenth century the arch-Copernican Galileo Galilei (1564–1642) met opposition from Catholic authorities in Rome," but fundamentally "their dispute focused on matters related to biblical interpretation, educational jurisdiction, and the threat Galileo represented to the entrenched 'scientific' authority of Aristotle, not on any supposed Copernican depreciation of the cosmic specialness or privilege of humankind." Importantly, historians have also pointed out that, from an Aristotelian-Catholic standpoint, the center of the cosmos was not actually a theologically esteemed locality. "If anything," writes Danielson, "Galileo and his fellow Copernicans were *raising* the status of earth and its inhabitants within the universe."[31]

Anthropology and sociology introductory materials also include references to Giordano Bruno amidst discussions of Copernicanism, the Enlightenment, and the Scientific Revolution. In these cases, Bruno is generally represented as a type of scientific martyr, condemned by the church for defending and promoting a scientifically generated heliocentric conception of the universe. These descriptions echo Andrew Dickson White's rejected nineteenth-century narrative of Bruno's immolation, and students in the social sciences are told that this episode helps demonstrate the historical reality of religion-science discord. For example, it is in this context that James P. Bonanno writes of the Copernican revolution: "The notion of an infinite universe, with untold number of planets, moons, and even suns was met with disdain by religious authorities, both Catholic and Protestant." In validation of this claim, he notes that the "Roman Inquisition sought to make an example of the Italian philosopher Giordano Bruno, a well-known supporter of heliocentrism who also elaborated on its precepts, by condemning him to death on the charge of heresy."[32] Anthropologist Jack D. Eller makes a similar assertion while outlining processes of secularization in the textbook *Introducing Anthropology of Religion: Culture to the Ultimate.* He details the following regarding Catholicism's early encounters with Copernicanism: "Received religious truth did not retire gracefully. The new theory was condemned by the Church, which was waging its wider war against heresy. In 1600, Giordano Bruno, a former Dominican brother who had been defrocked for his unorthodoxy in 1576, was burned at the stake for criticizing the traditional view and believing, beyond Copernicus, that the 'heavenly bodies' were merely distant suns with their own solar systems around them."[33] Nachman Ben-Yehuda also obliquely refers to Bruno while addressing the sociology of political deviance. "In today's liberal democratic societies, freedom of thought and speech are hailed as primary virtues," he notes, while in "other regimes, individuals who exercise freedom of thought, or challenge the 'order,' are liable to find themselves imprisoned or committed to an insane asylum."[34] To demonstrate his point, Ben-Yehuda provides the following vignette: "Science is no exception. Giordano Bruno died for challenging the Ptolemaic worldview and the morality which supported it; Galileo as well suffered because of this worldview."[35]

As with White's rendering of Bruno, these narratives neglect recent historical analyses of the legendary monk. For instance, modern studies emphasize that his brutal execution did not simply result from his acceptance of heliocentrism, or even a belief in an infinite universe, but from a series of views interpreted as being philosophical and theological heresies. In fact, scholarly consensus insists that Bruno's astronomical notions were, at most, a minor feature of his trial and conviction. Though acknowledging the complexities surrounding the monk's execution does not diminish its ghastliness, doing so does provide an important perspective on Bruno not detailed in anthropology and sociology texts, while reframing interpretations away from White's crude version of events. Consequently, such introductory materials blatantly neglect the intricacies associated with Bruno's murder, which demonstrate rather more than simply a collision between science and religion. At the same time, the textbooks and reference materials analyzed here also demonstrate a tendency to misrepresent early religious reactions to Charles Darwin's theory of evolution by natural selection.

Religious Reactions to Darwin

Introductory anthropology and sociology books feature numerous mentions of early religious reactions to evolutionary theory, as well as the persistent growth of science-religion conflict that followed Darwin's publication of *On the Origin of Species*. Such remarks can be found within a variety of contexts, including sections conveying histories of the social sciences, mechanisms of secularization, commentaries on evolutionary theory, and descriptions of religion, science, and society. Strikingly, in each case the pedagogical materials of both fields only describe negative responses to the theory of natural selection, which according to Peter Metcalf in *Anthropology: The Basics*, greatly "alarmed the established churches at the time."[36] As a result, the myriad of nineteenth-century religious reactions to evolution is distilled down to a straightforward fracas between two irreconcilable ways of viewing the world.

With comparable one-sidedness, Bonanno maintains in his overview of secularism that the "theory of evolution by means of natural selection presented a major crisis between science and religion," which added "tremendous fuel to a conflagration" of history's ongoing religion-science battle.[37] In similar fashion, the authors of *The Tapestry of Culture: An Introduction to Cultural Anthropology* discuss creationism, telling readers succinctly that following the publication of Darwin's seminal text, "For a time, both religious and scientific explanations of creation competed with one another."[38] Sociologist Amanda Rees provides students with a sketch of the concept of evolution and likewise stresses the apparent ramifications of Darwin's ideas, along with other nineteenth-century scientific discoveries: "The potential for conflict between the revealed truth of the Christian religion and the developing scientific worldview had existed since geological and paleontological research had indicated as far back as the eighteenth century that the earth might be far older than the Bible suggested, and the relationship between science and religion in the West had been seriously damaged by the publication of Darwin's theory of evolution."[39]

In *Sociology: Making Sense of Society*, a similar conflict narrative occurs under a section titled "The Crisis of the Modern Mind," which reviews the historical emergence of science and its sociological consequences. About science and religion, the authors first say that, while "one does not necessarily exclude the other," it was the case that "the popularity and prestige of scientific discovery was clearly seen as a threat to religious authority." And the text then comments, "Science not only offered different types of explanations for the workings of the natural world but also provided the tools for its conquest and exploitation. The discoveries of archaeologists and astronomers had long since struck at the heart of biblical truth, while Darwin's view of the origins of the human race was at the center of a growing conflict between science and religion."[40]

Associated with religious responses to Darwin, the notorious Huxley-Wilberforce debate is also presented to readers as being representative of this historic discord. "In 1860 at the University of Oxford, England," writes Birx while surveying the history of evolutionary theory, "the infamous Thomas Huxley and Samuel Wilberforce confrontation exemplified the intense conflict between the new evolution paradigm in science and an outmoded static worldview in religion." At the same time, Birx also points to the Scopes trial as yet another example of the long-standing conflict existing between religion, evolution, and science generally. "In 1925 at Dayton, Tennessee," he explains, "the infamous John Scopes 'Monkey Trial' had best represented this ongoing clash between science and religion over the factual theory of organic evolution." Essential to Birx's brief comments on evolution and religion is a clear-cut dichotomy comprised of two elementary options: "Religious Creationism or Scientific Evolutionism." Accordingly, he fails to discuss non-antievolutionist responses to biology and simply mentions intelligent design theory as a case of religion trying to harmonize itself with contemporary science.[41]

This friction between evolution and religious belief is often portrayed as being inescapable, owing to essential religion-and-science incompatibilities. Accordingly, Bonanno contends that "evolution remains an active source of debate in many societies due to the fundamental contradictions between religious interpretation and scientific investigation." This is also suggested via remarks which presume that evolution's materialistic account of nature necessarily violates primary religious claims. These include the acceptance of a creator God who designed the universe, as well as the belief in souls and supernatural forces. Bonanno, therefore, surmises, "Evolution and the principle of common descent demolished the scientific plausibility of creation and design for the universe."[42] Employing the same presuppositions in a discussion of Thomist philosophy, David A. Lukaszek first notes that through "common descent, multiplication of species, gradualism, and natural selection, Darwin provided an explanation for diverse life forms on this planet." He then insists, "The metaphysical implications are evident; the evidence for a God (designer), the soul, and afterlife are rejected in light of evidence and rational explanation."[43] In the *Companion Encyclopedia of Anthropology*, Tim Ingold expounds on interpretations of human nature, similarly pronouncing that Darwin's theory of evolution "of course had no place for mind or spirit except as the output of a material organ (the brain)."[44] Utilizing analogous reasoning in an overview of secularization, Eller explains to readers that Darwin's theory provided "a natural mechanism by which new species could arise." This achievement, he contends, helped to discredit religion in the face of scientific rationalism: "The one continuous insult to religion from the scientific program is that natural law seems to suffice to explain everything we see; as Richard Dawkins would say, 'the universe we observe has precisely the properties we should expect if there is, at bottom, no design, no purpose, no evil and no good, nothing but blind, pitiless indifference.' No matter where science looks, it does not find supernatural beings or forces. Even the Catholic Church has had to abandon its resistance to evolution."[45]

Aside from Eller's appeal to Dawkins, whom he describes as "the arch-evolutionist," what is particularly notable about his commentary is its reference to the Catholic Church.[46] This is because, if anything, the Catholic Church's response to evolution has historically been rather ambiguous. For instance, Richard G. Olson explains that the majority of Catholics have "had relatively little inclination to condemn Darwinian evolutionary theory."[47] Regarding recent history, Ronald Numbers has observed that Catholic creationism has failed to garner significant support, and even when divided on the issue of evolutionary theory in the United States, Catholics have rarely favored any sort of pro-creationist legislation.[48] Eller, therefore, seems unacquainted with historical interactions between Catholicism and evolution, resulting in a markedly counterfactual narrative of science-religion conflict. Similarly, descriptions of only negative nineteenth-century religious responses to evolution in anthropology and sociology texts skew the record through oversimplification. While there undoubtedly have been noteworthy negative reactions to Darwin's theory, as it challenged important theological and scientific concepts, many religious leaders were not hostile to the theory, and there were actually varied responses within the scientific community itself. Nonetheless, throughout such university-level introductory materials, accounts of religion during the Darwinian revolution are articulated without any indication of the diversity of reactions that occurred not only within Christianity but also in non-Christian religions and among scientists.

Science, Secularization, and Religion

Within sociology and anthropology textbooks, conflict model storylines associated with Copernicanism, the Scientific Revolution, and the theory of evolution occur most frequently alongside broad narratives of the history and mechanisms of secularization. These accounts suggest that, from the Enlightenment onward, science and its ostensibly counterreligious rationalism have directly led to secularization. This message is encapsulated in the words of Raymond Scupin, who explains, "Ever since the Renaissance and Enlightenment periods, Western industrial states have experienced extensive secularization, the historical decline in the influence of religion in society." Scupin then indicates that this has occurred because scientists "such as Galileo and Charles Darwin developed ideas that challenged theological doctrines."[49] Sociologist Robert J. Antonio seems to imply a similar story after first telling readers, "Atomistic materialism reemerged as a major cultural force during the Renaissance science revolution," as "Galileo and Newton again portrayed physical reality as ultimate particles moving in empty space." This led to the distinguishing of "science from metaphysical, aesthetic, or sociocultural thought," and ultimately, the "extraordinary success of Newtonian science contributed greatly to an extensive secularization of knowledge that reduced barriers to materialist approaches in human affairs."[50] Anthony Giddens makes an analogous tacit suggestion in the textbook *Sociology*, stating that because of the "development of science and the secularization of thought" it is now the case that we "no longer assume that customs or habits are acceptable merely because they have the age-old authority of tradition." Instead, because of the secularizing influence of science, "our ways of life increasingly require a rational 'basis.'"[51]

This message is conveyed plainly by Bonanno, who also delivers a sweeping account of secularization that involves the decline of religion, along with affiliated beliefs and dogma, resulting from the inexorable success of science. "The intellectual culture of skeptical inquiry that emerged during the 17th-century scientific revolution, and developed during the 18th-century Enlightenment," outlines Bonanno, "affected nearly every field of human thought." Consequently, belief in "supernatural explanations diminished as humankind developed a greater understanding of the universe," causing people "to accept the idea that science was the best means to understanding our world." While appraising the secularization process, he then insists that humankind's "greater understanding of the natural world has affected popular belief in supernatural phenomena at an inverse relationship."[52] Bruce, on the other hand, provides a more nuanced message in his overview of secularization that still bears some similar characteristics. Not only does he state that during the Middle Ages "the development of science was retarded by the church's imposition of orthodoxy on all fields of thought," he further explains: "Science was not easy for cultures that believed the world pervaded by unpredictable supernatural spirits and divinities. Systematic exploration of regularities in the behavior of matter required the assumption of regularities. The less God was implicated in the day-to-day operations of the universe, the freer people were to elaborate theories of its operations that paid only lip service to the creator." Our "increased reliance on effective secular means of securing this-worldly ends," resulted in a societal reduction of "reliance upon faith." Consequently, science and technology "have not made us atheists," but still cause us to be "less likely than our forebears to entertain the notion of the divine."[53] According to Bruce, therefore, from their inception onward, science, its technological products, and its rationalizing effects have all led to secularization.

Eller also frames the history of science as a sustained deposition of religion, ensuing from Newton's theories of gravity and celestial mechanics, as well as accumulated knowledge about electromagnetism and particle physics. In this context, he concludes that belief in the supernatural has historically been declining because "science does not seek nor accept supernatural explanations and so far has had no need to resort to them." In fact, according to Eller, from the nineteenth century onward, science has acted as a direct "alternative to religion for the average person." Additionally, modern physics has served to catalyze science's secularizing capacity by reducing the rationality of belief in God. "Even worse," Eller reasons, "quantum theory appeared to describe an almost incomprehensibly strange world that followed none of the rules of everyday reality. Worst of all, it is a *probabilistic* reality, not a rule-ordered one at all—not the kind of place where a reasonable god seems at home."[54]

In many ways, these secularization accounts reflect early social scientific theories of human progress and science articulated by the likes of Comte and Tylor, while also recapitulating central premises made by the secularization thesis's mid-twentieth-century founders. For instance, in 1966, the anthropologist Anthony F. C. Wallace stated confidently that "the evolutionary future of religion is extinction." He went on to predict, "Belief in supernatural beings and in supernatural forces that affect nature without obeying nature's laws will erode and become only an interesting historical memory. . . . But as a cultural trait, belief in supernatural powers is doomed to die out, all over the world, as a result of the increasing adequacy and diffusion of scientific knowledge and the realization by secular faiths that supernatural belief is not necessary to the effective use of ritual. The question of whether such denouement will be good or bad for humanity is irrelevant to the prediction; the process is inevitable."[55] Wallace acknowledged that this would most likely not occur within a century, but even so, in the decades following these forecasts, scholars began to doubt his prognostications. In particular, critics questioned the actual role that science was hypothesized to play in the secularization process. As Peter L. Berger, a former advocate of the secularization thesis, has more recently conceded, the modern world remains "massively religious," and it appears nothing like "the secularized world that had been predicted (whether joyfully or despondently) by so many analysts of modernity."[56] Charles Taylor has further contended that there seems to be no definitive link between science and secularization. "I'm not satisfied with this explanation of secularism: science refutes and hence crowds out religious belief," he notes in contrast with anthropology and sociology texts. "I don't see the cogency of the supposed arguments from, say, the findings of Darwin to the alleged refutations of religion."[57] It appears as though secularization is significantly more complicated than the arithmetic logic expressed within social scientific textbooks and reference materials, which simply suggests that the addition of science causes a societal subtraction of religion. For that reason, claims made by anthropology and sociology materials regarding science and secularization are, at the very least, readily contestable and disputed in contemporary academic literature.

Teaching Discord

Despite the expression of the conflict model anecdotes in sociology and anthropology introductory materials, it is apparent that not all textbooks and reference materials sketch the same combative picture. For instance, the textbooks *Anthropology: The Human Challenge* and *Anthropology* both deliver comparatively nuanced criticisms of the secularization thesis.[58] The authors of these works express uncertainty regarding science's true role within the process and critique the notion that religion has simply been displaced by science. *Discovering Anthropology*, published in 1992, also contains a small subsection examining the topic "Evolution versus Creationism." Here Daniel R. Gross questions whether evolutionary theory truly conflicts with religion, and remarks, "Many scientists who base their work on evolutionary theory are devoutly religious and accept the teachings of the Bible. Many of them feel that their scientific activities are perfectly compatible with their religious beliefs."[59] At the same time, even some of the introductory materials that generally perpetuate the conflict model also include some recognition of science-and-religion harmony. Illustrations of this can be found in the affiliated textbooks *Anthropology: A Global Perspective* and *Cultural Anthropology: A Global Perspective*, which have been cited previously throughout this chapter. Both of these texts contain near identical "Critical Perspective" pieces titled "Creationism, Intelligent Design, and Evolution,"[60] which not only criticize antievolutionists, but also concede that there are noncombative religious responses to evolutionary theory. Additionally, the aforementioned sociologists Kimmel and Aronson state, regarding religion and scientific practice, "Even though science and religion seek to do so many of the same things and often come to different conclusions, they are not necessarily rivals in society. Strong religious belief and deep scientific knowledge can coexist."[61]

Nonetheless, it is clear that narratives of science-religion conflict still persist within many contemporary anthropology and sociology texts. With this observation in mind, it is also important to consider some of the possible reasons why so many authors of social scientific introductory materials propagate the conflict model. The saliency of reflecting on this is underscored when social scientific pedagogical materials are contrasted with the introductory texts of other fields. For instance, religion-science legends associated with the advent of heliocentrism in Europe can be located in astronomy and physics textbooks. Hence, Chaisson McMillan, author of *Astronomy Today*, tells readers that Copernicanism was rejected "largely because it relegated the Earth to a noncentral and undistinguished place within the solar system and the universe," while he also links Bruno's immolation with "his heretical teaching that Earth orbited the Sun."[62] Another introductory work, *The Physics of Everyday Phenomena*, implies that before Galileo defended Copernicanism, numerous individuals had already been burned at the stake for espousing a heliocentric universe.[63] At the same time, historically doubtful claims associated with the Darwinian revolution can be found in biology textbooks, including insinuations that Darwin's contemporaries were predominantly six-day creationists.[64] Notably, however, when compared with social scientific introductory publications, such dubious conflict model claims appear far less frequently. Astronomy and physics pedagogical materials rarely mention the Galileo affair, and when they do, nuanced interpretations of events are often provided.[65] Accordingly, Douglas C. Giancoli tells readers in *Physics: Principles with Applications* that "Galileo's famous encounter with the Church" was in fact not a clash between science and religion, but an episode involving "politics, personality conflict, and authority."[66]

While many scientists, therefore, appear to maintain the conflict thesis (see chapter 15 in this volume), science introductory materials are comparatively less likely to include religion-science battle narratives. Certainly this contrast reflects the distinctive subject matter addressed in social scientific as opposed to scientific introductory materials. Though various topics presented in science textbooks provide potential opportunities for broaching religion-and-science matters, the themes raised throughout social science publications offer rife possibilities to comment on science-religion interactions. These include expositions of secularization, religious fundamentalism, and antievolutionism, as well as historical accounts of religion and the nascence of science and the social sciences. Moreover, the founders of the social sciences championed the conflict thesis, while discourses on science, religion, and secularization have endured as mainstays of social scientific inquiry. As a result, even if both scientists and social scientists adhere to the conflict model, social scientific research concerns, in addition to long-held conventions of upholding warfare narratives, are more contextually conducive to providing science-religion vignettes throughout introductory materials.

Notwithstanding whether characteristics of the social sciences lend themselves to the inclusion of conflict model anecdotes, it still seems rather troublesome that anthropology and sociology textbooks contain such narratives when there exists a preponderance of scholarly evidence refuting them. This is problematic because throughout many of the works analyzed here the authors contend that their disciplines are empirically based and rely on observable data as well as rigorous research methods. In spite of this, most conflict model sketches appearing within these introductory texts are unaccompanied by any sort of citations or pertinent bibliographic sources. As a result, readers are left wondering what academic articles, books, or authorities, if any, are actually being employed to derive the outmoded religion-and-science historiographies expressed throughout such publications. Perhaps even more disquieting is that in the few cases where in-text citations *are* provided alongside religion-science claims, the accredited sources often contradict the assertions being made. For instance, *Anthropology: A Global Perspective* and *Cultural Anthropology: A Global Perspective* both cite John Henry's *The Scientific Revolution and the Origins of Modern Science* in relation to conflict model allegations about the Renaissance, heliocentrism, and religion.[67] However, Henry's work forthrightly refutes the very conflict model proclamations articulated by Scupin and DeCorse in both textbooks.[68]

This resistance to, or disregard of, sound religion-science scholarship seems to undermine the textbook authors' stipulations that anthropology and sociology are professions founded on meticulous research. This is especially noteworthy because such introductory materials could quite possibly represent the most circulated images of both disciplines to the public. It is through such texts that social scientists define their vocations, explain to nonexperts what the purview and methods of their analyses are, while endeavoring to convince undergraduates of the inherent value associated with their craft. Consequently, in many ways these introductory materials represent idealized forms of the social sciences to the uninitiated. This is important because, while these idealized forms lack academic support for their conflict model assertions, they *do* remain true to the original science-religion discord plotlines expressed by the founders of these disciplines. Therefore, it may be that even though the conflict model is an outdated, academically rejected notion, and that contemporary researchers explicitly deny evolutionary accounts of social history, in the end both anthropology and sociology are invested as professional disciplines in the conflict narrative. This is because explanations of human progress and religion-science discord, pioneered by Comte, Tylor, and others, remain key elements in the origins story of the social sciences.

Moreover, the very genealogy of the social sciences is often linked with conflict legends such as those associated with the Enlightenment's separation of science and religion. It might then be the case that the persistence of the conflict model narratives within anthropology and sociology introductory materials is symptomatic of the difficulty social scientists face in divorcing themselves from the compelling specter of their fields' own genesis stories and associated progress narratives. Therefore, despite being an academically dubious supposition, the conflict model possibly survives in the social sciences, and is correspondingly propagated through introductory materials, because of its pivotal role in the founding narratives of both anthropology and sociology.

Conclusion

It is evident that university-level pedagogical materials from the social sciences, particularly sociology and anthropology, disseminate narratives of science-religion discord. These accounts include allegations that religion and science have remained foes since the Scientific Revolution and Enlightenment periods, typified by the Roman Catholic Church's assault on Copernicanism and by early religious responses to evolutionary theory. Readers are instructed that such occurrences result from fundamental science-religion irreconcilabilities, which have led to the usurpation of religious beliefs, as well as secularization from Galileo onward. It is rather sobering to learn that religion-science myths and chronicles of perennial science-religion warfare have been preserved within twenty-first-century textbooks and reference publications. As a result, the conflict model's plot is not merely a popular artifact, but it is also a premise kept alive in texts used to teach undergraduates on postsecondary campuses around the world. At least in part, therefore, it may be an idea that will not die because there are those in the academy still clinging to it and teaching its narratives to future generations.

NOTES

1. Auguste Comte, *Système de politique positive ou Traité de sociologie instituant la religion de l'humanité*, vol. 4, *Appendice* (1851–54; Paris: Au Siège de la Société Positiviste, 1929), 77. Translated and quoted in Mary Pickering, *Auguste Comte: An Intellectual Biography* (Cambridge: Cambridge University Press, 1993), 1:202.

2. George W. Stocking Jr., *Victorian Anthropology* (London: Free Press, 1987), 300.

3. John W. Draper, *History of the Conflict between Religion and Science* (1874; Cambridge: Cambridge University Press, 2009), vi.

4. Andrew Dickson White, "The Battle-Fields of Science," *New-York Daily Tribune*, December 18, 1869, 4.

5. John J. Macionis and Ken Plummer, *Sociology: A Global Introduction*, 5th ed. (Harlow, UK: Pearson, 2012), 678.

6. John J. Macionis, *Sociology* (Boston: Pearson, 2012), 458.

7. William Kornblum, *Sociology in a Changing World* (Belmont, CA: Thomson Wadsworth, 2008), 531.

8. Mark Erickson and Frank Webster, "Science and Technology: Now and in the Future," in George Ritzer, ed., *The Wiley-Blackwell Companion to Sociology* (Malden, MA: Blackwell, 2012), 621.

9. Leslie E. Sponsel, "Religion and Environment," in H. James Birx, ed., *Encyclopedia of Anthropology* (Thousand Oaks, CA: Sage, 2006), 2008.

10. James P. Bonanno, "Enlightenment and Secularism," in H. James Birx, ed., *21st Century Anthropology: A Reference Handbook* (Los Angeles: Sage, 2010), 465, 68.

11. Michael S. Kimmel and Amy Aronson, *Sociology Now* (Boston: Pearson, 2012), 487.

12. Macionis and Plummer, *Sociology: A Global Introduction*, 623.

13. Emil Visnovsky, "Enlightenment, Age of," in Birx, *Encyclopedia of Anthropology*, 817–18.

14. Visnovsky, "Enlightenment, Age of," in Birx, *Encyclopedia of Anthropology*, 819.

15. Diana Kendall, *Sociology in Our Times* (Belmont, CA: Wadsworth, 2011), 37.

16. Nigel Rapport and Joanna Overing, *Social and Cultural Anthropology: The Key Concepts* (London: Routledge, 2000), 172–73.

17. John H. Brooke, *Science and Religion: Some Historical Perspectives* (Cambridge: Cambridge University Press, 1991), 18.
18. Dan Edelstein, *The Enlightenment: A Genealogy* (Chicago: University of Chicago Press, 2010), 2–3.
19. Edelstein, *The Enlightenment*, 116.
20. Elisa Ruhl, "Aesthetic Appreciation," in Birx, *Encyclopedia of Anthropology*, 21.
21. Jon Shepard, *Cengage Advantage: Sociology*, 11th ed. (Belmont, CA: Wadsworth, 2013), 419.
22. Steve Bruce, "Science and Religion," in George Ritzer, ed., *The Blackwell Encyclopedia of Sociology* (Malden, MA: Blackwell, 2007), 4099; Joan Ferrante, *Sociology: A Global Perspective*, 7th ed. (Belmont, CA: Wadsworth, 2011), 470.
23. Macionis, *Sociology*, 459.
24. Robert J. Antonio, "Materialism," in Edgar F. Borgatta and Rhonda J. V. Montgomery, eds., *Encyclopedia of Sociology* (New York: Macmillan, 2000), 1781.
25. Abraham Rosman, Paula G. Rubel, and Maxine Weisgrau, *The Tapestry of Culture: An Introduction to Cultural Anthropology*, 9th ed. (Lanham, MD: AltaMira, 2009), 233.
26. Julie M. Albright, "Postmodernism," in *Blackwell Encyclopedia of Sociology*, 3572.
27. Raymond Scupin and Christopher R. DeCorse, *Anthropology: A Global Perspective*, 7th ed. (Boston: Pearson, 2012), 45–46.
28. Bill Cooke, "Naturalism," in Birx, *Encyclopedia of Anthropology*, 1695.
29. Ernan McMullin, "Galileo on Science and Scripture," in Peter K. Machamer, ed., *The Cambridge Companion to Galileo* (Cambridge: Cambridge University Press, 1998), 274, 275.
30. David C. Lindberg, "Galileo, the Church, and the Cosmos," in David C. Lindberg and Ronald L. Numbers, eds., *When Science and Christianity Meet* (Chicago: University of Chicago Press, 2003), 58.
31. Dennis R. Danielson, "That Copernicanism Demoted Humans from the Center of the Cosmos," in Ronald L. Numbers, ed., *Galileo Goes to Jail and Other Myths about Science and Religion* (Cambridge, MA: Harvard University Press, 2009), 51, 52.
32. Bonanno, "Enlightenment and Secularism," 465.
33. Jack D. Eller, *Introducing Anthropology of Religion: Culture to the Ultimate* (New York: Routledge, 2007), 258.
34. Nachman Ben-Yehuda, "Deviance: A Sociology of Unconventionalities," in Ritzer, *Wiley-Blackwell Companion to Sociology*, 221.
35. Ben-Yehuda, "Deviance," in Ritzer, *Wiley-Blackwell Companion to Sociology*, 221–22.
36. Peter Metcalf, *Anthropology: The Basics* (London: Routledge, 2005), 186.
37. Bonanno, "Enlightenment and Secularism," 468.
38. Rosman, Rubel, and Weisgrau, *Tapestry of Culture*, 234.
39. Amanda Rees, "Evolution," in Ritzer, *Blackwell Encyclopedia of Sociology*, 1510.
40. Ian Marsh and Mike Keating, eds., *Sociology: Making Sense of Society* (Harlow: Pearson, 2006), 15; Ian Marsh et al., *Sociology: Making of Society*, 4th ed. (Harlow, UK: Pearson, 2009), 18.
41. H. James Birx, "Evolution: Science, Anthropology, and Philosophy," in Birx, *21st Century Anthropology*, 596.
42. Bonanno, "Enlightenment and Secularism," 467.
43. David A. Lukaszek, "Aquinas, Thomas (1225–1274)," in Birx, *Encyclopedia of Anthropology*, 228.
44. Tim Ingold, "Humanity and Animality," in Tim Ingold, ed., *Companion Encyclopedia of Anthropology* (London: Routledge, 2003), 22.
45. Eller, *Introducing Anthropology of Religion*, 259–60.

46. Eller, *Introducing Anthropology of Religion*, 259.
47. Richard G. Olson, *Science and Religion, 1450–1900: From Copernicus to Darwin* (Baltimore, MD: Johns Hopkins University Press, 2006), 193.
48. Ronald L. Numbers, "The Creationists," in David C. Lindberg and Ronald L. Numbers, eds., *God and Nature: Historical Essays on the Encounter between Christianity and Science* (Berkeley: University of California Press, 1986), 397; Numbers, *The Creationists: From Scientific Creationism to Intelligent Design*, 2nd ed. (Cambridge, MA: Harvard University Press, 2006), 349.
49. Raymond Scupin, *Cultural Anthropology: A Global Perspective*, 8th ed. (Boston: Pearson, 2012), 260.
50. Antonio, "Materialism," 1781.
51. Anthony Giddens, *Sociology*, 6th ed. (Cambridge: Polity, 2009), 125.
52. Bonanno, "Enlightenment and Secularism," 466–67, 469.
53. Steve Bruce, "The Social Process of Secularization," in Richard K. Fenn, ed., *The Blackwell Companion to Sociology of Religion* (Malden, MA: Blackwell, 2003), 253–54, 255.
54. Eller, *Introducing Anthropology of Religion*, 258, 259.
55. Anthony F. C. Wallace, *Religion: An Anthropological View* (New York: Random House, 1966), 264, 265.
56. Peter L. Berger, "The Desecularization of the World: A Global Overview," in Peter L. Berger, ed., *The Desecularization of the World: Resurgent Religion and World Politics* (Grand Rapids, MI: Eerdmans, 1999), 9.
57. Charles Taylor, *A Secular Age* (Cambridge, MA: Harvard University Press, 2007), 4.
58. William A. Haviland et al., *Anthropology: The Human Challenge* (Belmont, CA: Wadsworth, 2011), 576–77; William A. Haviland, *Anthropology*, 9th ed. (Fort Worth, TX: Harcourt College Publishers, 2000), 692–94.
59. Daniel R. Gross, *Discovering Anthropology* (Mountain View, CA: Mayfield Publishing, 1992), 120.
60. Scupin and DeCorse, *Anthropology: A Global Perspective*, 64–65; Scupin, *Cultural Anthropology: A Global Perspective*, 26–27.
61. Kimmel and Aronson, *Sociology Now*, 487.
62. Chaisson McMillan, *Astronomy Today*, 7th ed. (San Francisco: Pearson, 2011), 38, 40.
63. W. Thomas Griffith and Juliet W. Brosing, *The Physics of Everyday Phenomena*, 6th ed. (New York: McGraw-Hill, 2009), 89.
64. James E. Bidlack and Shelley H. Jansky, *Stern's Introductory Plant Biology*, 12th ed. (New York: McGraw-Hill, 2011), 277.
65. John D. Fix, *Astronomy: Journey to the Cosmic Frontier*, 5th ed. (New York: McGraw-Hill, 2008), 74–75.
66. Douglas C. Giancoli, *Physics: Principles with Applications*, 7th ed. (New York: Pearson, 2014), 125.
67. Scupin and DeCorse, *Anthropology: A Global Perspective*, 46; Scupin, *Cultural Anthropology: A Global Perspective*, 21.
68. John Henry, *The Scientific Revolution and the Origins of Modern Science*, 2nd ed. (New York: Palgrave, 2002), 85–86.

CHAPTER SEVENTEEN

The View on the Street

JOHN H. EVANS

Prominent physicist and New Atheist activist Victor Stenger recently wrote an article claiming that the public was becoming more "antiscience" because of religion. Let us look for the point at which his argument moves from evidence to assumption based on the idea that would not die. He discusses religious objections to Darwinism and then turns to climate change, citing a congressman who said that "climate change is a myth because God told Noah he would never again destroy Earth by flood," and an evangelical organization—linked to the fossil fuel industry—that makes similar claims. He then links to a public opinion poll that shows conservative Protestants are much less likely than others to believe in global warming. He does not note that were that survey analysis to control for political affiliation, there would be no difference between conservative Protestants and anyone else regarding the evidence for human-caused climate change.[1] Instead, the survey fits with the assumption that religious knowledge is incompatible with scientific knowledge. He concludes that the way to counter skepticism about climate change is to eradicate religion, because religion is ultimately based on faith, and science on facts and observation. Religion "produces a frame of mind in which concepts are formulated with deep passion but without the slightest attention paid to the evidence that bears on the concept." Doubt about climate change "would not be possible except for the diametrically opposed world-views of science and religion," so what is needed is "the replacement of foolish faith and its vanities with something more sublime—knowledge and understanding that is securely based on observable reality."[2] I will return to his argument in the conclusion to illustrate the effect of the idea that will not die.

Public Opinion Studies and the Actual Conflict among the Public

In recent years, the empirical study of religion and science in the American public has made a distinction between epistemic and social/moral conflict. Epistemic conflict is the warfare narrative that academia in general is concerned with—that any conflict between science and religion occurs over truth claims about the natural world. To take the canonical case, did humans come into being as described by Genesis or by Darwinists?

But, an examination of case studies of conflicts between religion and science have revealed that conflicts in the public are typically not ultimately about epistemology.[3] Rather, the conflict is often over issues like whether religion or science will have more social authority. For example, I examined the case of public bioethical debates by looking at debates over human genetic engineering. I found that theologians and scientists were in conflict over which profession would have the authority to make statements about the ethics of manipulations of the human body. There was no disagreement over the facts of how genetics works.[4] In contrast to epistemic conflict, I call this social/moral conflict.[5]

The Limited Epistemic Conflict That Does Exist

There *is*, however, some epistemic conflict, and it needs to be identified as such. I conducted a formal test using opinion data of whether there was epistemic or social/moral conflict between contemporary religious people and science.[6] Many survey studies proceed by asking respondents a series of questions about "scientific knowledge," such as whether "the center of the earth is very hot" and "electrons are smaller than atoms." I think of these as noncontested "facts." There is no contrary religious claim about the size of electrons. Critically, for decades the epistemic conflict narrative was reinforced by the inclusion of "contested" facts with the uncontested, summing them all as "scientific knowledge." Contested facts include whether humans evolved from earlier species of animals and that the universe began with a huge explosion. Questions about evolution and the big bang have been found to be more measures of religious worldview than measures of scientific knowledge.[7] Respondents have the knowledge that scientists make these claims; they just do not agree with them.

Once the contested and uncontested facts are separated in analyses, there is no difference between any religious group and the not actively religious population on their knowledge of uncontested scientific facts, as well as their knowledge of scientific methods, claimed ability to understand science, the number of science courses they took in college, odds of having a scientific career, or being a science major in college. This includes the most conservative group that can be measured: high-attending, biblical literalist, conservative Protestants.[8]

Thus we see the actually quite limited epistemic conflict. Religious people are not opposed to scientific ways of understanding the world, and they are not less likely to participate in science. Rather, one religious group—conservative Protestants—dissents from the very few scientific claims for which they have a traditional religious counterclaim. The epistemic conflict is limited to beliefs about human origins for conservative Protestants. If you ask a conservative Protestant how we know that electrons are smaller than atoms he or she will describe the scientific method.

I will simply note at this point that scientists find conservative Protestants' selective use of the scientific method to be incomprehensible. Scientists presume that if you accept a methodological, naturalist explanation for one phenomenon, such as what an electron is, you by definition must reject fact claims that are not based on naturalism. But that is not the case.

Social/Moral Conflict with Science

Another type of conflict can be read between the lines of historical and sociological studies. This is social/moral conflict between science and religion, and it is in principle independent of epistemology. For example, a conservative Protestant may completely accept the mainstream scientific account of human origins but still be opposed to the social/moral influence of scientists' account of human origins.

Scientists tend to think of religion as a competitor in making fact statements about the natural world, and that science is morally neutral. However, religion is about much more than fact statements. That religion is not primarily about truth claims about nature is best stated by Andrew Buckser in a case study showing that secularization on a Danish island was not due to the introduction of science, but rather to the transformation of social relations brought on by agricultural mechanization. Citing the ideas of anthropologist Clifford Geertz, Buckser writes that secularization theory has defined religion as "a method of explaining the physical world through the supernatural. It generally attributes the decline of religious belief to science's superior explanatory power. But in any religion, explaining the physical world is only a subordinate task; it is explaining the social world, giving it meaning and moral value, which is religion's primary concern."[9]

Indeed, some recent studies show that learning more science does not lead to less faith. For example, it has been shown that having more education only leads to more belief in the truth of evolution for non–biblical literalists. For biblical literalists, more education does not change one's views.[10] Other studies show that attending college has no influence on preexisting belief about evolution and that taking classes in the natural sciences does not cause a greater decrease in religious belief compared with taking classes in other fields.[11] These studies suggest either that conservative Protestants studiously ignore the underlying content of their classes, or that views on religion and evolution are not primarily about anything you would learn from science.

It also seems likely that the public does not perceive science as engaged in the morally neutral pursuit of truth but instead sees it as having a moral agenda. In a number of public controversies, religion has struggled with science over which institution should influence the public's morality. For example, in debates over teaching evolution, while scientists talk about "scientific results, procedures, and verifications . . . , from the fundamentalists and evangelicals have come protests about the decline of Western morality."[12] Even William Jennings Bryan, of Scopes trial fame, was not only defending the Bible but thought that evolution "would, if generally adopted, destroy all sense of responsibility and menace the morals of the world."[13] Intelligent Design advocates today sound quite similar, arguing that Darwinism teaches moral relativism to society.[14]

Scientists are probably also seen as promoting a particular morality on issues like embryonic stem cell research. In public debates, scientists and their allies talk of the relief of suffering from disease, but, as those who believe in the sanctity of embryonic life often complain, largely ignore the life of the embryo. Religious leaders are quite ready to make a distinction between scientific knowledge and the ethical use of scientific accomplishments. For example, Pope John Paul II did "not hesitate to recognize the great achievements of modern science," even if he continued to "raise questions about their moral applications."[15]

I also tested the social/moral competition theory by examining if members of conservative religious groups, compared with the not actively religious, were less likely to want scientists to have influence in "deciding what to do about" global warming, stem cell research, and restricting the sale of genetically modified foods. Even controlling for how much the respondent thought scientists understood the issue at hand, which would be epistemic conflict, conservative Protestants were more likely to not want scientific influence in public debates.[16] Conservative Protestants may believe in the scientific method, and most scientific claims, but they are less likely to want scientists to be influential in public affairs.

I focused analysis on knowledge and social/moral conflict for the issue of global warming and found that there is no religious influence on the level of belief in scientists' claims about global warming (epistemic conflict). However, the biblical literalist, conservative Protestants, compared with the nonreligious, do not want scientists to be influential in public debates about what to do about climate change.[17]

I also examined who has confidence in the leaders of scientific institutions. I first showed that the survey question about confidence in scientists primarily measures opposition to scientists' moral influence in society, not opposition to the epistemology of scientists. It turns out that it is biblical literalist, conservative Protestants who have the least confidence in scientists, and their level of confidence decreased from 1984 to 2010.[18] Later studies generally concur. Making a different set of comparisons, David Johnson and his coauthors found that the more religious their respondents were, and the more they attended religious services, the less confident they were in scientists.[19] Again, this suggests a social/moral not epistemic conflict.

I used the time line of debates in the public sphere to speculate about what constituted the social/moral conflict. In the 1950s and 1960s, the primary sociopolitical concern of conservative Protestants was the threat of communism.[20] In the late 1970s, they joined Catholics in the antiabortion movement and then moved more broadly toward the religious right, which started making claims about the ethics of the human body on issues such as euthanasia, reproductive technology, embryonic research, and sexual ethics. Prior to the conservative Protestants' shift of focus to the ethics of the human body, they would not have seen themselves as moral competitors with scientists. The ethics of the body would prove contentious because at the same time conservative Protestants became interested in the subject, so too did scientists. Scientists were engaged in technologies such as birth control, human genetic engineering, organ transplantation, extended life support of people with brain damage, and so on.[21] Starting in the late 1970s, conservative Protestants would have increasingly seen themselves in social/moral competition with scientists in debates in the public sphere. Social/moral conflict seems chiefly concerned with technologies of the human body.

Studies of social determinants of abstract faith in science also support the thesis that the primary contemporary conflict is social/moral and not epistemic. I conducted a study of the change over time in three types of faith in science.[22] For the study, faith was defined as a belief in that for which there is no evidence. So the first type of faith is that science can provide meaning for society, that it can be the cognitive basis for everything in society, and it was measured by the degree of agreement with the statement: "We believe too often in science and not enough in feelings and faith." In the context of this item, "believing in," "feelings," and "faith" reference nonmaterialist, nonrational reasons for doing anything. The second type of faith is related to fact claims about nature, and thus to faith in the epistemology of science, and to the ability of science to solve general problems of society. This was measured by the degree of agreement with the statement: "Overall, modern science does more harm than good." The third type of faith is more narrowly focused, as belief in science's ability to solve problems in the physical world. It is measured by the degree of agreement with the statement: "Modern science will solve our environmental problems with little change to our way of life."

I found that in general the religious respondents have less faith than the nonreligious in science providing meaning. This is akin to social/moral conflict, because it is not really linked to fact claims about the natural world. Faith in science solving general problems was also less likely to be found among the religious. Critically, for faith in science solving concrete problems in the physical world, which depends on scientific facts, there were no differences between the religious and the nonreligious. The upshot of all of this is that religious people have faith in science when limited to concrete manipulations of the physical world to solve problems—which is quite close to epistemic, nonmoral questions like whether the inside of the earth is hot. They *are* in conflict with science when science is about producing meaning for society—which is conceptually close to social/moral conflict.

In sum, studies of public opinion in the contemporary American public sphere show that as long as "science" concerns making fact claims about the natural world, then there is no conflict between religion and science. The one exception is the few fact claims for which there has traditionally been a conflicting claim from conservative Protestantism. This is essentially only about human origins, and it appears that a properly measured attitude about evolution would result in this narrow conflict being limited to conservative Protestants. However, there is evidence of social/moral conflict between conservative Protestants and science that is broader than human origins. While the conflict over evolution does not "spread" to other fact claims, it does appear that the social/moral conflict, whatever its origins, has "spread" to distrusting scientists' involvement with any issue in the public sphere, even global warming. This means a more widespread and consequential conflict.

Why Are Social Elites Believers in the Epistemic Conflict Narrative?

I now turn to what I see as the two sociology of knowledge questions of this volume: (1) Why does belief among elites in the generalized epistemic conflict narrative survive, despite the public not being in such a conflict and specialized academics showing that it is false? (2) Why do these elites not see social/moral conflict? The persistence of these two myths is what explains, in Jon Roberts's terms, "the idea that would not die."[23] I am not aware of any systematic sociological research into why the myths continue, although there certainly should be. Below I offer a number of hypotheses that should be examined more systematically.

The first reason for the myths is not very academically interesting, but it is probably part of the explanation. The survey results are clear that to the extent there is any epistemic or social/moral conflict between religion and science, it really only involves conservative Protestants. Despite the fact that conservative Protestants make up only 25 percent of the population of the United States, the general American public is disproportionately exposed to their views, and for the uninformed American, religion is conservative Protestantism. Michael Evans has found that ordinary citizens think that all religion discussed in the public sphere is going to be conservative.[24]

One reason for this public perception is that the media tends to focus on conservative Protestantism because it is more interesting than other Christian religions. "The calling card for entering the news is conflict, and any group that is able to create such situations typically gains access to media through the reports of hungry journalists," writes Quentin Schultze. The media only reports when there is conflict, particularly between or within religious groups.[25] So, writing a story about the mainline Protestant clergy who are generally believers in Darwinian evolution is boring, because most reporters also believe in Darwinian evolution. But, interviewing the few conservative Protestant geocentrists would be much more interesting. Conflict sells newspapers or, in this day and age, generates clicks.

This process not only pulls attention from certain traditions, like mainline Protestantism, that have little or no conflict, but it also focuses on the more conflictual parts of conservative Protestantism. The few dozen members of the Westboro Baptist Church who picket military funerals claiming that the death was justified because of America's acceptance of homosexuality probably receive more press than does the entire mainline Protestant tradition.

So, one reason why the myth of conflict *with all religions* survives is that the media focuses on the one religious tradition that has some limited conflict with science. Despite all this attention, the media may not even teach us an accurate account of the epistemic conflict for conservative Protestants, because most articles and broadcasts do not have time to explain that conservative Protestants' opposition to Darwin does not imply they are opposed to modern chemistry. On the other hand, the media may well effectively teach us that conservative Protestants tend to have a social/moral conflict with science.

The second reason why the myths may continue is that scientists get to define the topic of any dialogue. In the many exchanges between religion and science, the scientists view the process and purpose of their work as generating fact claims about nature, exclusively. Epistemic conflict is then the only topic to talk about, given that scientists think their work is only about fact claims and not about ethics. The constricted nature of the discourse is evident from the fact that, among all of the "religion-and-science-dialogue" events that have occurred over the decades, I am unaware of any that were not primarily about epistemology. Of course, some religions do make fact claims about nature, but this is a very small part of what religions are about. Focusing on the tiny overlap between the two gives the impression that religion is only about making fact claims about nature.

Sociologists are actually part of the problem, wrapping the epistemic conflict narrative into sociological definitions of religion. Social science was born in the nineteenth century under the influence of methodological naturalism, which is "a disciplinary method that says nothing about God's existence." It also assumed metaphysical naturalism, "which denies the existence of a transcendent God."[26] Consistent with the Enlightenment assumptions that gave birth to sociology, religion itself has been defined by sociologists as that which is not verifiable by the methods of science.

For example, Durkheim's definition of religion depends on splitting the world into the sacred and the profane. The profane world is that which is explainable by human reason and thus able to be observed. The sacred world is that which operates outside of the ability of rationality and science to explain it. This is usually called supernatural—that which is above or beyond "nature," and "nature" is that which is explainable through reason. By the definition of one sociology textbook, "[R]eligion can be defined as a system of beliefs and practices by which a group of people interprets and responds to what they feel is sacred and, usually, supernatural as well."[27] In theological terms, sociologists tend to define religion in terms of a "God of the gaps," where God exists in the phenomena that science cannot (yet) explain.[28] To better see social/moral conflict with science—instead of just looking for the limited cases of epistemic conflict—will require sociologists to create a definition of religion in which the nonepistemological is given at least equal standing.

The third possible reason the myths continue is that sociologists who study the contemporary public have not been able to explain why there is conflict at all. This makes any attempts to debunk the myths less convincing. For example, social scientists have found there to be epistemic conflict for conservative Protestants on a few scientific claims (for example, evolution) but do not have an explanation for why there is not conflict on others. Moreover, what exactly do conservative Protestants perceive as the social/moral conflict with science?

I will leave social/moral hypotheses for later work, but I can suggest a few explanations for the epistemological divide that could be tested. The most obvious to the sociological eye is basic cultural production theory which black-boxes how much sense a claim makes, assuming that any idea with enough resources could become influential (consider, for example, astrology). There are innumerable epistemic conflicts that come to the mind of individual religious people and scientists, but only in a few cases do individuals on both sides of the same claim obtain enough resources to publicize their ideas. Darwin was able to get the resources to make his claims central to the minds of scientists. Proponents of young-earth creationism were able to do the same. The result is conflict—explained not by the ideas themselves but by the ability to gather the resources to promote the ideas. Moreover, in this perspective, both sides must obtain the requisite resources. Physicists may point to geocentrism as a conflict, but the few geocentrists have never been able to obtain the resources in the religious community to promote their ideas. Analogously, but from the other direction, theologians may say that the Christian resurrection is a conflict with science, but no scientist cares about this claim. Therefore, there is no actual conflict.

Another possible explanation for sociologists would be that epistemic conflict depends on the extent to which the claim is linked to other important claims in religion or science. For example, one of the reasons why elites in the Church of England in the early twentieth century were reluctant to accept Darwin was not that this contradicted a literalist Genesis narrative. They were not believers in that anyway. The deeper problem was that Darwin logically negated the Fall, and once the Fall is gone, so is the atonement by Christ on the cross for the sin of the Fall—which would then require a massive reconception of Christian theology and belief.[29] Similarly, conservative Protestants claim that if the Genesis narrative is not literally true, then none of the Bible is true, which would also require a reconceptualization of the faith. From the other side, scientists get concerned about evolution because it is seen as the key to modern biology. Presumably, they would be less concerned about something with fewer dependencies. Like my previous hypothesis, in this explanation it would not be the claim itself that indicates it will be defended against an opposing claim, but the number and importance of the other claims that depend on it.

Or, sociologists should consider, in order to improve their explanations, whether some scientific claims are accepted and others not because conservative Protestants are Baconian, Scottish commonsense realists, while others have moved on to a notion of science different from this dominant nineteenth-century standard. Nineteenth-century Baconian science was opposed to abstraction and thought that "the things worth understanding were not particularly opaque." Things were as they appeared to be, and therefore "theories, hypotheses, metaphysical thoughts, and other mental complications were unnecessary."[30] People could use their common sense to observe nature and build up generalized understandings from these observations. In general, you should trust your observations, and not theories or models.

There is evidence that contemporary conservative Protestants have this Baconian view. The Intelligent Design textbook *Of Pandas and People* essentially makes the distinction between observations and theories in explaining when you should believe science and when you should not.[31] Since this view of science was no longer supported by mainstream science, and would not be taught in high schools, this epistemic stance could be incubated as a general intellectual orientation that could later be applied to the conservative Protestant epistemic stance toward the "other" book of God besides "nature," which is the Bible. Historian George Marsden notes that while inerrancy was not invented by Baconianism or Scottish commonsense realism, it contributed to this hermeneutic.[32] God's truth in nature and the Bible were revealed in the same way: "In nineteenth-century America, Baconianism meant simply looking carefully at the evidence, determining what were 'the facts,' and carefully classifying these facts. One might scrupulously generalize from the facts, but good Baconianism avoided speculative hypotheses. The interpretation of Scripture, accordingly, involved careful determination of what the facts were—what the words mean. Once this was settled the facts revealed in Scripture could be known as surely and as clearly as the facts discovered by the natural scientist."[33]

If contemporary conservative Protestants think that true knowledge of the Bible is uncomplicated, transparent, and available via a commonsense reading, they may continue to think that other knowledge—like knowledge of nature—is similarly uncomplicated, transparent, and available via common sense. If so, then conservative Protestants will believe institutional science claims that can be immediately observed (the average temperature of the earth) and not claims based on "theories" like climate models and speculations about primates from millions of years ago. Moreover, human evolution and the big bang cannot be observed and thus are based on speculations, but that the earth goes around the sun can be observed.

My fourth reason for the continuation of the epistemic myth may be the most important. The elites may see conflict when the masses are not in conflict owing to different forms of reasoning in the two groups. In sociological terms, certainly every person with a PhD is a member of the societal elite. The greatest contribution to the dialogue between religion and science would come from demonstrating to the elites that they are using a form of reasoning different from that used by the public, and that their reasoning—and thus the conflicts it posits—cannot be necessarily extrapolated to the public.

This elite reasoning can best be described as an "ideology" or "worldview." In these outlooks, ideas are "organized in a hierarchical fashion, in which more specific attitudes interact with attitudes toward the more general class of objects in which the specific object is seen to belong."[34] Ideologies also "assume that causation flows from the abstract to the specific, so that when an individual is faced with the question of what government should do in a given instance, his or her preference will be based, in part, on more general principles. This model thus assumes "a degree of deductive political reasoning, from abstract beliefs to more specific political preferences."[35] Note that "constraint" and "deductive political reasoning" in this literature refer to what would be called "logic" in regular conversation. If you have a strong ideology or worldview, if you believe in a higher-level abstract idea like "materialism," "methodological naturalism," or "God controls the world," you will then, owing to logic, reach consistent conclusions with a number of more concrete claims, such as "Darwin was correct." Those with a strong ideology or worldview would not hold inconsistent ideas. Despite that most academics believe in this consistency themselves, and sociologists have long assumed that the general public also displays this consistency, in recent years sociologists have tried to not assume coherence but to treat coherence as a measured trait that must be explained.[36]

The survival of the epistemic conflict myth is then partly the result of elites who write about the conflict assuming that the public has a tight ideology—that the public is as logically consistent as they are. I am mainly thinking here of the scientist-theologians who come up with ingenious ways to make new scientific developments logically consistent with perhaps an only slightly modified theology. But, the public has much less logical consistency than do elites.[37] This is not an insult but a matter of sociological realism. The reason for the difference is that the only people who have the time and motivation to develop airtight, logically consistent beliefs all the way back to first principles are those who are rewarded for doing so. Academics are rewarded for this with tenure—analytic philosophers are an extreme case. If you paid the man or woman on the street to come up with a logically consistent ideology, they could do so, but until then, the difference remains.

So, while a philosopher may say that it is inconsistent to believe in a materialist explanation for sound waves but a nonmaterialist explanation for the Eucharist, ordinary people do not bother to think back to the principles beneath beliefs and thus have no problem with "incompatible" beliefs. For the average person, no scientific question that has any religious valence matters to their everyday life enough to take the effort to synthesize it with their other beliefs. What would it matter to the everyday life of someone that the method that generated claims of human evolution is incompatible with the method that generated claims about the resurrection?

In regard to religion and science, the ideology or worldview that academics probably have, but probably mistakenly assume the public also has, is that there is an abstract, higher-level belief called "materialism," which drives conclusions about all sorts of lower-level claims. So, under the presumption of logic embedded in an ideology, if someone believes scientists' claims derived from materialist scientific methodology that the earth goes around the sun, you must apply materialist reasoning to all other claims about nature. This is the standard that would be used for tenure cases in a philosophy department, but not among ordinary citizens.

So, elites see an epistemic conflict between scientific and religious claims because, if you have an ideology that is consistent back to a first principle like "materialism," there *is* conflict. But, since the public generally does not create ideologies like this, it has no epistemic conflict. This explains my findings that there is epistemic conflict over fact statements about human origins but no other scientific claims. Conservative Protestants may just see these as conflicting statements—"I learned in church that God created humans, and my biology teacher says that humans evolved from primates." For religious people to also disagree with scientists about radio waves, the distance to the moon, or why flowers open in the morning requires that people logically reason back from these statements to higher-level principles like "materialism." People do not do that, so there is only limited conflict over lower-level fact claims. So, the myth continues because academics extrapolate the reasoning style they are rewarded for to the general public.

The fifth and final possible explanation for why the myth continues is that historians may be inadvertently undermining proper understanding of conflict. Ronald L. Numbers writes that historians have had little to say about the public's views of religion and science, and he even regrets that two of his own edited volumes have neglected the views of the public.[38] As you would expect, one reason for this neglect is that the common folk "left little evidence of their thoughts, and much of what we have is filtered through the writings of those who observed them."[39] Historians then focus on the elite debate because there are no other data, but also because the debate before the twentieth century *was* largely between elites, owing to the education and resources necessary to participate in it. To take the obvious case, when Newton was writing, very few of his fellow countrymen and women were literate, to say nothing of literate in Latin, so obviously this debate did not involve the masses. The religion and science debate has historically been a very elite affair, and the types of arguments that the elites make reflect this.

These ideas really mattered to the people historians study, and they were socially rewarded for having a strong, logically consistent ideology. Wilberforce knew he would have to respond directly to Huxley's criticisms, and William Jennings Bryan knew he would be forced to defend his ideas against Clarence Darrow. They had incentive to make their ideas logically iron-clad, which usually results in building ideological systems back to first principles or value spheres that are simply untestable assumptions.

As an example, consider the data for historian Peter Bowler's encyclopedic review of "the debate" between science and religion in early-twentieth-century Britain, which, without explicitly saying so, is almost exclusively from elites. The book shows heroic attempts at iron-clad logical consistency by the elites of the time. This debate was largely about materialism and whether materialism as a high-level principle could be altered to make room for closer-to-the-ground traditional religious belief. For example, some thought that "matter itself was mysterious, and thus offered no suitable foundation for the kind of materialism that sought to eliminate mind and purpose from nature."[40] In this era there were many attempts to modify deep scientific constructs like materialism to make them logically compatible with traditional religious views, as well as attempts to modify deep religious principles to make them compatible with the science of the time.

Most historical studies do not explicitly say that they are examining the ideas only of the elites or note that the reasoning of the elites and the public may be different. In the same way that the reasoning of the evangelical director of the National Institutes of Health, Francis Collins, is probably quite different from that of a contemporary randomly selected evangelical, Bishop Wilberforce's reasoning was probably quite different from a randomly selected Victorian-era Anglican. Without identifying the elite/public distinction, people may think that Wilberforce's reasons reflect on what a contemporary evangelical might think. People who read the historical literature may think that it tells us something about the contemporary public debate about religion and science—but if I am right, it does not. By focusing on elites who have tight ideologies, without explicitly saying that it is only elites who have these ideologies, historians may be undermining the sociological explanation that there is little to no epistemic conflict in the public.

Conclusion

Stenger's article, described in the beginning of this chapter, makes sense in light of the difference between epistemic and social/moral conflict, as well as the reasons why the epistemic myth continues. Stenger is making claims about the average members of the public who are religious, but he is using the statements of elites to do so. He assumes that belief in God is incompatible with on-the-ground statements of facts from scientists because people who believe in God cannot believe in the methodology of science—they must just believe in faith and biblical exegesis like the congressman he cites. Scientists assume that regular citizens have iron-clad systems of justification back to first principles. Moreover, they assume that any conflict with them must be over fact claims, since in the self-image of science, all science does is discover facts. Stenger the scientist cannot see that a conflict with science might not be just about fact claims at all, but may be social or moral. Stenger is repeating the myth, and to the extent his readers are persuaded by it, they would be less likely to bother reaching out to religious people to help pursue Stenger's own stated goal of mitigating climate change.

Social science shows a very limited type of conflict where one religious group—conservative Protestantism—has an epistemic conflict with science over a few of the claims of scientists but not more. This same group seems to be in a social/moral conflict with science. I think that the mythical parts of the conflict claims will recede if we pay attention to the social causes of the conflict narrative itself that I outline above.

NOTES

1. John H. Evans and Justin Feng, "Conservative Protestantism and Skepticism of Scientists Studying Climate Change," *Climatic Change* 121 (2013): 595–608.
2. Victor Stenger, "The Rising Antiscience," *Huffington Post*, September 27, 2013.
3. John H. Evans and Michael S. Evans, "Religion and Science: Beyond the Epistemological Conflict Narrative," *Annual Review of Sociology* 34 (2008): 87–105.
4. John H. Evans, *Playing God? Human Genetic Engineering and the Rationalization of Public Bioethical Debate* (Chicago: University of Chicago Press, 2002).
5. For an elaborated version of the argument in this chapter, see John H. Evans, *Morals Not Knowledge: Recasting the Contemporary U.S. Conflict between Religion and Science* (Berkeley: University of California Press, 2018).
6. John H. Evans, "Epistemological and Moral Conflict between Religion and Science," *Journal for the Scientific Study of Religion* 50 (2011): 707–27.
7. J. Micah Roos, "Measuring Science or Religion? A Measurement Analysis of the National Science Foundation Sponsored Science Literacy Scale 2006–2010," *Public Understanding of Science* 23 (2014): 797–813.
8. Using more precise methods, this finding was confirmed by David R. Johnson, Christopher P. Scheitle, and Elaine Howard Ecklund, "Individual Religiosity and Orientation towards Science: Reformulating Relationships," *Sociological Science* 2 (2015): 106–24.
9. Andrew Buckser, "Religion, Science, and Secularization Theory on a Danish Island," *Journal for the Scientific Study of Religion* 35 (1996): 439.

10. Joseph O. Baker, "Acceptance of Evolution and Support for Teaching Creationism in Public Schools: The Conditional Impact of Educational Attainment," *Journal for the Scientific Study of Religion* 52 (2013): 216–28.

11. Jonathan P. Hill, "Rejecting Evolution: The Role of Religion, Education, and Social Networks," *Journal for the Scientific Study of Religion* 53 (2014): 575–94; Christopher P. Scheitle, "U.S. College Students' Perception of Religion and Science: Conflict, Collaboration, or Independence? A Research Note," *Journal for the Scientific Study of Religion* 50 (2011): 175–86.

12. Mark A. Noll, "Evangelicalism and Fundamentalism," in Gary B. Ferngren, ed., *Science and Religion: A Historical Introduction* (Baltimore, MD: Johns Hopkins University Press, 2002), 274.

13. Noll, "Evangelicalism and Fundamentalism," 275.

14. Discovery Institute, "The Wedge," n.d., http://www.antievolution.org/features/wedge.pdf.

15. James L. Heft, "Catholicism and Science: Renewing the Conversation," *Journal of Ecumenical Studies* 39 (2002): 376.

16. Evans, "Epistemological and Moral Conflict between Religion and Science."

17. Evans and Feng, "Conservative Protestantism and Skepticism of Scientists Studying Climate Change."

18. John H. Evans, "The Growing Social and Moral Conflict between Conservative Protestantism and Science," *Journal for the Scientific Study of Religion* 52 (2013): 368–85.

19. Johnson, Scheitle, and Ecklund, "Individual Religiosity and Orientation towards Science."

20. Robert Horwitz, *America's Right: Anti-Establishment Conservatism from Goldwater to the Tea Party* (New York: Polity Press, 2013).

21. John H. Evans, *The History and Future of Bioethics: A Sociological View* (New York: Oxford University Press, 2012).

22. John H. Evans, "Faith in Science in Global Perspective: Implications for Transhumanism," *Public Understanding of Science* 23 (2014): 814–32.

23. Jon H. Roberts, "'The Idea That Wouldn't Die': The Warfare between Science and Christianity," *Historically Speaking* 48 (2003): 21–24.

24. Michael S. Evans, *Seeking Good Debate: Religion, Science, and Conflict in American Public Life* (Berkeley: University of California Press, 2016).

25. John Dart and Jimmy Allen, *Bridging the Gap: Religion and the News Media* (Nashville, TN: Freedom Forum First Amendment Center, 1993), 18 (Schultze quotation), 15. See also Ronald L. Numbers, "Aggressors, Victims, and Peacemakers: Historical Actors in the Drama of Science and Religion," in Harold W. Attridge, ed., *The Religion and Science Debate: Why Does It Continue?* (New Haven, CT: Yale University Press, 2009), 50ff.

26. Evans, *Morals Not Knowledge.*

27. Ronald L. Johnstone, *Religion in Society: A Sociology of Religion* (Upper Saddle River, NJ: Prentice Hall, 1997), 13.

28. Allen Verhey, "'Playing God' and Invoking a Perspective," *Journal of Medicine and Philosophy* 20 (1995): 347–64.

29. Peter J. Bowler, *Reconciling Science and Religion: The Debate in Early-Twentieth-Century Britain* (Chicago: University of Chicago Press, 2001), 226.

30. Christopher P. Toumey, *God's Own Scientists: Creationists in a Secular World* (New Brunswick, NJ: Rutgers University Press, 1994), 16.

31. Percival Davis and Dean H. Kenyon, *Of Pandas and People: The Central Question of Biological Origins* (Dallas, TX: Haughton Publishing, 1989).

32. George M. Marsden, "Everyone One's Own Interpreter? The Bible, Science, and Authority in Mid-Nineteenth-Century America," in Nathan O. Hatch and Mark A. Noll, eds., *The Bible in America: Essays in Cultural History* (New York: Oxford University Press, 1982), 90–91.

33. Marsden, "Everyone One's Own Interpreter?," 83.

34. John Gerring, "Ideology: A Definitional Analysis," *Political Research Quarterly* 50 (1997): 969, 975.

35. Mark A. Peffley and Jon Hurwitz, "A Hierarchical Model of Attitude Constraint," *American Journal of Political Science* 29 (1985): 876–77.

36. William H. Sewell Jr., "The Concept(s) of Culture," in Victoria E. Bonnell and Lynn Hunt, eds., *Beyond the Cultural Turn: New Directions in the Study of Society and Culture* (Berkeley: University of California Press, 1999), 58.

37. Philip E. Converse, "The Nature of Belief Systems in Mass Publics," in David E. Apter, ed., *Ideology and Discontent* (New York: Free Press, 1964), 206–61; Paul DiMaggio, "Culture and Cognition," *Annual Review of Sociology* 23 (1997): 263–87.

38. Ronald L. Numbers, *Science and Christianity in Pulpit and Pew* (New York: Oxford University Press, 2007), 9, 142.

39. Numbers, *Science and Christianity in Pulpit and Pew*, 12.

40. Bowler, *Reconciling Science and Religion*, 87.

THOMAS H. AECHTNER is a senior lecturer in religion and science at the University of Queensland. Among his published articles is "Galileo Still Goes to Jail: Conflict Model Persistence within Introductory Anthropology Materials," *Zygon* 50 (2015): 209–26.

RONALD A. BINZLEY is an independent scholar living in Madison, Wisconsin. He holds a PhD in history from the University of Wisconsin–Madison and is the author of "American Catholicism's Science Crisis and the Albertus Magnus Guild, 1953–1969," *Isis* 98 no. 4 (December 2007): 695–723.

JOHN HEDLEY BROOKE served from 1999 to 2006 as the inaugural Andreas Idreos Professor of Science and Religion at the University of Oxford. He is the author of the landmark *Science and Religion: Some Historical Perspectives* (1991) and other works on the history of science and religion.

ELAINE HOWARD ECKLUND is the Herbert S. Autrey Chair in Social Sciences, professor of sociology, and founding director of the Religion and Public Life Program at Rice University. Her publications include *Science vs. Religion: What Scientists Really Think* (2010).

NOAH EFRON teaches the history and philosophy of science at Bar Ilan University in Israel and has served on the city council of Tel Aviv–Jaffa. He is the author of *Judaism and Science: An Historical Introduction* (2007) and *A Chosen Calling: Jews in Science in the Twentieth Century* (2014).

JOHN H. EVANS is professor of sociology, associate dean of social sciences, and codirector of the Institute for Practical Ethics at the University of California, San Diego. His most recent books are *The History and Future of Bioethics: A Sociological View* (2012) and *What Is a Human? What the Answers Mean for Human Rights* (2016).

MAURICE A. FINOCCHIARO, distinguished professor of philosophy emeritus at the University of Nevada–Las Vegas, is author of several books on Galileo, including *The Galileo Affair: A Documentary History* (1989).

FREDERICK GREGORY is professor emeritus of the history of science and European history at the University of Florida and past president of the History of Science Society. His several books include *Nature Lost? Natural Science and the German Theological Traditions of the Nineteenth Century* (1992).

BRADLEY J. GUNDLACH is professor of history and director of the Division of Humanities at Trinity International University. He is the author of *Process and Providence: The Evolution Question at Princeton, 1845–1929* (2013).

MONTE HARRELL HAMPTON teaches intermittently at North Carolina State University and the University of North Carolina at Chapel Hill. He is the author of *Storm of Words: Science, Religion, and Evolution in the Civil War Era* (2014) and editor, with Regina D. Sullivan, of *Varieties of Southern Religious History: Essays in Honor of Donald G. Mathews* (2015).

JEFF HARDIN is the Raymond E. Keller Professor and chair of the Department of Integrative Biology at the University of Wisconsin–Madison, where he has taught since 1991. He is the author or coauthor of more than fifty peer-reviewed articles and a dozen book chapters in his area of scientific research, which focuses on embryonic development. He currently serves as chair of the board of the BioLogos Foundation.

PETER HARRISON is an Australian Laureate Fellow and director of the Institute for Advanced Studies in the Humanities at the University of Queensland, having formerly occupied the Andreas Idreos Professorship of Science and Religion at the University of Oxford. His latest book is *The Territories of Science and Religion* (2015).

BERNARD LIGHTMAN, professor of humanities at York University, specializes in the relationship between Victorian science and unbelief. A former editor of *Isis* and the president-elect of the History of Science Society, he is the general coeditor of *The Correspondence of John Tyndall* and author of a forthcoming biography of Tyndall.

DAVID N. LIVINGSTONE is a professor of geography and intellectual history at Queen's University, Belfast. He is the author of *Dealing with Darwin: Place, Politics, and Rhetoric in Religious Engagements with Evolution* (2014).

DAVID MISLIN is an assistant professor in the Intellectual Heritage Program at Temple University and the author of *Saving Faith: Making Religious Pluralism an American Value at the Dawn of the Secular Age* (2015).

EFTHYMIOS NICOLAIDIS is director of the Institute for Neohellenic Research at the National Hellenic Research Foundation in Athens, Greece, and former president of the International Union of History and Philosophy of Science and Technology / Division of History of Science and Technology. He is the author of *Science and Eastern Orthodoxy* (2011).

MARK A. NOLL is the Francis A. McAnaney Professor of History at the University of Notre Dame and author of numerous books, including *America's God: From Jonathan Edwards to Abraham Lincoln* (2002) and *In the Beginning Was the Word: The Bible in American Public Life, 1492–1783* (2015).

RONALD L. NUMBERS is the Hilldale Professor Emeritus of the History of Science and Medicine and of Religious Studies at the University of Wisconsin–Madison, where he taught for nearly four decades. He has written or edited more than two dozen books, including *Galileo Goes to Jail and Other Myths about Science and Religion* (2009).

LAWRENCE M. PRINCIPE, Drew Professor of the Humanities and director of the Singleton Center for the Study of Premodern Europe at Johns Hopkins University, has published extensively on the history of alchemy and on the history of science and religion. He is the author of *Science and Religion* in the Teaching Company's Great Courses.

JON H. ROBERTS is the Tomorrow Foundation Professor of American Intellectual History at Boston University. He is the author of *Darwinism and the Divine in America: Protestant Intellectuals and Organic Evolution, 1859–1900* (1988) and coauthor, with James Turner, of *The Sacred and the Secular University* (2001).

CHRISTOPHER P. SCHEITLE is an assistant professor in the Department of Sociology and Anthropology at West Virginia University. His publications include *Places of Faith: A Road Trip across America's Religious Landscape* (2012).

M. ALPER YALÇINKAYA is an assistant professor of sociology and anthropology at Ohio Wesleyan University. He is the author of *Learned Patriots: Debating Science, State, and Society in the Nineteenth-Century Ottoman Empire* (2014).

INDEX

Abbott, Lymann, 241
Abduh, Muhammad, 216–17
Abdulhamid II, Sultan, 215
Aeterni Patris, 111
Afghani, Jamaluddin al-, 209–10, 215, 217
Agassiz, Louis, 58
Aggelou, Alkis, 139
Albright, Julie M., 309
alchemy, 131, 135
Alexander, Denis, 230, 239, 247–49, 254. *See also* neo-harmonists
Alexander, James W., 54, 56
American Institute of the City of New York, 9
American Jewish Committee, 191, 195
American Jews: belief in the harmony between Judaism and science, 186–87, 189–90, 197; conflicts between traditionalists and reformers among, 188–89; opposition to antievolution campaign of, 193–94, 195; opposition to Protestant hegemony of, 191–92, 195–97. *See also* Christianity, relationship between science and; conflict thesis
American Scientific Affiliation (ASA), 165, 177, 179–80
Anaplasis (Greek Orthodox Christian association), 129; *Anaplasis* journal of, 129, 133–34
ancient Greece, 65, 128, 130, 132
Andalusia, 208
Anglicanism, 262
anthropology, 58, 302–3, 320–21. *See also* textbooks, introductory level for the social sciences
Anti-Defamation League, 191
antievolution campaign, 174–76, 195
Antin, Mary, 193
Antonio, Robert J., 308, 315
Answers in Genesis (AiG), 233
Apostolides, Nikolaos, 129–30, 132
Appleton, William H., 21, 73
Aquinas, Thomas, 111–13
Arabs, 20, 206–10
Aristotle, 34, 65, 111, 310
Ark Encounter, 233
Arnold, Matthew, 153
Aronson, Amy, 305, 318
atheism, 94–95, 220–21, 251, 270, 304. *See also* New Atheists
Atkins, Peter, 232, 235n11
Atran, Scot, 235n11
Atwater, Lyman, 167–68
Augustine, 12, 34, 111–13, 119
authenticity, 204
Averroes, 20, 206, 208, 216
Avicenna (Ibn Sina), 206–7
Ayala, Francisco, 253

Bacon, Francis, 49, 53–54, 59. *See also* Baconianism
Bacon, Roger, 190
Baconianism: as factor in post-Darwinian debates, 261; influence on 19th-century American Protestants, 53–58, 61n17, 62nn31–32; influence on 20th- and 21st-century American conservative Protestants, 333–34
Bainbridge, William Sims, 242
Bakewell, Robert, 51
Baldwin, James Mark, 168

Barberini, Cardinal Maffeo, 29. *See also* Urban VIII, Pope
Barbour, Ian, 253; typology of science-religion relations of, 80, 83n65, 257n43, 267
Barth, Karl, 87, 99
Battershall, Walton, 154
Bavinck, Herman, 179
Beecher, Henry Ward, 73, 241
behaviorism, 157
Bellarmine, Cardinal Robert, 29–30, 37, 41
Bellows, Henry Whitney, 73
Benedict XIV, Pope, 30
Ben-Yehuda, Nachman, 311
Berger, Peter L., 317
Berlin, 118
Biagoli, Mario, 237n26
biblical inerrancy, 144, 155, 165, 171, 334
biblical interpretation: approach of American liberal Protestants 148–49; approach of conservative Protestants, 228, 333–34; approach of 19th-century American Protestants, 47–49, 54–55; approach of Roman Catholics, 108–10; and the Galileo Affair 28–30, 32–34; and geology, 49–52, 57–59, 109–10; and rationalism in Germany, 4, 87–88
BioLogos, 230–31, 250–51
Birx, H. James, 304, 313
Bitsakis, Efthychis, 123
Bixby, James, 144
Bloom, I. Mortimer, 194
Bonanno, James P., 310, 312–13, 315
Bourdieu, Pierre, 198
Bowler, Peter, 269, 337
Bowne, Borden Parker, 151
Boyer, Pascal, 235n11
Boyle, Robert, 7
Bracken, Joseph, 253
Brandeis, Louis Dembitz, 194
Brecht, Bertolt, 309
Brewster, David, 37
Briggs, Charles, 171
"brights," 223
British Association for the Advancement of Science, 3, 65, 67, 74
British Humanist Association, 229
Brooke, John Hedley, 225, 232, 236n15, 237n26, 238n32, 306; *Science and Religion* of, 241, 264–65
Brown, C. Mackenzie, 281
Brown, Frank Burch, 270
Brown, William Adams, 155
Bruce, Steve, 307
Bruno, Giordano, 65, 137, 184, 190, 310–11
Bryan, William Jennings, 174, 176, 327, 337
Bucaille, Maurice, 217
Büchner, Ludwig, 90, 215
Buckland, William 50, 61n18
Buckle, Thomas, 16–17, 21
Buckley, Michael, 264
Buckser, Andrew, 326
Bultmann, Rudolf, 99
Bumstead, Henry, 193
Byzantium, 126, 127, 130, 135–36, 139

Calvinism, and science 261. *See also* Princeton Seminary and College
Campanella, Tommaso, 33
Campbell, Alexander, 52
Cantor, Geoffrey, 2, 83n64, 165, 237n26, 265–66, 272–73
capitalism, 205
Carpenter, William, 266
Carrier, Richard, 225–26
Castelli, Benedetto, 29, 33
catastrophism, 50
Catherine II, Empress of Russia, 126
Catholicism. *See* Roman Catholic Church
CERN (European Council for Nuclear Research), 282
Cevdet, Abdullah, 217
Chamberlain, Neville, 223, 227, 235n8, 254
Chambers, Robert, 16–17, 25n36
Chatard, Francis, 112
Christian Union of Scientists, 138
Christianity, relationship between science and: American Jews on, 185–87, 197; historiography of, 258, 262–63, 268–70; modern Greek intellectuals on, 131–32, 137; modern Muslim intellectuals on, 204, 215–16; New Atheists on, 80, 223, 225–27, 236n61. *See also* liberal Protestants; Protestant evangelicals; Roman Catholic Church
Christina of Lorraine, 29
Christomanos, Anastasios, 129–33
City College (New York, NY), 196
Clarke, William Newton, 151

Clayton, Phillip, 253
Cobb, John B., 253
Coffee, Rudolph, 190
Collins, Francis, 230–31, 239, 245, 250–52, 254
colonialism, 205–6, 210, 272
Columbia Theological Seminary (Columbia, SC), 47, 53, 58
Columbia University, 196
communism, 94, 137, 177, 328
complexity thesis, 39, 43, 164, 225, 265
Comte, Auguste, 22, 25n29, 306, 316, 320; "Law of Three Stages" of, 14–15, 302
conflict model, 279, 302. *See also* conflict thesis
conflict thesis, 2, 20, 123, 229; American Fundamentalist Protestant responses to, 174–77; American Jewish responses to, 186–90, 197–98; American-evangelical Protestant responses to, 165–74, 177–80; American liberal Protestant responses to, 144–46, 157; diversity in formulations of, 79–80, 83n64, 163–65, 203–4; Greek Orthodox Christian responses to, 135–36, 138; Wilhelm Herrmann's criticism of, 97–99; historiography of, 258–59, 264, 271–72; Islamic responses to, 207–11, 215–17; New Atheists' implicit acceptance of, 221–26; origins of, in United States, 46–59; origins of, in Germany, 87–95; origins of, in modern Greece, 136–37, 139; prehistory of, 6–8; reasons for persistence of, 42–43, 99, 120, 198–99, 303–17, 331–35; Roman Catholic responses to, 107–12, 115–19; scientists' views on, 280–81, 289–90, 298–99; Otto Zöckler's criticism of, 96–97
conservative Protestants, 326–30, 333–34, 336
contested facts, 337
Conybeare, William 50
Cooke, Bill, 309
Cope, Edward Drinker, 73
Copernicanism: in Galileo Affair, 27–30, 32–33, 35, 39–41; in social science textbooks 307–11, 314, 318
Copernicus, Nicolaus, 27–29, 37–38, 65, 77, 136, 140, 305, 307–11
Cornell, Ezra, 10–11
Cornell University, 9–12
Cosmas Indicopleustes, 59, 123
Cosmos (PBS television show), 221
Cousin, Victor, 206–7
Coyne, Jerry A., 223–24, 227, 230, 233, 252, 254, 279
creationism, 177–79, 227, 232–33, 244–45, 248, 272, 312–14, 332
Creation Museum, 233
Crick, Francis, 221, 224
cultural myths, 38–39, 303, 307. *See also* Galileo Affair
Czolbe, Heinrich, 89

D'Alembert, Jean Le Rond, 7–8
Danielson, Dennis R., 310
Darwin, Charles, 16, 65, 67, 70, 112, 117, 118, 167, 176, 269, 270, 311, 332; *On the Origin of Species*, 17, 58, 67, 91, 93, 180, 248, 312
Darwinian fundamentalists, 226
Darwinism, 2–4, 129, 260, 267, 269, 318, 324, 327, 333; and American Evangelical Protestants, 174, 176; and American Jews, 202n64; and Calvinist theology, 261, 270–71; and German monism, 93–95; and Greek Orthodox Christians, 129, 132–34; and neo-harmonists, 242–43, 245, 247–48; and New Atheists, 226–27; and Ottoman public discourse, 215; in social science textbooks 303, 311–14
Daum, Andreas, 88, 91, 94–95
Davy, Humphry, 130
Dawkins, Richard, 80, 167, 220, 222–24, 226–34, 236n12, 239, 244, 279, 314
Dearborn Independent, 194, 196
de deification of nature, 263
deduction, 53, 62n33
deism, 47, 49, 68,78–79, 167, 170
Dembski, William, 232
democratization, 56
Democritus, 65
Dennett, Daniel C., 220, 221, 223, 226, 227, 228, 231
Descartes, René, 65, 89, 111, 142n20, 262, 263
Dick, Thomas 47
dictatorship in Greece (1967–74), 138–39
Diderot, Denis, 7
Dillenberger, John, 264
Diner, Hasia, 188, 191
dissonance, in relation to methods of harmonization, 259, 265, 269, 273
diversity, in science-religion discourse, 158, 262, 268

divine immanence, 147–48
Dixon, Thomas, 272
Dixey, Frederick, 269
Dobzhansky, Theodosius, 271
Dorlodot, Henry de, 119
Draper, Elizabeth, 18
Draper, John William, 25n40, 73, 110, 125, 134, 168, 171, 180, 185, 197, 224, 227, 228, 231; advocate for the conflict thesis, 30, 93, 165–66, 171, 186, 203, 220, 244, 303; at annual meeting of the British Association for the Advancement of Science (1860), 25n40, 67–68; connection with Edward Youmans, 21–22; correspondence with John Tyndall, 68–70, 77; criticisms of, 96, 143, 190, 225, 261–62, 269; on Eastern Orthodoxy, 123–24; *History of the Conflict between Religion and Science,* 6, 9, 13, 17–22, 66–67, 77–79, 96, 103, 104, 123, 143, 184, 186, 189, 216, 262, 303; *History of the Intellectual Development of Europe,* 13, 68, 75, 77–78, 96; on Islam, 19–20, 216, 272; origin of his conflict thesis, 12–21; on Protestantism, 18, 124; on Roman Catholicism, 18–19, 78–79, 103–4, 124. *See also* conflict thesis
Du Bois-Reymond, Emil, 94–95, 97
Duffield, John T., 169
Duhem, Pierre, 37

Eastern Orthodoxy. *See* Orthodox Church
Ecklund, Elaine Howard, 281
Eddington, Arthur, 271
Edelstein, Dan, 306–7
Efron, Noah, 265, 273–74
Eginites, Demetrios, 129
Einstein, Albert, 30, 36–37, 224, 271
Eller, Jack D., 310–11, 314, 316
Ellis, George, 253
Elzas, Barnett, 190
Emerson, Ralph Waldo, 72
empiricism, 48, 54–55
empiricist-biblicist paradigm, 54–55, 57–59. *See also* biblical interpretation
England, Richard, 269
Enlightenment, the, 7–8, 34–35, 46, 170, 231, 331; and 19th-century American Protestants' conceptualization of the relationship between science and religion, 47–48, 53–55, 60n5; and the propagation of the conflict thesis in social science textbooks, 303–10, 315, 320–21
Epicurus, 65
epistemic conflict, 325–26, 331–37. *See also* social/moral conflict
epistemology, 53–55, 259, 273, 325–26, 329, 333–34. *See also* epistemic conflict
Erickson, Mark, 304
eugenics, 117–18
evangelicals. *See* Protestant evangelicals
Evangelische Kirche, 86
Evans, Michael, 330
evolution, theory of, 70–71, 91, 134, 146, 157, 166, 232, 261, 318, 321; controversy over teaching in public schools, 175, 194, 198, 227, 279; and epistemic conflict, 329–30; Islamic responses to, 272; and neo-harmonists, 244, 246, 248–49, 251, 253; and Princeton Seminary, 166, 168, 169–71, 174; and Roman Catholic apologetics, 105, 108–12, 117–19, 228–29; Roman Catholic debates over, 112–15. *See also* Darwinism

Fahri, Mehmed, 217
Farabi al-, 207
Faraday, Michael, 66, 70, 265–66, 274–75
fatalism, 20, 216
Ferber, Edna, 194
Feuerbach, Ludwig, 87, 90
Feyerabend, Paul, 37
Feynman, Richard, 224
Finke, Roger, 243
Fisher, Ronald, 271
Fiske, John, 20–21, 241
flat earth, 123, 227
Flückiger, Felix, 85
Fodor, Jerry, 2–3
Ford, Henry, 194–96
Foscarini, Paolo Antonio, 29, 33
Fosdick, Harry Emerson, 155
Free Religion movement, 87–89, 92
Freudianism, 157, 231
Frisch, Ephraim, 190
Fuad, Beşir, 214–15
Fuller Theological Seminary, 165, 177, 180
fundamentalist Protestants, 157, 165, 172–79, 183n40, 193, 197, 201n45, 228, 327

Galileo, Galilei, 7, 56, 59, 65, 136, 137, 140, 259, 265, 282, 307; *Dialogue on the Two Chief World Systems, Ptolemaic and Copernican,* 27, 29, 30,

35–36, 41; *Letter to the Grand Duchess Christina,* 29, 35, 265; origins of trial before the Roman Inquisition, 27–30. *See also* Galileo Affair
Galileo Affair, 7, 91, 96, 136, 231, 247, 259, 260, 267, 231; as conflict between conservation and innovation, 33–34; deep structure versus surface structure of, 31–34, 40–42; interpreted as demonstrating complexity thesis, 43, 227, 266; interpreted as demonstrating conflict between science and religion, 30, 35–37, 143, 307–10; interpreted as demonstrating harmony between science and religion, 31–32, 37; as source of cultural myths, 36–38, 41–42; treated in Catholic apologetics, 105, 107–8, 119; used in anti-Catholic polemics, 7, 103, 105, 131
Gassendi, Pierre, 65, 262
Geertz, Clifford, 326
Genesis: astronomy and interpretation of, 52; evolution and interpretation of, 169, 325, 333; geology and interpretation of, 50–52, 57, 61n18, 63n44, 109. *See also* biblical interpretation
geology, 50–52, 109–10
German Romanticism, 72, 76, 86–87
Giancoli, Douglas C., 318–19
Gibbon, Edward, 206–7
Giberson, Karl, 231, 250
Giddens, Anthony, 315
globalization, 272, 274
Gmeiner, John, 108, 112
"God of the gaps," 332
Goethe, Johann Wolfgang von, 86–87, 92
Gompers, Samuel, 194
Goren, Arthur, 192
Gould, Stephen J., 72, 224, 226, 227, 268
Graham, Billy, 177
Gray, Asa, 241
Greek Enlightenment, 128, 136, 140
Greek Orthodox Church. *See* Orthodox Church
Greeley, Horace, 9
Greene, William Brenton, 171
Gregorian calendar, 133–34
Griffin, David Ray, 253
Gruner, Rolf, 263

habitus, 212–13
Haeckel, Ernst, 89, 91–95, 97–98
Ham, Ken, 233
Hameed, Salman, 272
Hammerstein, Wilhelm Freiherr von, 94
Harding, Fred F., 192
harmony thesis: and American Jews 189; and American liberal Protestants, 157; and American Protestant evangelicals, 164, 179; and confessional apologetics, 263; and Galileo Affair, 31, 34–35, 39, 43; and Muslim intellectuals 204, 207–9, 215–17; and neo-harmonists, 241–42, 247, 252; and 19th-century American Protestants, 51; and Orthodox intellectuals, 136
Harnack, Adolf von, 155
Harris, Sam, 220, 222, 224, 230, 236n12, 239, 252
Harrison, Peter, 226, 259, 267–68, 272–73
Hartshorne, Charles, 253
Haught, John F., 228–29, 253
Harvard University, 196
Hawking, Stephen, 221, 224
Haytham, Ibn al-, 207
Heilbron, John, 225, 269
heliocentrism. *See* Copernicanism
Hellenic Open University, 139
Henry, Carl F. H., 165
Henry, John, 319–20
Henry, Joseph, 52, 54
hermeneutics, 30, 47, 267. *See also* biblical interpretation
Herrmann, Wilhelm, 97–99
Hesychasm, 127–28, 138, 140
Hewit, Augustine, 108–9
Hirst, Thomas, 72, 74–76
"historian myth buster," 241, 244–46, 248, 254
historicism, 144
Hitchcock, Edward, 51, 53, 57
Hitchens, Christopher, 220, 222–23, 224, 239
history of science, 259–60, 275, 316; and Greek intellectuals, 135; and Muslim intellectuals, 207
Hodge, Alexander Archibald, 164
Hodge, Charles, 53, 144, 164, 167, 169, 181n18
Hollinger, David A., 191, 239, 240, 241, 246
Hooker, Joseph Dalton, 66–67, 74
Hooykaas, Reijer, 253, 263, 267
Hovenkamp, Herbert, 165
Howe, George, 58–59
Humani generis, 119
Humboldt, Alexander von, 88–89, 206
Hunter College, 196

Hutton, James, 50
Huxley, Julian, 271
Huxley, Thomas Henry, 2, 21, 66–67, 70, 73, 103, 168–69, 269, 274–75; advocate of conflict thesis, 104–5, 110, 143, 167, 169–70, 171; correspondence with John Tyndall, 71, 74–75; debate with Bishop Samuel Wilberforce (1860), 2, 16, 67, 271, 313, 337

idealism, 79, 89, 150–51, 167
Iliou, Philippos, 139
induction, 54, 62n33, 181n18. *See also* Baconianism
Ingold, Tim 313–14
Institute for Creation Research, 233
Intelligent Design, 221, 231–32, 245, 248, 249, 327
International Social Survey Programme (ISSP), 284
interpretation, of science and religious ideas, 262, 264, 265, 268
Isaacs, Abram, 185
Ishak, Harputlu, 214, 217
Islam, 131, 211; Muslim intellectuals' belief in harmony between science and, 204, 207–9, 215–17; as obstacle to scientific progress, 206, 209; as promoter of scientific progress, 207–8, 210, 216
Islamic state in Spain, 206–8
Iverach, James, 241

Jaki, Stanley, 225–26, 253, 263
Jewish Protective Association, 192
Jews, see American Jews
Jisr, Hussein al-, 215, 217
John Paul II, Pope, 30–31, 34, 327
Johnson, Phillip, 221, 231, 232
Judaism, 131; and science, 186–89, 274. *See also* American Jews

Kant, Immanuel, 76, 85–87, 89
Karas, Yannis, 139
Keith, Arthur, 2
Kemal, Namik, 208, 210
Khan, Sayyid Ahmad, 211, 215
Khazini al-, 272
Kimmel, Michael, 305, 318
Kingsley, Charles, 241
Kitsikis, Nikolaos, 136–38
Klaaren, Eugene, 253
Kohler, Kaufman, 188–89
Kras, Nathan, 190
Kroto, Harold, 1–2
Küçük, B. Harun, 272
Kuhn, Thomas, 259–60
Ku Klux Klan, 195–96
Kuyper, Abraham, 179

Lamar, James S., 54
Lamarckian evolution, 55, 214
Lange, Albert, *History of Materialism and Critique of Its Present Significance* of, 74–75
Langer, William, 89
Laplace, Pierre Simon, 52, 223
law, of nature, 14–17, 53, 55, 68, 90, 146, 274, 314
Le Conte, Joseph, 52, 241
Leeser, Isaac, 186–87
Leo XIII, Pope, 108. See also *Providentissimus Deus* and *Aterni Patris*
Leroy, Maurice Dalmas, 112–13, 115, 120
Leuba, James, 280–81
Lewis, C. S., 250
Lewis, Tayler, 57
liberal Protestants, 164, 169, 170, 197, 261; openness to harmonizing Christianity with advances in scientific knowledge, 144–45; revisions to Christian theology of, 150–53; revisions to concept of revelation of; 148–49; Protestant opponents of, 172–74; on relationship between science and theology, 154–57
Libri, Guglielmo, 36
Liebig, Justus, 90
Lightman, Bernard, 269
Lindberg, David C., 236n15, 240, 241, 262, 269–70, 310
Livingstone, David N., 4, 237n26, 270
Locke, John, 273
Lord, David, 57
Lord, Eleazar, 57
Lucretius, 65
Lukaszek, David A., 313
Lyell, Charles, 50

Machen, J. Gresham, 165, 172–74
Macintosh, Douglas Clyde, 152
Macionis, John J., 304, 307–8
Macloskie, George, 168, 170–71

mainline Protestants, 144, 158n10, 330–31
Malled du Pan, Jacques, 37
Marsden, George, 174–75, 334
Marshall, Louis, 195
materialism, 65, 75–77, 89–91, 132, 138, 270; critiques of 19, 96–97, 129, 133–34, 214–15
Mathews, Shailer, 152
Maximus the Confessor, 128, 138
Maxwell, James Clerk, 266, 274–75
Mazlsih, Bruce, 259
McCabe, Joseph, 220
McCosh, James, 164, 167–70, 174, 241
McGiffert, A. C., 171
McGrath, Alister, 229, 253
McMillan, Chaisson, 318
McMullin, Ernan, 309–10
Mencken, H. L., 174
men of science, Muslim, 211–12, 216, 218
Merton, Robert, 258, 267
metaphysical naturalism, 331
metaphysical realism, 152–53
Metcalf, Peter, 312
methodological naturalism, 52–53, 55, 99, 146–47, 231, 331, 335
Midgley, Mary, 234
Midhat, Ahmed, 214–17
military metaphor, 8, 11, 163–64, 177; and American Protestant Fundamentalists, 175–76; and Princeton Seminary, 165–68, 171–72
Mill, John Stuart, 167
Miller, Glenn T., 165
Miller, Kenneth, 230, 254
Miller, Perry, 258
Miller, Samuel, 48–49, 53–54
millet system, 125–126
Millikan, Robert A., 156
Milton, John, 7, 36, 41
Minton, Henry Collin, 171
Misteli, Hermann, 90
Mitchell, Elisha, 56
Mitchell, Ormsby MacKnight, 52
Mitsopoulos, Constantine, 129, 132–33. See also *Prometheus*
Mivart, St. George Jackson, 105, 111–14, 115–16, 120, 266–67
modernism (movement within Roman Catholic Church), 105, 115–16
Mohammedanism. *See* Islam
Moleschott, Jakob, 90
Molloy, Gerald, 109
monism, 91–93, 95; idealistic versus naturalistic, 89
Moore, Aubrey L., 241, 269
Moore, James R., 19, 66, 163, 222, 241; *The Post-Darwinian Controversies* of, 4, 158n2, 240, 261. *See also* military metaphor
Moors, see Islamic state in Spain
Morrill Land-Grant Act of 1862, 10
Morris, Henry M., 178, 233
Morris, Simon Conway, 253
motivation, 262
Müller, Karl, 91
Murphy, Nancey, 253
"muscular Christianity," 175
Muslims. *See* Islam
Myers, PZ, 223–24, 254
myths, concerning the relationship between science and religion, 2, 244–46, 248, 271, 303, 307, 309, 321, 330–32, 334–38. *See also* cultural myths

Nagel, Thomas, 3
Napoleon Bonaparte, 86, 223
National Association of Evangelicals, 177
National Institutes of Health (NIH), 230, 250
natural selection, 2–4, 25n40, 91, 117–19, 222, 226, 232, 249, 253, 261, 267, 269, 311–13
natural theology, 8, 50, 85–87, 99, 260–61, 263–64, 266–67
Naturphilosophie, 86, 88
nebular hypothesis, 52
Nelson, Paul, 235n8
neo-evangelicalism, 177–80
neo-harmonists, 220, 239–52
Neo-Orthodox (movement within Greek Orthodoxy), 139
Neo-Orthodoxy (movement within Protestant theology), 156–57
New Atheists, 220–34, 239, 242, 248, 252
Newman, Louis, 195–96
New Theology, 144, 151
Newton, Isaac, 87, 222, 263, 271, 315, 336
Newtonian physics, 126, 305
New York University, 196
Nietzsche, Friedrich, 176, 220
Noll, Mark, 165, 228

"noncontradiction," doctrine of, 58
NOMA (non-overlapping magisteria), 72, 226–27, 268
Numbers, Ronald L., 4, 179, 223, 235n8, 235n10, 236n15, 240, 241, 247, 262, 269–70, 314, 336
Nursi, Said-i, 217

Obama, Barack, 250, 252
Olson, Richard G., 99, 314
Open University (OU) distance-learning course, 260–61
Orfanidis, Theodor, 129
orientalism, 210–11
Orthodox Church, 124, 125, 130, 133, 135, 136; and 18th-century debate in Greece on teaching of science, 125–27, 139; opposition to Gregorian calendar of, 133–34; opposition of left-wing Greek intellectuals to, 123, 137; relationship to modern Greek state of, 128–29, 134, 138; and voluntary religious associations in Greece, 129
Osborn, Henry Fairfield, 168
Ottoman Empire, 125–26, 128, 136, 206–8, 212–13
Overing, Joanna, 306

Paine, Thomas, 49
Palamas, Gregory, 127–28, 138
Palmer, Benjamin Morgan, 58
paradigms, 260. *See also* empiricist-biblicist paradigm
Pascendi dominici gregis, 115
Pasha, Ziya, 207, 213
Patton, Francis L., 172
Patton, George S., 171–72
Paul, Apostle, 229
Paul V, Pope, 29, 41
Peacocke, Arthur, 253
People's College (Havana, New York), 10
Perkins Professorship of Natural Science in Connexion with Revealed Religion, 47, 58, 60n4
personal idealism, 150–51
philology, 46–47, 57, 63n44, 259, 273
physico-theology, 85, 268
Piatelli-Palmarini, Massimo, 2
Pinker, Steven, 224, 252
Pittsburgh Conference, 188
Pius IX, Pope, 18–19, 25n46; *Syllabus of Errors* of, 78, 104
Pius X, Pope, 115
Pius XII, Pope, 119
Plantinga, Alvin, 3, 45n23, 232
Plummer, Ken, 304
Polkinghorne, John, 253
Polygenism, 4, 58–59
Popper, Karl, 30
popularization of natural science in Germany, 88–89
Porter, John W., 176
Portland Deliverance, 171
positivism, 14–15, 19, 68, 259, 263, 268
Poulton, Edward, 269
preadamism, 58
Presbyterian Church, U.S.A., 172
Presbyterians: and controversy over higher criticism of the Bible, 171; and fundamentalism, 174; in 19th-century American south, 47, 58–59; reception of Darwinism in different countries and regions by, 270. *See also* Princeton Seminary and College
Priestley, Joseph, 259
Princeton Review, 164–67
Princeton Seminary and College, 164–66. *See also* military metaphor
Princeton Theology. *See* Princeton Seminary and College
Princeton University, 196
Principe, Lawrence M., 78, 225, 236n15
privatization of religion, 264
progress: social linked to scientific, 125, 131–32, 137, 190, 205–8, 210, 217, 316, 320; in thought of John William Draper, 14, 68–69
Progressive Orthodoxy. *See* New Theology
progressivism, 186, 192–93, 198
Prometheus, 132–35
Protestant evangelicals, 47, 49, 164, 254, 327; and fundamentalist movement, 172, 174–76; and neo-evangelicalism, 177–80; Princeton Seminary as exemplar, 165, 168–69; use of military metaphor by, 166
Protestantism, 36, 187, 228; John William Draper's views on, 18, 78, 80, 124; and science, 52–54, 56, 58, 267–68. *See also* conservative Protestants; fundamentalist Protestants; liberal Protestants; mainline Protestants; Protestant evangelicals
Protocols of the Elders of Zion, 194

Providentissimus Deus, 30, 108–10, 114, 120
Provine, William B., 221
Ptolemy, 34
public schools, 175, 192, 194–95, 198, 227, 279
Pupin, Michael Idvorsky, 138
Puritanism, and science, 258, 262
Purves, George T., 171

Quakers, 266
quota system, and American higher education, 196
Qur'an, scientific exegesis of, 216–17

racism, 191, 209–10
Ralston, Thomas, 52
Ramberg, Peter, 87–88
Ramm, Bernard, 165, 177–79
Rapport, Nigel, 306
rationalism, 47–50, 55–59, 87–88
Rau, Herbert, 89
Ravenscroft, John Stark, 56
Reimarus, Hermann, 85
Reiss, Michael, 1
relationship between science and religion: revisionist historiographies, 240–41, 244–45, 258, 260, 261, 274. *See also* complexity thesis; conflict thesis; harmony thesis
Religion among Scientists in International Context (RASIC), 283–88, 290–92
religious experience, 149, 156. *See also* biblical interpretation
religious feeling, 147
religious societies (or associations) in Greece, 128–29, 138
Remzi, Huseyin, 217
Renan, Ernest, 208–10
Rees, Amanda, 312
revelation, 85, 88, 148–50, 154, 174
Richards, Robert, 91–93
Riley, William Bell, 174–76
Rimmer, Harry, 175–76
Ritschl, Albrecht, 154–55
Ritter, Henry 130
Roberts, Jon, 263–64, 330
Roberts, Richard, 1
Roman Catholic Church, 75, 97, 143, 178, 225, 264, 269, 279, 282, 328; condemnation of modernism by, 115; Congregation of the Index of, 29, 113–15; Council of Trent, 34; John William Draper's opposition to, 18–19, 78–79, 103–4, 124; First Vatican Ecumenical Council of, 19, 20, 78, 104; and Galileo Affair, 31, 35, 39, 40–41, 59, 103, 105, 107–8, 307–8, 310–11; Greek intellectuals' views on, 125, 131–32, 134, 137; opposition to eugenics, 117; response to conflict thesis by intellectuals, 107–12, 115–19; Roman Inquisition of, 8, 27, 29, 32–33, 36, 39, 136, 184, 310; Second Vatican Ecumenical Council of, 31, 119; and theory of evolution, 108, 112–13, 115, 117–19, 314; John Tyndall's views on, 69, 76
Rosenberg, Maximilian Theodore, 184–86, 197
Ross, Alexander, 37
Roßmäßler, Emil, 91
Royal Institution, 65, 66, 69, 74
Rudwick, Martin, 263
Rupke, Nicolaas, 271
Ruse, Michael, 226, 227, 228, 254
Russell, Bertrand, 30, 220
Russell, Colin A., 233, 237n26, 260–261, 268
Russell, Robert J., 253
Rutty, John, 272

Sacks, Jonathan, 229–230
Sagan, Carl, 221, 222, 224
Sage, Henry, 11
Said, Edward, 210
Sami, Mustafa, 207
Sandeen, Ernest, 165
Sandemanians, 265
Savage, Mino Judson, 241
Schelling, Friedrich, 86–88, 92
Schleicher, August, 92
Schleiermacher, Friedrich, 147
Scholasticism, 110–11. *See also* Thomist philosophy
Schönborn, Cardinal Christoph, 2
Schultze, Quentin, 330
Schwartz, Angela, 88
Schweitzer, Albert, 168
scientific materialism, 89–91
scientific naturalism, 66, 274–75. *See also* methodological naturalism
Scientific Revolution, 262, 303, 305–6, 311, 315, 321
scientists, and secularization, 280–81
Scopes, John, 196
Scopes trial, 174, 193–96, 198, 272, 313

Scott, Eugenie C., 254
Scott, William Berryman, 167–68
Scottish commonsense realism, 53–55, 59, 333–34
scriptural geologists, 57, 63n42
Scupin, Raymond, 315
secularization, 280–81, 304, 311, 312, 314–17, 319, 326
Sedgwick, Adam, 50, 266
September 11, 2001, 220, 222
Seward, William H., 165
Shank, Michael H., 225
Shapin, Steven, 1
Shapiro, Adam, 272
Shepard, Jon, 307
Shipley, Maynard, 178
Shumayyil, Shibli, 215
Silliman, Benjamin, 51, 54, 57
Skaltsounis, Ioannis, 129
Smith, Henry Preserved, 171
Smith, Samuel Stanhope, 55
Smith, Wilbur, 180
Smith, William Robertson, 270
Smyth, Newman, 147, 241
social/moral conflict, 326–30. *See also* epistemic conflict
sociology, 302–5, 310, 320–21, 331. *See also* textbooks, introductory level for the social sciences
"Southern Reformation," 18, 19, 216
Spalding, John, 111–12
Spencer, Herbert, 17, 20–21, 66, 73, 167
Spencer, Nick, 247
Stanley, Matthew, 271, 274–75
Stark, Rodney, 225–26, 239, 242–47, 248
State Agricultural College (Ovid, NY), 10
Stearns, Lewis French, 147
Stenger, Victor J., 224–25, 324–25, 337–38
Stephanides, Michael, 135–36
Strauß (Strauss), David Friedrich, 87, 167
Stroumpos, Dimitrios, 129–32
Stuart, Moses, 58
Suarez, Francisco 110
Sunday, Billy, 175
surveys: attitudes of public toward religion-science conflict, 139–40, 325–30; attitudes of scientists toward religion, 280–81, 283–90
Swetlitz, Marc, 188

Tahtawi, Rifa'a al-, 207, 218n9
Tappan, Henry P., 11
Taylor, Charles, 317
telescope, 28
Templeton Foundation, 227
textbooks, introductory level for the social sciences, 302–3; advocating the conflict thesis, 303–7, 317–21; presenting heleocentrism as an example of science/religion conflict, 307–11; presenting post-Darwinian debates as an example of science/religion conflict, 311–14; presenting science as a driver of secularization, 314–17
Thackray, Arnold, 260
Thomist philosophy, 313
Tillich, Paul, 99
Turner, Frank M., 66, 237n26, 258, 269, 271–72
Turner, Stephen, 84
"two books" theology, 58–59, 61n17, 242, 247, 333–34
Tylor, Edward Burnett, 302, 306, 316, 320
Tyndall, John, 3, 9, 21, 66, 96, 266; American trip (1872–73), 71–73; correspondence with John William Draper, 68–70, 77; presidential address of, 3, 65–66, 73–76, 270; understanding of relationship between science and religion, 75–76, 78–80, 270. *See also* conflict thesis
Tyndall Centre for Climate Change Research, 66

Ule, Otto, 91
uncontested facts, 325–26
uniformitarianism, 50
Unitarianism, and science 260
University of Athens, 129, 139
University of Chicago, 196
University of Padua, 126
Urban VIII, Pope, 37, 41, 91

Vaughan, Herbert, 114
Vestiges of the Natural History of Creation, 16–17, 55
Viardot, Louis, 207–8
Virchow, Rudolf, 93–95
Visnovsky, Emil, 306
Vogt, Karl, 89–90
Voltaire, Francois Marie Arout de, 36, 41

Wallace, Anthony F. C., 316–17
Walsh, James Joseph, 107, 116
Ward, Keith, 253
warfare thesis. *See* conflict thesis
Warfield, Benjamin Breckinridge, 164, 169–170, 174
Wasmann, Erich, 118
Watson, James, 224, 250
Weber, Max, 280
Webster, Charles, 262
Webster, Frank, 304
Wegscheider, Julius, 87
Weinberg, Steven, 224, 236n15
Weir, Todd, 87–89
Weldon, Stephen P., 221
Westboro Baptist Church, 331
Westfall, Richard, 263
Westminster Theological Seminary, 179
Whewell, William, 206
White, Andrew Dickson, 73, 123, 134, 155, 178, 185, 203, 224; advocate for conflict thesis, 30, 46, 59, 66, 93, 104, 153, 166, 168–69, 171, 180, 220, 225, 228, 231; connection with Edward Youmans, 21–22; criticisms of, 12, 96, 116, 153–54, 169–70, 176, 225, 232, 244; on Eastern Orthodoxy, 124–25; *A History of the Warfare of Science with Theology in Christendom*, 6, 9–12, 67, 96, 123, 153, 168; origin of his conflict thesis, 9–12; on Protestantism, 153, 163, 169; on Roman Catholicism, 103–4. *See also* conflict thesis
Whitehead, A. N., 253
Wilberforce, Samuel, 2, 16, 67, 261, 271, 313, 337
Wilkins, John, 37
Wilson, David B., 268
Wilson, David Sloan, 227–28
Wilson, Edward O., 224, 271
Wilson, Woodrow, 173, 176
Winchell, Alexander, 241
Windle, Bertram, 107, 115–18, 120
Wise, Isaac Mayer, 187–88
Wise, Stephen, 196
wissenschaft, 84–85, 87–88, 96–97
Witherspoon, John, 55
Woodrow, James, 47, 53, 58–59
World's Christian Fundamentals Association, 174
World Values Survey (WVS), 284
World War I, 117, 174, 177
World War II, 137–38, 177, 199
Worthen, Molly, 165
Wright, George Frederick, 241

Yale University, 196
Youmans, Edward L., 21–22, 71, 73, 77, 143
Young Ottomans, 207–8

Zahm, John Augustine, 113, 115–16, 120
Zöckler, Otto, 95–98
Zoe (Greek Orthodox religious association), 138. *See also* religious societies